THE STERNBERG FAMILY
OF FOSSIL-HUNTERS

Charles Hazelius Sternberg at McKittrick
Courtesy Scientific American (Sept. 1928 issue)

THE STERNBERG FAMILY
OF FOSSIL-HUNTERS

Martin O. Riser

The Edwin Mellen Press
Lewiston/Queenston/Lampeter

Library of Congress Cataloging-in-Publication Data

Riser, Martin O.
   The Sternberg family of fossil-hunters / Martin O. Riser.
      p.   cm.
   Includes bibliographical references and index.
   ISBN 0-7734-8985-1 (acid-free)
    1.   Sternberg, Charles H. (Charles Hazelius), b. 1850.
  2. Sternberg family. 3. Paleontologists--United States--Biography.
I. Title.
QE707.S8R57   1995
560.9--dc20
[B]   94-48127
                                                                                                              CIP

A CIP catalog record for this book is available from the British Library.

Copyright © 1995 Martin O. Riser

All rights reserved. For information contact

The Edwin Mellen Press	The Edwin Mellen Press
Box 450	Box 67
Lewiston, New York	Queenston, Ontario
USA 14092-0450	CANADA L0S 1L0

The Edwin Mellen Press, Ltd.
Lampeter, Dyfed, Wales
UNITED KINGDOM SA48 7DY

Printed in the United States of America

This book is dedicated to my mother, Imelda Catherine (Murray) (Riser) Gunn, with all love, as well as to Doctors Ritterman, Yuen, Valentine, McCarthy, Tobler and Stinson, and the talented nurses of Stanford Medical Center without whom this book would have never been finished.

There is a beauty about fossil bones...that transcends their basic quality as the remains of ancient life, a beauty involving the forms of the fossils and the texture of their surfaces. The minerals included in the process of fossilization whereby the bony material is infiltrated and replaced and thereby petrified-- that is, turned to stone--often lend color to the fossil that makes it esthetically more interesting to look at than if it had the colorization of the original bone.

Edwin Colbert, *Digging into the Past*

TABLE OF CONTENTS

	Introduction	i
One	The Beginning	1
Two	Bone Venture	25
Three	Life With Cope	41
Four	Badlands In Balance (1876)	59
Five	More Kansas	73
Six	Oregon Desert & The John Day	83
Seven	Fossil Hunter Etiquette	105
Eight	Lawrence, Kansas	115
Nine	Texas Permain & The Red Beds	133
Ten	The Bone-Hunter As A Type	159
Eleven	Cope-Marsh Feud	171
Twelve	Sternberg The Professional	191
Thirteen	Ex-Patriot American	205
Fourteen	American Museum Mystique	229
Fifteen	Torpedo & Cyclone	239
Sixteen	George Miller Sternberg	249
Seventeen	The Family	267
Eighteen	California Sojourn	301
Nineteen	Sternbergs In The San Juan	331
Twenty	Christianity	341
Twenty-One	Number One Son: The Early Years	371
Twenty-Two	The Younger Sons: Early Years	379
Twenty-Three	Charles As Author & Lecturer	389
Twenty-Four	Sidelights	405
Twenty-Five	George Makes It On His Own	413
Twenty-Six	The Boys In The North	437
Twenty-Seven	The Sternberg Legacy	443
Appendix A	Date Sequencing	449
Appendix B	Sternberg Family Tree	457
Bibliography	Sternbergs	461
Bibliography	Other	473
Index		485

AUTHOR'S NOTE

As with any worthwhile endeavor there are always those who promote by influence or help in ways that cannot be easily expressed. No historical book of any variety can be written without helps from myriad people. This book is certainly no different, and I want to acknowledge the many and varied helps of a number of special people.

Initial encouragement was provided by a librarian at the United States National Museum, Carolyn Hahn. Her kindness in giving extra prompted me to surge forward at a time when the book could easily have gone by the wayside. An associate of mine, Susan Schmalbach, was also of help at that time by taking out time to humor a fellow worker.

My brother, Patrick J. Gunn, and Margaret Tuft of the Merced County Library have provided vast helps by searching for articles and books I couldn't get locally. The staffs of the Berkeley Public Library, the Richmond Public Library, the Earth Sciences Library and Main Campus Library for the University of California at Berkeley have been praiseworthy. The special library service BAYLIS for the city of Oakland is most appreciated for the information they were able to obtain for me on William Sternberg.

Mae Thornton, Curator of the Ft. Harker Museum in Kanopolis, Kansas, provided personal information on the family, and William McKale of the United States Cavalry Museum at Ft. Riley was also helpful.

Margaret Hanna, wife of G. Dallas Hanna, who is still active at the California Academy of Science in San Francisco, graciously provided valuable correspondence between her late husband and Charles Sternberg that allowed me to fill in many gaps.

Special thanks should also be given to Carol Barsi, librarian at the San Diego Natural History Museum, and her able staff of volunteers (Mrs. Harwell and Mrs. Voss) who were of immeasurable help in providing document copies and other helps with a fervor that is catching.

Thanks are due as well to Dr. Richard J. Zakrzewski of the Sternberg

Memorial Museum in Hays, Kansas for his assistance in providing George Sternberg's scrap book information and other helpful documents.

The British Museum (of Natural History) kindly provided photocopies of a substantial number of letters between Charles Sternberg and Dr. A. Smith Woodward. These were provided thanks to the intercession of Clive Reynard, and the kindness of Dorothy Norman, both to whom I am endebted.

Special thanks are also due James Armistead for his encouragement and support, as well as for his assistance in explaining the book publishing business and (along with Michael Hildenbrand) in finding an appropriate publishing house.

My sister-in-law, Laura Wethern, provided special assistance in providing a photographic plate of the Sternberg letterhead logos, and Jim McCarthy of Fairhaven Bible Chapel was most helpful in critiquing chapter 20.

No work would be complete without special thanks to those that help in less obvious, but equally valuable ways. Thanks to my dad, John Patrick Gunn, for the trip we took together through the midwest, and to my wife Lila, daughter Julie, and son Michael for their forbearance in putting up with my being away, or sequestered, at times when I could have been more help to them. My daughter Julie, a writer herself, provided great assistance by editing most of the chapters.

Tim Tokaryk of the Saskatchewan Museum of Natural History provided substantial informations well beyond my wildest expectations. He provided fieldnotes for Charles and each of the sons, as well as providing invaluable family information. One box of documents he kindly sent actually weighed in at 7.46 kgs!

A.E. David Spaulding has also provided valuable helps and support. A Sternberg expert in his own right he graciously assisted in a number of different ways, including arranging an introduction with Raymond Martin, son of Charles M. Sternberg.

Other contributors of great assistance have been:
Shelley Wallace, Hartwick College, Oneonta, NY
Jeanne Mithen, Riley County Historical Society & Museum
Dr. John Stolz, MD, a special Hartwick alumnus & archivist

Mrs. Kenneth (Betty) L. Main, Central Iowa Genealogical Society
Kathie Willson, Historical Society of Marshall County
The Ellsworth County Historical Society, Ellsworth, Kansas
William R. Massa Jr., Yale University Library
National Archives, Washington, DC & San Bruno, California
John Chorn, the University of Kansas
Robert W. Hook, Austin, Texas
Diane Szabo, Palatine House (Lutheran Parsonage), Schoharie
The Old Stone Fort, Schoharie, NY
Jens L. Franzen, Senckenberg Museum, Frankfurt, Germany
Royal Ontario Museum, Toronto, Ontario, Canada
Field Museum of Natural History, Chicago, IL
Tyrrell Museum, Drumheller, Alberta, Canada
Ellen Sulser, State Historical Society of Iowa, Des Moines
Elizabeth Barbour Love, Riley County Historical Society
Janet Johannes, Ellis County Historical Society, Hays, KS
Barbara Gill, Berks County Historical Society, Reading, PA
Mrs. Orloff & Mrs. Sternfeld, American Museum of Natural Hist
New York Historical Society Library
Sunnyvale Patent Info Clearinghouse, Sunnyvale Public Library
John Mark Lambertson, Kansas State Historical Society
Sutro Library, San Francisco, California
Pastor David Brooks, Ganson Street Baptist Church, Jackson, MI
Thomas J. & Ruth Tonkovich, La Mesa, California
Raymond M. Martin, Victoria, BC, Canada

TEXTUAL NOTES

Throughout the book there was a necessity to use some abbreviations that could help to keep the books length down. The following abbreviations were frequently used:

CHS - Charles Hazelius Sternberg

CMS - Charles (Charlie) Mortram Sternberg
GFS - George Fryer Sternberg
LS - Levi Sternberg (son of CHS)
AMNH - American Museum of Natural History, New York
BMNH - British Museum (Natural History), London, England
CAS - California Academy of Science, San Francisco, CA
KAS - Kansas Academy of Science, Manhattan, KS
MCZ - Museum of Comparative Zoology, Cambridge MA
SDNHM - San Diego Natural History Museum, San Diego, CA
USNM - United States National Museum (Smithsonian), Wash, DC

Relating to Charles Hazelius Sternberg's Books:

Life - *The Life of a Fossil Hunter*
Red Deer - *Hunting Dinosaurs on the Red Deer River Alberta, Canada*
Romance - *A Story of the Past or the Romace of Science*

INTRODUCTION

I begin this biography with the simple premise that biographies are only of interest when there is something tangible and worthwhile that can be extracted from reading about how another individual lived their life. What a massive waste of time it would prove reading about a person if it were not so. Unless the life in question has meaning for today of what value is it to us? The value I see in the life of Charles Sternberg should be self-evident if you choose to read these pages. To me, he and his family are indicative of the type of men and women who made America great, for they show the same qualities of hope, hard work and persistence that we like to think are the elements responsible for making this country into a world power, and that will keep our country both free and a force for freedom for many years to come. Our Constitution, which has stood virtually undaunted after over 200 years of onslaught, is perhaps the symbol of our freedom, but it is the men and women who have given of themselves in great sacrifice who are the true essence of this country's greatness.

You might well ask "why write a biography about Charles H. Sternberg?" After all, didn't he write his own biographies, *The Life of a Fossil Hunter* and *Hunting Dinosaurs in the Badlands of the Red Deer River Alberta, Canada*? What else is there to tell? Or more likely, you might ask (unless you're an inveterate fossil hunter), "who is Charles H. Sternberg? Couldn't be much since I've never heard of him!"

I hope in the course of this book to both show that there is much yet of great worth that can be written about this man and his sons, and that if you have never heard of him it is simply the result of the oddities of our scientific community. He and his sons certainly deserve better. For it does not always follow that just because a person is not well known in general scientific circles that he has really been relegated to a lesser position because his science wasn't good enough, or lacked originality. The same sort of logic applies when we realize that the scientists who have been famous in their own day do not always have the same kind of following or notoriety just a few years later. The science of paleontology has certainly proved to be just such a fickle science, and it is subject to the same winds that blow unpredictably wherever man can be found.

I initially became interested in Charles H. Sternberg as a result of my interest in the history of dinosaur-hunting, and because I wanted to learn more about the methods and techniques of hunting and preserving bone. I found all this and much more in his writings. He tells us where to hunt bone specifically, tells us about the area geology, tells us what the animals were most probably like when they were alive, and he tells us about the lives and foibles of the famous men who used to hunt the dinosaurs; both friend and foe. His writings glory in the telling and description of bone, and one can tell from his eloquently phrased sentences that he had a genuine love for these vestiges of a former life, and that his life was truly given over to their capture.

But these are not the only things that bring me back to his writings, and make him of continued interest to us today. It is Charles H. Sternberg the man who is ultimately the force that brings me back again, and again. By the very nature of his relationship to his fine sons we are led to want to know more about them as well. They were shaped into men of real prominence in great part by their father, and had the same admirable qualities in their makeup also.

Charles Sternberg was an interesting man; a man of history, a man of complexity, yet a man of simplicity, a man with true zest for life, a man in tune with the world he lived in, a man in touch with God, and a courageous man of science.

Charles Sternberg will also always be remembered for his place in history; he was indeed a man in the right spot at the right time. Although dinosaurs were known earlier in the century in America the avalanche in hunting had not yet arrived. Sternberg was blessed with perfect timing for his arrival in Kansas. This then allowed him to play his great part in the "Great Dinosaur Rush" that was to so dismay the world. This Rush was to turn the scientific world upside-down with the discoveries of these unbelievably immense and curious creatures. America needed his special kind of services right then and he was equal to the task. The famous rivalry that was to develop between Edward Drinker Cope and Othniel C. Marsh, and the integral part he played in that majestic competition, also helps to keep Sternberg's name before the fossil and history-reading public.

Sternberg is of interest to the history-minded because of the importance of the transitional time in which he lived. The West was just beginning to truly be opened to the white man when he entered the fray, and was just as surely being closed, finally conquered, as his life ended. As a youngster he'd met the famous Wild Bill Hickok, he'd marvelled at the thundering buffalo herds that are no more, and he'd even heated himself at fires drawn from buffalo chips. He knew the rough and tumble side of life, was educated in the free-for-all atmosphere of the frontier towns of the West, and he knew what it was to fear being caught by marauding Indians. He was to live in sod-houses and dug-outs and even in tents, and yet for all this rough exterior he was to always remain a classic gentleman.

For anyone immersed in the lore of early American paleontology the names of those giants that Sternberg had rubbed shoulders with on the fossil fields are immortal: John Bell Hatcher, Benjamin F. Mudge, Barnum Brown, Othniel C. Marsh, Samuel W. Williston, Edward Drinker Cope, J.L. Wortman and many others. He knew them all, and was their equal as a collector and fossil-hunter. In some cases he could more aptly be called their teacher as he tutored some of these men in collecting skills. Part of his interest as a man is to be found in the fact of his not being so well-known, nor as well-credited, as these others. This is one instance perhaps where being less well known can be thought of as a benefit, for one can hardly fail to be original if you're writing about a lesser known individual who sparks real interest.

It is peculiar, but not uncommon, to find these instances where the teacher fails to get credit for the successes of his pupils. Such a situation applies to Sternberg. Often the great in a profession are molded into their field by parents who force them into certain paths of behavior. The parents, although they spend countless hours and suffer great deprivations rarely receive any real credit, and are more often than not viewed in an unpleasant light by biographers. In a sense at least, Sternberg was a parent to some of these young men who worked with and for him, and he certainly has not been given the credit due him for forging them into men of greatness. Even those following, who didn't work directly under him, were to benefit from his legacy to the field. But those who would give him his due, even today, have been few. I hope to change that.

In order to begin to understand the context in which Sternberg lived we must better understand the innate differences between the original founding fathers of the science, and what the professional paleontologist has become. One of the ways to get a glimpse of life back then is to look at the early camera images that portrayed life on the fossil fields. The textbook pictures of these men of the early years show them to be rag-tag in appearance; perhaps even unkempt by some standards. Ragged hair pushed into slouched hats, many sporting handlebar mustaches or heavy beards, they looked decidedly older and more experienced than their young years. Suspenders generally held heavy-bagged pants and long-sleeved shirts were the order of the day even in the hotter climates. The appearance, however, could not be considered complete without the ever-present side arm or rifle needed for protection against the stray war party or for supplying the next meal. The other hand most often held a standard digging pick, or in later years a geologist's (or Marsh) pick. This has always been the true symbol of the trade.

These pictures cannot be fully appreciated without looking at the same men in later years in their "modern" refinery indicative of later esteemed position in life. Their new "work" clothes consisted of suits and ties, more in keeping with the museum or geological survey positions they had attained in their more advanced age. This to me is one of the more obvious indications that these men were indeed a cut above and that their field capabilities were not without a high degree of intelligence and drive. These men were successful because they were good at what they did, learned and developed from others around them, and most decidedly worked hard for their living. They were truly scientific in their approach to this business of bone-collecting and clearly surpassed the humdrum by seriously studying their finds to the point that we today can better understand what the world and its creatures were like during the primal era.

The telling of the coming and the going of the true "bug-man," or bone-hunter is a saga as wide-ranging and regal as that found in the movie *Giant*. The country changed so much in just a few short years, and the science most certainly did. Begun with men of questionable credentials and even less expertise the science bloomed with the dinosaurs and became a true and sophisticated science second to none. Sternberg spanned the whole of this period, and if it were for that only he is worthy of special notice.

Not only is Charles Sternberg known for his own merits as a collector and preparator, but he managed to raise three famous sons: George Fryer Sternberg, Charles Mortram Sternberg and Levi Sternberg. Sternberg trained each of his sons well in the intricacies of his trade, and each was to become famous in their own right and contributed greatly to the advancement of paleontological knowledge. This bond between father and sons, and this occupation-sharing (so foreign to our own day), is of special interest. It says something to us today about what he contributed as a father, and how there does not have to exist contention and division, or even competition, between father and son.

Interesting also is the fact that Sternberg lived in a world foreign to us in the sense of it being a man's world. He shared in what might be considered the last hurrah in regards to that special camaraderie between men that is known only infrequently today, and then generally only in the contexts of war or sports. Although this might not be fashionable, it is certainly of great historical interest to those of us today who are interested in the "way it was," and I expect that most men long secretly for a return to a day more like when Sternberg lived. A man could definitely "prove himself a man" in those days, and that romantic ideal is not without its appeal. One need not necessarily be ranked among the chauvinists in order to appreciate this male kinship and how it lacks in real expression in our modern age.

The adventurer, and what he came to mean to the American youth that were to follow, was generally to prove to be the more visible Roy Chapman Andrews and men of that mien, but Charles Sternberg could certainly have filled those same shoes, and would have done so without so much self-interest being shown. It is hard for us today to understand that same sense of adventure when we can easily get from one point to another in short order, and without the same fears of being threatened by highwaymen or ravenous beasts. Adventuring has become a cold and calculated business these days, more designed for the accountant than for the brave.

The bone-hunters were a lonely bunch, often showing distrust of others because of the possibilities of someone stealing their latest discovery. It was very similar to the industrial secret stealing or espionage we hear of today. On the rare occasions when they let down their guard and got together with others of their trade, or when they sat back in advanced age after leaving the rigors of the field, they shared an amiable bond of kinship that was sustained through the plying of their trade and the knowledge they had amassed. This "man's only" existence perhaps best finds voice in the fact that not once in his 287 page autobiography, *The Life of a Fossil Hunter,* does Sternberg mention the wife of his children. The family men (his sons), however, are presented with frequency and clear and evident pride. While some might incorrectly label this as a sign of his not giving proper due to his wife and women in general, we will show that this was not the case, and that he was really portraying the "way it was" in those rough and tumble days, and that he acted in accord with other men writing at the time. The fact that he was to celebrate his 58th wedding anniversary certainly must say something about the depth of the relationship between he and Anna.

Besides the breadth of Sternberg's actual collections he will long be known for his contributions to the methods of collecting and preservation of bone. Although there will always be scoffers who might dispute his being the first to use the modern technique of wrapping fossils in bandages (in order to prevent their being damaged in removal and transportation), there is literally no one who can dispute that he was at least one of the first to develop aspects of this technique to a science. I hope to provide additional insight into this area and prove that it was his innovation that led to the modern method of plaster and bandage preparation.

There continue to be those who refuse for whatever reason to give him the credit he deserves, and that belittle his tremendous efforts. Douglas J. Preston in his *Dinosaurs in The Attic* ballyhoos the Roy Chapman Andrews' of this world (and of the American Museum) and yet both scoff at Sternberg as being eccentric while praising his great mummified dinosaur find. While it is true enough that Sternberg is a man "whom almost no one has ever heard of," Preston seems to imply that this is deservedly the case. He also impugns Sternberg further by calling him a "hardworking but unimaginative compiler of data." Let me assure you that Preston's book, which has sections of real merit and a subject of great interest, shows a lack of firm research as far as Charles Sternberg is concerned, for Sternberg was a brilliant man in many ways and this fact would be obvious to one thoroughly acquainted with the subject. While he was hardworking and persistent and dogged in his pursuit of a fossil, he was at the same time humorous, loving, an excellent writer, and had a solid grasp of geology that he had picked up on the trail. I call these traits memorable, and not pedestrian as Preston implies.

Sternberg was also a diverse man in regards to his various occupations. He had an abiding interest in mechanical things and always wanted to use the latest device to assist him in the latest endeavor. He was to personally patent a fuel composition, investigate various uses for clays and silicates, manufacture various types of hand soaps, and was involved in a dutch cleanser-type production among other things. He also farmed and raised stock at various times.

We also find much to interest us in the rest of the Sternberg family, outside the field of paleontology. Charles' brother, George Miller Sternberg, was Surgeon-General of the United States Army, and he will long be remembered for his contributions to the early medical history of the United States. He has been aptly called the "American Pasteur," and was an acclaimed pioneer in preventing the spread of communicable diseases in this country. It was he who commissioned Dr. Walter Reed to pursue the cause of yellow fever, and everyone knows of Reed's success.

There were also other Sternberg family members of note such as brother William who became a very prominent member of society in Tacoma, Washington, and Frank, another brother who made his mark in San Francisco. The Reynolds branch and the Miller family were also people of substance, and their stories will be woven in as well.

Yet it still remains Charles Sternberg's character traits that we find so inspiring. His vocal love for nature and the creations of God are very evident, and find voice in the beauty he is able to animate in his writings. He speaks, for instance, of being taught by Edward Drinker Cope not to harm any life placed on this earth by the Creator, and explains how after some thought on the matter he made this his lifelong credo also. His visionary scenes of past life lend credence to the stark beauty that he is able to instill in the dry bones and the parched terrain where he earned his living, and we cannot help but be moved by the simple joy he takes in his surroundings. Although some might question his technical poetic abilities, few would not find sheer beauty in the pictures he draws of the life that passed before us. Is this after all not the essence of a good writer?

One of the most interesting aspects of this man is that he was able to do what many other Christians could not manage to do in the past, and cannot do today: reconcile or balance the Biblical magnificence of God with the vision of the past as found in their interpretations of bones and earth. Such men of talent as LeConte, Agassiz, Cuvier, King, Dawson, Silliman and Gray all grappled with this problem, yet Sternberg did not seem to be as troubled in his acceptance of aspects of science and religion as did the others and he managed to live both fully and faithfully. This ease of life, walking in both seemingly disparate worlds, is worthy of further attention.

Charles Sternberg's bravery and fortitude were tested rather formidably on numerous occasions, but perhaps the most exemplary aspect of his field-life was his loyalty under duress. He was always very loyal to the person paying his expenses, even when it was at personal cost in either money or time. His loyalty knew the right bounds too, for he never would stoop to the dirty tricks so prevalent in the Marsh parties in destroying fossil material so as to prevent it from falling into "enemy hands." Long hours and hard work under the most adverse conditions were frequently mentioned in his autobiographies, and by those who wrote of him.

Charles Sternberg lived to be 93 years old, and he was still in the rugged back-area fossil fields when in his late 70's. He was the original "iron man." There has never been a field collector like him in terms of longevity and character and love for the trade, and we most certainly want to recognize him as a man to learn from, and a man worth knowing. This hard-won seniority made him a well-respected man of science, even though he lacked the credentials of the others.

His sons bear the same impressive marks of excellence as he, had the same loyalty to trade and tradition, loved the land as much, and maintained the same independence as their father, even when faced with financial ruin. Paleontology was their life, and they led that life to its fullest. Each of them will be long remembered for their individual contributions both in the United States and in Canada. They indeed were dinosaur-hunters in the truest sense, and much of our knowledge today of those magnificent beasts of old is attributable to the work that Sternbergs performed in remote areas of our country. We owe them a great debt. This book is an attempt to pay back some of the lack of deserved recognition that is long overdue.

One of the real difficulties I've had with this book is in trying to ascertain what kind of book to write. Initially it was mainly the paleontological aspects that were of interest to me, and this was followed by the equally interesting character trait study of Sternberg himself. Beyond this, however, I found myself becoming even more interested in the times in which these men lived, as well as the growth of the place that was to become known as the state of Kansas. When one begins to see the puzzle of history begin to take shape in relation to different segments of a life, one cannot fail but to be interested even further; often to the point of distraction. An example in my case would be the interest spurred in steamboats and their history in westward development as a result of Sternberg's casual mentioning of Cope's trip on the steamer *Josephine*. My first desire was to pack in a bunch of interesting facts that I'd learned (but they proved to be irrelevant to the story), so I had to force myself to cut back and trim. How much more fun (but how much less readable) it would be to write the history of Kansas by taking each subject and dwelling upon the intricacies of its affect upon Kansas and the world today. How sad it is that life is so short, and money in so short supply, that one cannot on a whim take upon such a voluminous task.

My final decision therefore was to write a book about the Sternbergs, all of them, with a decided emphasis on their geologic contributions, but yet with a heavy-handed slant to the history that was unfolding all around them. I simply felt that the times were too decisive, and of too much interest to go unheralded, and lent too much to the lives of these people to react in a half-hearted attempt. The family was also too noteworthy to go unnoticed and hence I have elaborated here also.

I hope, dear reader (as Charles would say), that you will forgive my meanderings if you find them distracting or unrelated, as I have written this chronicle in the hopes that it will reawaken your sense of the noble man in America, and will provide you with a desire to learn more about this family and the times from which they grew. If I even begin to approach a fraction of that end I will be happy, and the eight years I spent writing this will not have been in vain.

CHAPTER ONE
THE BEGINNING

Charles Hazelius Sternberg was born June 15, 1850 in the small town of Middleburgh, New York. He was one of twin boys born to the Reverend Doctor Levi Sternberg and his wife, Margaret Levering Miller. The twins, Charles and Edward, were mid-siblings in a large family comprised mainly of boys. The family was to prove tenacious and the children were to uphold the family name through their thrift and common sense. A number of the Sternberg children were to become "people of substance" from a worldly perspective, and most notably Charles.

Information on the family is sketchy to almost non-existent in some cases. Some of the children cannot be traced much beyond their name, and the difficulties of tracing is heightened by the fact that the family was to spread out early over the entire breadth of the nation. Surprisingly, in the course of two biographies on Charles' oldest and most famous brother, George Miller Sternberg, and in the two autobiographical works Charles produced himself, there is little mention of the rest of the family members. Of the few that are named there is little fleshing out of anything beyond the merest scraps of data as to occupation or locale. More frequently the terms "my brother" or "sister" is used in place of the family member's name. Although Charles certainly deserves better, his sister-in-law does not even mention him by name in the work she wrote about Charles' more famous brother, George, even though the brothers were quite close.

Charles himself mentions only his brothers George, Edward and Theodore, leaves his "Los Angeles sister" (Rosina) nameless, and mentions none of the other nuclear family at all in either of his lengthy works. Although one might construe this as a sign the family was not close, such an assumption would be mistaken since they were a close-knit family, even when they were separated by countless miles. There is also no evidence of any animosity or strife between the various family members, and they seemed to have a healthy respect for one another, regardless of their individual depth or lack of accomplishments.

This failure to mention names seems to be similar to that whereby John Bell Hatcher, who in his summary of his Patagonian fossil trip, recounts that he took along with him his brother-in-law O.A. Peterson, but could not stretch himself to do more than refer to his in-law as "Mr. Peterson." He didn't even refer to the relationship between them at all. This reticence is amusing to us now, but was apparently natural to that day.

During Charles' formative years his father Levi was principal at Hartwick College (1850-1865) in Otsego County, New York. This small Lutheran Seminary was located about four or five miles below Cooperstown on the road to Milford in the Valley of the Susquehanna. As originally built, the seminary (or Academy as it was often called), consisted of a two-storied brick structure that housed eight classrooms, a kitchen and dining facilities. The rooms themselves were relatively small as the entire building was only 48 feet long and 35 feet wide. The edifice was topped off with a cupola that held the official school bell. Initially there was a 13-room home built for the professor, and then in 1839 an adjacent church was built to serve as a chapel and community church. This became known as the Evangelical Church of Hartwick. In 1841 two wings were added onto the original building, and the building stayed in this manner through Levi's tenure as principal before it was changed further.

Hartwick is the oldest Lutheran institution of higher learning in the United States, and played a substantial role in setting standards of education for this country. It was, for instance, one of the first institutions to allow women as students (1851) and as staff members (1852). Additionally, the seminary was to play a major role in early Lutheran history in the United States, and much of that church's far-flung staff would pass through its doors. One cannot also overlook the major role this fine seminary played directly upon the lives of the Sternberg family, since the children learned and taught here as did their father and their grandfather.

Hartwick, both home and school to the Sternberg children as they grew up, was founded in 1761 by the venerable but odd John Christopher Hartwick. It had been Hartwick's endeavor to found a religious community that would be guided by its staunch emphasis on education. As one of the first white men into this area he had taken out a patent upon the land around the lake with the hopes of founding this community. Hartwick, the town, was begun in this way, and its early direction was entirely at the hands of Hartwick himself. It remained so until 1791 when he finally conceded what a monumental task this was and turned the administrative chores over to William Cooper, for whom the town of Cooperstown was to get its name. Cooper was given authority, as agent, to dispose of, at his discretion, the major portion of Hartwick's holdings, with the exception of about 3,000 acres which were held back under Hartwick's special instructions.

Cooper, the father of James Fenimore Cooper, and Hartwick were unfortunately not to get along, and after a falling out that ended in lengthy litigation Cooper was to divest himself from Hartwick, although a good portion of the property was to come into his possession.[1] Cooper's famous son, often called the first truly international American author, was to write exciting tales about the area and its people. Reading a number of Cooper's books gives one a good feel for the pristine nature of this area when it was first settled.

Christopher Hartwick was to die in 1796, although his legacy of the seminary was not actually to come to fruition until 1815. This was largely due to Hartwick's having named Jesus Christ as his heir in his will. There was much legal maneuvering necessary in order to honor the spirit of the will, since they were not able to physically contact the heir. It was at this prestigious institution that the boys' father Levi was to teach and to administer.

The general area in upstate New York around Middleburgh (including Cobbleskill, Oneonta, Canajoharie, Schoharie and Cooperstown) is all fairly similar in terrain. One is immediately struck with the greenness of things. Although the summer weather can be quite hot and muggy, the ground stays sufficiently moist from residual snow water so as not to turn brown. As such it provides both a beautiful and pleasant backdrop for the already magnificent countryside.

The greenness is accentuated by the open patches of meadow on the hillsides and in the valleys, and by the large accumulation of trees. Trees dominate the countryside and line the roads. Maple, birch, oak, weeping willow, pine and mountain ash are apparently the most prolific varieties, but there are others as well. Trees are closely packed in a manner reminiscent of tree growth in northern Germany. This may account in part for the special bond the German emigrants felt with Schoharie. The expanse of green so evident in the countryside translates to the villages as well, as there is often ample green lawn between dwellings.

Abundant fauna continues to be a major component of the countryside even today throughout the whole of the area, and side roads especially are littered each morning with the carcasses of small animals. Raccoons, possums, and what appeared to be muskrats were particularly evident on a recent trip to the area, and deer can be seen roaming the area at dusk, and often fall prey to road traffic.

Rooftops are all high-sloped in order to accommodate the heavy snows (which can exceed 50 or more inches in a season), and the house designs retain that Germanic look about them. There is a neatness about the villages that adds to their attractiveness. Although many of the old barns show major damage, they are often still quite serviceable and are painted a spritely red. The Germanic love of bright colors is embedded in the community.

If one ignores the telephone lines, the inevitable video stores, the modern convenience stores and gas stations, one can easily imagine they have been transported back 100 or more years in time. This would have been the same Middleburgh as Charles Sternberg walked and played in as a child.

Although the area thrives mainly on its dairies, agriculture remains important as well. Corn, both for food and fodder, appears to be a mainstay although other crops are noticeable as well. It is more than likely that the Sternberg family grew crops right on their own property in order to supplement their income and provide added food for the family.

Geologically one can see that there is much that would have appealed to young Sternberg. Particularly as one heads south from Albany there are a profusion of rock exposures that entice the fossil-hunter's eye. Near Cobbleskill and Schoharie are a number of limestone caves. Howe Caverns, the most prominent of the caves, was undoubtedly well-known to Sternberg. Since it was discovered in 1842, Charles may have even had the opportunity to explore its depths before the modern elevators were added to facilitate traffic.

The present site of the Petrified Creatures Museum, north of Cooperstown, was most likely another area well-known to the Sternberg family. The site today provides a good outlet for children who like the hands-on experience of digging for fossils. This fossil-rich site is perhaps one of those that gave Charles his early taste for the craft.

Charles' recollections in his autobiography, *The Life of a Fossil Hunter*, places great emphasis on his love for the outdoors, and most particularly the countryside of New York. Trees, lakes, flowers and the other marvels of nature dominate his writings, and they help us to understand why Charles finally chose the lifestyle he settled upon. It also points to one of the clearest reasons why Sternberg was often referred to as a naturalist, since his writings transcend the mere geologic and often bask in a real appreciation of the flora and fauna he encountered.

His lack of formal training in geology (which seems to eliminate him from being labeled as a geologist or paleontologist by modern standards) may account for the label of naturalist also, but the most obvious reason still seems to be the very evident love for wildlife and countryside that permeates his writings. One suspects that this love for the outdoors oddly comes from his mother more than from his father. For, although his father was to become a large rancher and livestock owner, in addition to his call as a preacher, he never seemed to display the same true love for the land as did his son, and he could never find much in his son's choice of occupations to please him.

Margaret Sternberg, however, was to always prove more ethereal in her approach to such things, and undoubtedly found her son's choice a pleasant one. She could see beyond the mere practical aspects of earning a living, while Levi could not. There is certainly irony here if one considers that Levi himself chose a very impractical occupation for himself (at least by the standards of many) that generally never translated into goods wages or an easy life. He would have measured his success in terms of saved souls in much the same way that Charles would have measured his in found fossils.

Charles' mother Margaret was a refined woman of culture, who was not accustomed to the rigors of western life. She was a strong woman in the mythic sense of the western woman, evidenced by the fact that she learned to cope with the harshness of life in Kansas and even managed to bring some finesse to the areas where they were to live. The daughter of a clergyman, she had been versed in all of the arts. She was exceptionally talented with languages, having varying degrees of skill in Latin, German, Spanish and French. She was also an accomplished musician, and played the church organ at Levi's churches in her later years. In many ways she was to prove similar to Martha Pattison Sternberg, second wife of Margaret's son George Miller Sternberg, since they were both brought up in clerical families, and both had mastery over the arts and languages. It is not hard to see why son George would have picked such a wife, since she was so similar to his mother.

The gentler side of Charles was apparently derived from his mother. It was she who schooled him in the art forms of poetry and prose that would serve him so well in later years. Her schooling was naturally enhanced by the teaching at Hartwick since poetry was an integral part of the curriculum there and Charles undoubtedly was a member of the Philophronean Society, just as were his brother George and their father before them. The naturalist's love of animated beauty that is so evident in Charles' writings must have been inculcated and fostered by his mother at a very early age, for it evidenced itself early in his life. Charles also was to gain from his mother her quietness of spirit.

Levi, on the other hand, should not be ignored for he too had much to offer to a growing son. He would have instilled the more practical values in his sons, and would have administered strongly the word of God upon his offspring. His strength of character, that made him a commanding presence among other men and caused him to be a well-respected man in many areas, was passed onto his children as well.

In Charles' various writings there is also a benevolence and an openness that is perhaps not so easily understood today, but of such merit that one can only hope man might come back to that state in the future. This gentleness is one of the most commendable of Sternberg's many character traits, and it provides for us real insight into the heart of this man:

> How I love flowers! I carried to my mother the first crocus bloom that showed its head above the melting snow, the trailing arbutus, and the tender foliage of the wintergreen. Later in the season I gathered for her the yellow cowslip and the fragrant water-lily; and when autumn frosts had tinged the leaves with crimson and gold I filled her arms with a glorious wealth of color.[2]

This joy of expression is what we see and feel in looking back at old-time movies, with their clearly espoused innocence. This sentimentality seems so alien to us today because our modern books and movies mock all that smacks of conservatism, and liberality unfortunately reigns.

There is a freedom of expression in this poetical passage that transcends the bounds of modern man's usual strictures of emotion, and its sensitivity inspires. It also helps us to begin to flesh in the shadows of the man so that we can begin to see who this Charles Sternberg really was. Although it might appear difficult to reconcile this kind of gentle nature with the inevitable picture of the grizzled and hardened bone-hunter we are led to expect, we will find that neither can be separated from the other. It is this complexity that makes him of such interest to us today. To call a man both simple and complex in the same breath seems a dichotomy, but in reality he was both. It is evident that his parents played a major role in the formation of this simple/complex concept as we shall see later.

Charles' having brought flowers to his mother was both practical and thoughtful, since his mother "with advancing years and fewer cares...began a study of the wild flowers, which she painted separately and in bouquets, to impart to her Eastern friends an idea of the beauty and interest of the prairies." [3] Not only did she instill this interest in friends and neighbors, but in her sons as well, for they all seemed to have developed a love for blooms. This may account in part for Charles' success as a collector of fossil leaf impressions, since this would have given him a good background in botany.

Charles' father Levi, a Lutheran clergyman, spent 14 years (1850-1865) at Hartwick Seminary as Principal, after serving at various congregations in Danville, Buffalo and Middleburgh, New York. He had been a student at Hartwick for five years (1828-1832) before completing his education at Union College. He then returned to Hartwick as an assistant teacher from 1836 to 1839.

A picture of Levi taken in his mid-to-late 40's shows him to have been a very stern man in appearance. He much resembled Abraham Lincoln because of his gaunt appearance and similarity of beard. His eyes, however, were his most prominent feature, and are perhaps best described as "piercing." Although a stern disciplinarian, he was at the same time a very energetic man known for the power of his words.

It was at Hartwick that the boys received their primary education. George, the oldest and best known of the sons, not only received his early schooling there, but also was allowed to use his talents as teacher there (1854-1857). In his last years at the institution he taught mathematics, chemistry and natural philosophy. While the intellectual climate and teaching credentials were vastly different from that of our own time, there is every reason to believe that all of the essentials of an excellent education were met, and that many higher schools of today would be hard pressed to provide a better foundation.

A listing of course titles at Hartwick shows just how extensive their education was. Besides theological classes one could choose from bookkeeping, surveying, chemistry, solid geometry, astronomy, physiology, botany, geology, agriculture, hydrostatics, magnetism, electricity, mechanics, optics, civil engineering, Latin, Greek, Hebrew, German and French.

The years at Hartwick were good years for the family, but they also proved to be lean years. Then, as now, even the most advanced and prestigious positions of service to the church paid but paltry salaries, and those with large families were particularly hard hit. George's choice to leave the family at an early age in order to participate in the family bread-winning was in part caused by this plight.

At the same time, however, both George and Charles were to describe these early years in glowing terms, and there is no question as to the value they placed upon their education and time of life while at this institution. As with so many happy families, where deprivation is felt and hardships encountered, the true response of the family to the adversity can actually prove uplifting if the family operates as a team. The Sternberg family was just such a unit, and because of it they thrived. Both George Miller Sternberg and Charles were to return to visit Hartwick in later years, and George was to even include the school in his last will and testament with a substantial bequeathment.

The former site of Hartwick Seminary, tucked away within the languishing hills just south of Cooperstown is in a particularly beautiful area. The archaic countryside retains all the best of its former days, and because of its remoteness, it provides a very attractive setting, unspoiled even today. It is not hard to understand why Christopher Hartwick would have been enchanted with the land.

Hartwick College today has been relocated to the nearby town of Oneonta. The college remains an attractive campus, and still provides a fine education for today's student. The red brick buildings of the campus form a pleasant contrast to the green of the hills at Oneonta. Nestled on the low sections of the hills, Hartwick is laid out below the campus of S.U.N.Y. which stands on the next level up.

The already strong religious upbringing fostered by schooling at Hartwick was further bolstered by the fact that Charles' grandfather on his mother's side, Dr. George B. Miller, was professor of Theology at Hartwick Seminary. Miller was on staff at Hartwick during most of the period between 1827 and 1869. He served as principal of the institution between 1830 and 1839, and then again between 1848 and 1850. Miller used to spend considerable amounts of time with his grandchildren.

Levi's oldest son George, was named after his father-in-law, and in later years when Charles had his own family he was to name his oldest son after his grandfather as well. These are certainly marks of the admiration the family showed for this man and his teachings. George Miller played an even greater part in the life of his grandson George, since George was to actually live with the Millers for an extended period. Dr. Miller was principal of Hartwick himself just prior to his son-in-law, and actually gave up the reins at his own insistence. He also was always there to provide whatever help he could to both the family and to Levi during this time-frame.

Levi's having chosen to name his son Charles Hazelius Sternberg was not without special meaning. The Rev. Dr. Ernst L. Hazelius was the first administrator of Hartwick Seminary, taking that position in 1815. Hazelius, a strong and able administrator, did much to further Hartwick and the Lutheran church. It was for he that Charles was to get his middle name. Hazelius was also a relative of Charles' on his mother's side. Hazelius was an uncle by marriage to George B. Miller's wife, Delia B. Snyder Miller.

It is perhaps a measure of the importance of the faiths of Levi Sternberg and his father-in-law George B. Miller that they had their differences. Levi, when nearing the end of his principalship at Hartwick, became disenchanted with the course of things at the seminary, and began to espouse what was then known as the "New Measure Movement." This put him at odds with Miller since Levi felt there was error infiltrating the theological department, and Miller headed that department. After various confrontations Levi Sternberg decided it was best that he resign, and he did so in 1865. It is also a measure of the man that he was asked to again serve as principal in 1870 (after Miller's death), but he declined because he was then well-settled in Kansas.[4]

Following Levi's tenure at Hartwick College he left to take over the principalship of Iowa Lutheran College in 1865. This small college was located in Marshall County, at the town of Albion in central Iowa. Iowa Lutheran College was a very young college at the time, having only been founded some six years earlier. Oddly enough, the seminary was founded upon public funds (swamp land funds) which the populace had voted to be released from the county for that purpose. Care for the property had been turned over to the Lutheran church, who were to form a coalition with the local Presbyterians. In 1861 it became a Lutheran institution and in 1865 Levi Sternberg took over as both principal of the seminary and as pastor of the connected church. A few years after Levi left for Ft. Harker, the institution was managed by the Methodist church for a period, and then was sold to the public school district and was turned into a high school facility. The various changes in church management may suggest that the institution had its difficulties right along, and this perhaps helps to explain Levi's departure for Kansas and another lifestyle.

This move to Iowa was certainly a good one for young Charles as it gave him his first view of the midwest and its rolling hills, lush grass, and vast plains. There can be no doubt that this expanded vision of the world played a great part in Charles' later avocation, and his desires to move on to different fossil fields.

At the same time, however, Charles was only too acutely aware of what

Iowa lacked as well. In an article of 1903 he was to state "...my eyes were opened to a new region of treeless plains." He found it disappointing because there were "no mountains to explore, or beds of fossiliferous rocks."[5]

In 1867, at the tender age of 17, Charles moved, along with his twin brother Edward, (and at the behest of his older brother George), to a ranch south of Ft. Harker. This ranch was located in Kansas on prime river bottom-land, in Ellsworth County, near present-day Kanopolis. It was at this place that Charles really began to develop his interest in geology and fossils. The ranch itself had been owned by his brother George, who had early realized the potential in the land while serving at Ft. Harker. Charles apparently was afforded a position of some authority on the ranch, in light of his young years, for he describes in his autobiography that he was "in charge" of the ranch in the year 1872. The ranch was eventually given to Levi Sternberg by George, since Levi had expressed an evident love for the land, and for farming, when he made a visit to the ranch on his son's invitation. It is a measure of the devotion that George displayed to his father in this regard, for he turned the ownership over to his father with some evident trepidation. Martha Pattison, wife and biographer for George M. Sternberg was to comment that George had real fears in allowing his refined and cultured mother the necessity of settlement in the "wilds" of Kansas. It is also a measure of his mother's strength that she was to actually excel in the move, and bring her culture with her rather than psychologically leave it behind. And Kansas was to prove the better for it.

Charles was to be a man in the right place at the right time in a number of different ways. Since he was born in 1850 he was too young at the outset of the Civil War to serve (being only 11 years of age) and hence he did not have to leave the family farm at that point. He also found himself in the right situation to get maximum value as a collector. If he'd been born earlier he would have predated the "dinosaur rush," and if later, his lack of credentials would probably have forced him into a different occupation. In addition to all this he came to live in fossil country at the right time as well.

Important to consider in this move to a new state is that Kansas at the time was much more feral than was Iowa, and the Kanopolis area was on the more primitive edge of what was considered to be populated Kansas. This move to a more pristine area, this encounter with primal nature, while being fraught with a kind of immediate danger from a number of possible perils, also provided the thing Charles most needed: unadulterated beauty. This land had not yet been ravaged by the invading white man and his infernal machines, wire had not yet been strung up to fence off the land from one's neighbors, and the railroads had not yet caused their swath of destruction across the land. This stark and beautiful place called Kansas lay as yet unchallenged before Charles, and to him it was more than alluring.

Willa Cather in her novel, *My Antonia*, describes the kind of country that the brothers met with in their move to Kansas:

> We were talking about what it is like to spend one's childhood in little towns like these, buried in wheat and corn, under stimulating extremes of climate; burning summers when the world lies green and billowy beneath a brilliant sky, when one is fairly stifled in vegetation in the colour and smell of strong weeds and heavy-harvests; blustery winters with little snow, when the whole country is stripped bare and grey as sheet-iron.[6]

Kansas itself was to receive its name from one of the more populous Indian tribes in the territory, the Kansas Indians. The tribal name translates as "people of the south wind." The name, now changed substantially, was originally to be spelled or pronounced as "Kausau."[7]

A glance at a map of Kansas shows additionally why this move was such a beneficial one for the young, fledgling fossil hunter. Kanopolis is almost dead center in the state of Kansas, and is on a straight line with the "bad-landed" Logan and Gove Counties, and the prominent towns of Manhattan and Lawrence. These places were all to play significant roles in Charles' fossil education, and no better places could be designed by a Creator were He to set a man down with intentions of making him into a fossil hunter. Charles himself had no doubt that

God had ordained him for work in fossils. And one cannot ignore the fact that Kansas is to fossils what oranges are to Florida, or potatoes are to Idaho. So it was to be the Kansas chalk that made Charles Sternberg, and Charles Sternberg that made the Kansas chalk.

Perhaps this is a good place to interject with a description of "chalk," since it was to play such a major role in Sternberg's life and that of his son George. Chalk is a white mineral soft enough to be sliced through with a knife. When acid or strong vinegar is poured over its surface it effervesces in small, rapid bubbling as does limestone and marl. This frothing is somewhat analogous to the fizzing of a carbonated drink. Chalk is a compound of carbon dioxide and calcium. Under microscopic examination it shows itself to be formed of minute shells or other fragments of ancient sea life. It obviously then is part of the strata that is laid down on an ocean bottom, and in many spots in the world can be found to cover many miles in area. It is also particularly great fossil- bearing material since fossils are best found where there are factors that contribute well to an immediate sort of preservation of animal life, and what better place to be found but in an area enveloped in watery and sandy soil?

The Kansas chalk equates with what is commonly called the Cretaceous period of geologic life. The word "Cretaceous" comes from the Latin word meaning chalk, although the Cretaceous stratas of the world are often not composed of any chalk at all. The Cretaceous, although it is often pictured as having its culmination in the White Cliffs of Dover, finds its real prominence in Kansas. The Cretaceous, or chalk, in Kansas stretches almost from one end of the state to the other. The major rivers in Kansas, the Smoky Hill, the Arkansas and the Cimmaron all criss-cross the chalk beds, and cause the eroding out of the fossils that were to make Charles Sternberg famous.

Kansas in the 1870's was wild and savage by today's standards, and it was then in its infancy in terms of its population. The eastern part of the state was only very sparsely populated, and the western portion had little beyond the occasional Indian band to traverse its many wonders. Ellsworth (near Kanopolis) was on the western edge of the sparsely populated "east" part of Kansas, although it is really in the center of the Kansas of today. As such, Ellsworth was particularly primitive, and provided a primeval natural countryside that made it an ideal place for an adventurous man to make a living. Since Sternberg was from the outset one of the major collectors in Kansas he had the benefit of traveling over virgin territory that had not yet been stripped of its fossil wealth. In fact, there were few men at that point who even knew that Kansas had the potential for fossils at all. But they would know in short order, for Kansas was to become a major pawn on the great dinosaur battlefield.

The western "frontier-line" in 1860 centered on the eastern edges of Nebraska and Kansas and split Texas in two. West of this line were the western wilds composed of Indians and spacious open and unfenced lands, with almost no white settlers at all. Moving northward to Canada the "wilds" were even more pronounced in that they were more barren and less settled. The unsettled nature of the area was a boon to the adventurous like Sternberg, but it was far too much a wasteland for those who could not do without the social graces of the east.

The Kansan towns of the day were notably rootless, as might befit a savage land. Sternberg's description of the typical frontier town was that it had "streets paved with playing-cards, and whiskey for sale in open saloons and groceries."[8] Most certainly the average Kansas citizen would not have portrayed their towns in quite the same graphic manner, but imports from the east would certainly have agreed that there was a major difference between the comforts of the east and the Kansas frontier. Sternberg's aversion to drinking and gambling made his comments somewhat predictable, but in his defense one has to admit that the debauchery and untamed atmosphere of Ellsworth and some of the other wild towns in Kansas is a matter of historical record that cannot be denied. Ellsworth was perhaps the worst of these, and might be characterized as the archetypal western town during this time period.

Between 1860 and 1890, however, there was a major explosion in the population of the plains states that did much to civilize these Kansas towns.

Kansas was to grow from about 100,000 people to about 1,400,000 in 1890. Surrounding plains states were also able to show phenomenal growth in the period, but Kansas growth was the most striking. This mushrooming of the population was the result of a number of factors, of which the largest was certainly the added accessibility to the area that the railroads were to provide.

Another major factor relating to this phenomenal growth has been attributed to the fact that the military men who had been sent to the west after the Civil War to control Indian trouble, found Kansas much to their liking. Being adventurous by nature they enjoyed the wide open spaces where fences and people were few and far between, and therefore decided to stay on after their term of service was completed. In the five years after the Civil War's close over 75,000 ex-soldiers were to settle in Kansas. Since these men were the real heartwood of the nation, this made Kansas a place of real opportunity, and its leadership was the strongest around.[9]

Kansas had only been granted statehood a few years before Sternberg moved to Kansas (1861), and technology was only then beginning to enter the state. The first train had not crossed the borders into Kansas until 1860, and it would be another six years before the tracks would reach to Manhattan and Ellsworth, the area around which Charles and his family were to live during his formative years. When the trains came the technology seemed to grow by leaps and bounds.

Kansas is and was literally the center of it all. Not only is it the geographic center of the conterminous United States, but it is also the acknowledged center of the North American continent, as far as land surveying is concerned. Kansas is also the embodiment of the "bread basket," and it is hard to imagine that there could ever have been a time when Kansas did not supply the major wheat crop for the entire United States, and much of the world. At this time, however, Kansas was just beginning to discover its own riches, and farming was just taking root.

Kansas is also the archetype of the plains state. It shows little noticeable variation in relief from one end of the state to the other. It contains 82,200 square miles of sameness. Its rolling plains of a burnt russet color roll on for endless miles, uninterrupted by any other natural visual annoyance. It is partially bordered on the northeast by the Missouri River and on the southeast by the Ozark plateaus. Very few trees are to be found in the whole of the state, and not much water. What it does have, however, is a stark beauty that is beyond description, and fossils; a plentitude of fossils that has defied many an imagination.

Frances Parkman Jr. was to describe the plains area rather eloquently when he wrote:

> For league after league, a plain as level as a lake was outspread beneath us; here and there the Platte, divided into a dozen thread-like sluices, was traversing it, and an occasional clump of wood, rising in the midst like a shadowy island, relieved the monotony of the waste. No living thing was moving throughout the vast landscape, except the lizards that darted over the sand and through the dank grass and prickly pear at our feet.[10]

Yet in the plains with their sometimes dreary and monotonous sameness there can be found a peculiar kind of vivid beauty. But there are as many who can't see the beauty (or won't see it) as there are who enjoy it. I think that much has to be attributed to where one grows up, for there does seem to be a connection between beauty and familiarity. There is no question, however, that all of the Sternbergs seemed to thrive in their new home state, and they loved the land.

Charles' first encounter with Kansas fossils was with fossil leaf impressions that he managed to garner from the hills near his home. He described these Dakota group fossils of the Cretaceous period as coming from rocks consisting of "red, white and brown sandstone, with interlaid beds of variously-colored clays."[11] These initial forays into the Kansas countryside netted him some fine examples of various sassafras leaves from the area that he was to label "Sassafras Hollow," and these whetted his appetite for what he now knew he could find and collect without a great deal of effort. Of the many sassafras varieties he was to discover, one of the finest was the three-lobed *Sassafras cretaceum* newberry.

Much has been made about the "mystical" dream-sequence experience he describes in his autobiography, and a number of commentators seem to want to play up the experience well beyond what Sternberg intended. Douglas J. Preston for instance, has extrapolated and intimates Sternberg often found fossils by dreams. Preston says "he [Sternberg] once wrote that he discovered these localities through dreams."[12] One expects that were Charles living today he might well have chosen to simply drop the incident altogether from his autobiography as it seems to place him in the negative light of the fanatic, which he certainly was not.

The dream that he describes, and that has caused the controversy, takes place with him walking up a ravine in the Smoky Hill bluffs southeast of Ft. Harker where he was to find:

> On either slope were many chunks of rock, which the frost had loosened from the ledges above. The spaces left vacant in these rocks by the decayed leaves had accumulated moisture, and this moisture, when it froze, had enough expansive power to split the rock apart and display the impressions of the leaves.
>
> Other masses of rock had broken in such a way that the spaces once filled by the midribs and stems of the leaves admitted grass roots; and their rootlets, seeking the tiny channels left by the ribs and veins of the leaves, had, with the power of growing plants, opened the doors of these prisoners, shut up in the heart of the rock for millions of years.
>
> I went to the place and found everything just as it had been in my dream.[13]

Sternberg himself explains the dream as being most likely the result of his having seen the specimens before subconsciously in some other context, "while chasing an antelope or a stray cow," and felt that he saw but did not really recognize (except with his subconscious) what he had seen until it came time for him to dream. Under no circumstances could one construe him intimating at anything supernatural here, although many have seemed to take that tack.

Although Sternberg was clearly interested in dreams, and perhaps had an affinity for remembering his, he did not assign any mystical significance to them.

Charles forwarded his numerous collected leaf samples to Professor Spencer F. Baird of the Smithsonian Institution in 1870. The specimens were eventually forwarded from Baird to Dr. John Strong Newberry at Columbia University who would later include them in his volume, *Later Flora of North America,* which was not to be published until after Newberry's death. These early finds still rank as among the finest of their kind ever discovered, enough so to place Sternberg within the pages of most textbooks in Paleobotany.

Sternberg was to describe three different major collecting sites from which he was able to collect leaf impressions. Sassafras Hollow, his most prolific site, was said to be about a mile southeast of the Thompson Creek schoolhouse near the property owned by the Hudson brothers. Lesquereux himself was to collect there in 1872 and was to name one of the large leafed specimens after Charles, *Protophyllum sternbergii.*

A second locality, the "Betulites Locality," was named for its abundance of birch specimens. This was not a site discovered by Sternberg, but by the famous Judge E.P. West. This site was a much larger collecting area since he claimed it was about a mile in length and topped the highest hills in the county.

The third locality was the "dream sequence location" just described. From this location he notes finding two very large-leafed examples of Kansas flora, *Aspidophyllum trilobatum* and a three-lobed example of sassafras that was over a foot in diameter, *Sassafras dissectum.* This he felt was a location known only to him, and that others had never collected there before.

In 1872 Sternberg had the good fortune to be able to meet with and talk privately to Dr. Leo Lesquereux, who was at that time one of the most prominent paleobotanists in the United States. Lesquereux had been imported from Europe under the direct influence of Louis Agassiz. Agassiz was the founder of the Museum of Comparative Zoology at Harvard in Cambridge, and he had also brought over such prominent men as Girard and Guyot who had been fellow students with Lesquereux.

Lesquereux had received through a friend of Sternbergs a number of duplicate specimens, and was apparently impressed with Charles and with his enthusiasm. Dr. Lesquereux, because of his fame as a botanist, was visiting Ft. Harker as the personal guest of the fort commander. Charles must have learned late of the visit for he records that he made a mad dash there by horse at the last minute. Although the meeting itself occurred during a party given for Lesquereux, and was therefore rushed, Charles described the limited time he spent with the great botanist as "golden moments."[14] It seems obvious that this brief encounter helped to cement Sternberg's future perhaps more so than any single event, with the possible exception of his first communication with Cope.

According to Sternberg, Dr. Lesquereux was simply delighted with the rough sketches that Charles had produced a few years before and had brought with him, and subsequently Lesquereux was able to recognize unnamed material from the crude drawings. This brief visit was to be the cause of Lesquereux later describing much of Charles' material, and provided additionally an avid correspondent for Sternberg to write to for technical advice.

To understand just how prolific Sternberg's collection of fossil plant impressions was, we have only to note that he indicates in his autobiography that in the year 1888 alone he sent over 3,000 leaf impressions from the Dakota formation of Kansas to Lesquereux. Many of these especially fine samples of early Kansan flora can be seen even today in the National Museum in Washington, DC. Other specimens are to be found in the major museums of the world, but the primary collection still remains at the Smithsonian. The sheer volume of material collected by Sternberg in such a short time attests both to Sternberg's skill as a collector and to the bountiful nature of fossil remains to be found in Kansas.

Sternberg includes with great pride in his book *The Life of a Fossil Hunter* a letter from Lesquereux, dated Apr. 14, 1875, that was written from Columbus, Ohio and that Charles was to label his "lodestone." It is clear that they had discussed the possibility of Charles studying medicine, for he says:

> I much approve of your purpose of studying medicine. Your taste for natural history will help you much and encourage you. But allow me still to say to you as a friend would do that you cannot expect to become useful to others and to yourself in science except by hardwork, pursued with patience and a fixed purpose.
>
> Science is a high mountain. To go up to its top or at least high enough to gain free atmosphere and wide horizon necessitates hard climbing, through brushes thickets rocks etc. (sic) Those who from the beginning look around for commotion, and soft paths merely enter the gloom of the wood at the base. They are seen from nobody and see nothing but indistinct forms and because their horizon is thus limited to darkness they think there is nothing else and nothing more to see from high above toward the top of the mountain.
>
> Moreover there is not a true hard step in science or in life which does not give its reward in one way or another. While we have not a single moment of laziness of unmerited, comfortable rest, which does not bring us some kind of disappointment and has not to be paid by a little more trouble and work.[15]

This is good and useful advice, and it remains as relevant today as it was then. At this time in his life Charles must have considered Lesquereux a special friend and mentor, although Cope would come along shortly and he was to usurp all former allegiances and become the true and lasting mentor. Additionally one might note that the clear merit of such a letter falls squarely upon the fact that it has the end result of supercharging an individual with the grandeur of the lifestyle that is dedicated to science. This was true fodder to Charles, as he felt this way himself long before Lesquereux said so.

One of the other special things that Lesquereux and Sternberg shared was their hearing impairment. Charles had one bad ear in which he was totally deaf and Lesquereux was for all intensive purposes deaf and had to read lips as a result of a childhood accident. This similarity of disability perhaps helped to fortify their special relationship, and perhaps explains why men with a 40 year age difference would get along so well.

If any could doubt the true scientific fervor with which Charles Sternberg approached the subject of natural history, and his feelings as to his calling to the profession, they need only read a few lines from any of his writings. He most certainly felt himself much more than a mere collector, and felt that his contributions were of lasting value. He was to state, for instance:

> I have never kept a single specimen [fossil leaves] for myself, although I love them dearly, and it has often been hard to give them up. But the object of my life has been to advance human knowledge, and that could not be accomplished if I kept my best specimens to gratify myself. They had to go, and they went...[16]

As an amateur collector myself I can certainly attest to the difficulty one has in giving up a hard-won fossil, since the effort one must often put into the finding, extraction and cleaning of a fossil is substantial and can only be measured in a scale of many hours. It is perhaps analogous to the heart-wrenching difficulty with which the artist allows his work of art to be torn from his control. This relinquishing of a treasured article is difficult no matter what the profession, but the complete artist or scientist must learn to rise above that. Sternberg certainly did.

So as we have seen, Charles Sternberg got his training as a fossil hunter and collector all on his own (except for the urgings of Lesquereux and Cope), and he developed those skills through plain hard work. Since this art (and it was an art in his hands) was not practiced by many, and the United States was itself very new to the science, there wouldn't have really been anyone around to teach him too much anyway, at least as it related to the job of hunting and extrication.

Considering this fact, it is really amazing that Sternberg was to accomplish so much on his own, as he had no one to push him or give him the firm direction a student might have had. His limited but prolific forays into the neighboring Kansas chalk to take out his fossil leaves gains much more significance when we consider what motivation he might have had. When the only obvious motivation comes from within the person themself (and this clearly seems to have been the case with Sternberg), and the grandeur, gain or glory for such an effort made is intangible or does not exist, it strikes home how marvelously strong that individual must be to shoulder work under those circumstances. In light of all the negative factors (poor wages, danger from occupation and Indian perils, being away from loved ones, and suffering from the many ills the business could produce) it is amazing that the early bone hunters were as durable as they were. Sternberg himself even had to overcome the admonitions of his strong and persuasive father, and yet he managed to scrape up a plentitude of fossils with a few good tools, perseverance and trust. Certainly there is something fine here, and something that we can learn from. And in the process how can we fail but to be better for it?

NOTES

1. Rev P.A. Strobel, *Biographical Sketches*, Memorial Volume, Centennial Anniversary of the Hartwick Lutheran Synod of the State of New York, Lutheran Pub Soc, Philadelphia, 1881, Pg. 413

2. Charles H. Sternberg, *The Life of a Fossil Hunter*, Henry Holt & Co, NY, 1909, Pg. 2

3. Martha L. Pattison Sternberg, *George Miller Sternberg*, American Medical Association, Chicago, 1920, Pg. 19

4. Hardy Heins, *Throughout all the Years*: Hartwick 1746-1946, Lutheran Pub House, Blair, Neb, Pg. 54

5. Charles H. Sternberg, "The Life of a Fossil Hunter," *The American Inventor*, #10, June 15, 1903, Pg. 1 [Not to be confused with the book of same title]

6. Willa Cather, *My Antonia*, Houghton Mifflin Company, Boston, The Riverside Press, Cambridge, 1954, Pg. 1 of Introduction

7. Chas. P. Beebe, *Kansas Facts*, Topeka, Chas. P. Beebe, Pub, 1929, Pg. 52

8. CHS, *Life*, Pg. 64

9. *Kansas Historical Collections*, Vol XII, 1911-1912, Wm. A. Calderhead, "The Service of the Army in Civil Life After the War," State Printing Ofc, Topeka, Pg. 16

10. Francis Parkman Jr., *The Oregon Trail,* Caxton House, Inc., NY, 1910, Pg. 48

11. CHS, *Life,* Pg. 15

12. Douglas J. Preston, "Sternberg and the Dinosaur Mummy," *Natural History,* Vol 92, Jan 1982, Pg. 89

13. CHS, *Life,* Pg. 19

14. Ibid, Pg. 19

15. Ibid, Pg. 25

16. Ibid, Pg. 30

CHAPTER TWO
BONE VENTURE

Although Sternberg's actual tenure as a bone hunter had not yet begun, and although he had not yet had any practical experience in collecting bone, it was clearly his intention to pursue that course. In the winter of 1875-76 he was a student at the newly established Kansas State Agricultural College in Manhattan, where he was apparently in pursuit of a furthered background in the geological sciences. At the time there were 238 students in the entire college (much fewer in the science section) and although the college had been in existence for 10 years it had graduated only a total of 27 students to that date. This limited winter training period was to prove to be the total extent of Charles' formal geological training. Everything other than that was strictly in the way of "hands on training," or in being schooled through reading or conversing with schooled paleontologists and other men of related sciences.[1]

Kansas State Agricultural College (which had been founded as Blue Mont College and has since become Kansas State) had been founded by Methodists in 1864. This school was to spawn some of the finest early pioneers in Paleontology, and both Sternberg and Samuel W. Williston got their initial educations in science there. The major reason why the college had gained such prominence in the field of the geological sciences was the direct result of the efforts of Benjamin F. Mudge. Mudge began his tenure at the college in December 1865, and shortly after became professor of natural history and higher mathematics. It is a measure of this man's talents that he was to teach a myriad of different subjects over the years; subjects as wide ranging as astronomy, botany, geology, physical geography, meteorology, zoology, entomology, physiology and others.

Samuel Williston, a giant in early Kansas history also, was to develop the same kind of hero worship for his teacher Mudge that Sternberg was to develop for Cope. In a brief history of Mudge's life which he wrote for the *American Geologist*, he was to picture Mudge as a "kindly faced, plain old gentleman, as ready to talk with the uncouth farmer as with the aristocrat." He called Mudge an "enthusiastic and able lecturer...loved and revered by the people of Kansas."[2] He went on to say that it was Mudge's teaching alone that was responsible for fostering within him the love of science, and that this exuberant teaching made all the difference, since the college had no equipment from which to teach. Apparently the sum total of scientific equipment at the college consisted of an "electrical machine, three leyden jars and six test tubes."[3] Mudge was indeed a teacher and motivator in a manner very similar to Edward Drinker Cope and the incomparable Louis Agassiz.

In 1876, Professor Benjamin Mudge, described by Sternberg as the "enthusiastic state geologist" and the "popular professor," secured a party to explore in the western part of Kansas. Mudge had previously been in that general area as early as 1871, and was already a long-standing disciple of O.C. Marsh for whom he was to have loyalty to for a number of years. Mudge had left the college under adverse circumstances in February of 1874, just prior to Charles' attending the college. Although there is no record of which I am aware, it is likely that Sternberg's reason for not continuing at the college may have related in part to the fact that the program of geological sciences (so carefully cultivated by Mudge), fell on immediate and permanent hard times upon his departure. With him gone there would clearly have been no educational benefit from Charles' point of view, for Mudge was undoubtedly the attraction.

The story of Mudge's departure is in itself interesting because of the parts played by its principals, and because the effect on Charles' future life was immediate. Mudge (who at the time was the second most senior professor on campus), along with professors H.J. Detmers and Fred E. Miller were summarily dismissed by the Board of Regents of the college because of their "insubordination and gross misconduct."[4] This extreme action was carried out because the three professors had boldly traveled to Topeka to lobby against the ratification of certain members of the to-be-reconstituted Board of Regents, that had been appointed but not yet ratified by the Senate. Most prominent of the men whom they were lobbying against was Major Nathaniel A. Adams, who was a prominent churchman and politician who had some direct ties to college President John A. Anderson. It was Adams who had pushed for Anderson's nomination.[5] Anderson was ratified by the regents in July 1873 and took office on Sept. 1, 1873. The new Board of Regents, however, had not been confirmed by the legislature since they were not in session at the time. It was not therefore until late January of 1874 that Mudge and his fellows attempted and failed to block the nomination of Adams and company.

Just why Mudge was so anti-Adams is unclear, but the fact that the professors attempted to block the nominee by traveling to Topeka speaks for itself. They certainly must have felt strongly about the issue as they clearly must have recognized the inherent dangers in such a move, and what failure might bring. According to Julius T. Willard, historian for the college, they failed blocking the confirmations by only one vote. If Mudge had stayed there seems little doubt that Charles might have become a credentialed paleontologist, rather than being limited to the less prestigious titles of naturalist or collector.

Of some irony here, considering Williston's close ties to his teacher Mudge, is the fact that Williston took his Bachelor of Science degree from the college in 1872, and it was he who actually took over some of Mudge's classes for a period. By Richard Swann Lull's assessment this was due to Williston "not understanding the situation" and that the "apparent injustice of the matter...made him so strong a partisan of Prof. Mudge that his manifestation thereof proved more than the authorities could stand and he was asked to leave."[6]

But, I have failed to point out one of the interesting twists in all this: Charles Reynolds, Charles' future father-in-law was one of the Board of Regents at the time this decision was made, and had known Adams personally for a number of years as well as President Anderson (since they, both Anderson and Adams, were ministers in Manhattan). More interesting still is the fact that Charles Sternberg's father Levi had just left the Board of Regents himself! One wonders how Charles' life might have been different if Mudge had stayed at the college, and his affiliation had been with Mudge and Marsh rather than with Cope.

Charles had apparently attempted to secure a place on Mudge's expedition, and had failed. His description, claiming he had "made every effort" seems to imply that he felt some astonishment, and perhaps even hurt and resentment, at not being allowed on the expedition. The reason given him was that it was full and therefore could not accommodate further members. Since this expedition was not under the auspices of the college itself one can easily see how it might have been filled before Charles had full realization as to what was happening, so there is no reason to believe he was treated unfairly.

Many prior commentators have continued to foster the error that Sternberg was taught by Mudge at the college. This was simply not the case as has been shown since Mudge was gone from the college before Charles' limited attendance. Nor did Mudge have any official connection with the college at the time that the expedition referred to was being set up. In fact his presence anywhere around the college would probably have been viewed negatively by the administration. Although Mudge may have recruited members somehow through the college, it was without any official sanction.

Having failed to become a member of the Mudge expedition it was "almost with despair" that Charles took the step that was to eventually lead to his making a lifelong commitment to the pursuit of fossils: that step being the writing of a letter to Edward Drinker Cope asking to be considered for a position as an independent collector in the Kansas chalk. Apparently he had communicated previously with Cope and perhaps had even sent him some fossils already. It seems almost sure that his brother George had already started a relationship with Cope prior to this date, so Cope was most likely prepped for this request.

It is not hard to see how Cope could not have been anything but impressed by Charles' epistle, considering the very persuasive nature of Charles' eloquent writing, and Cope's own love for the dedicated and loyal individual. It certainly had to be Charles' exuberance, for he lacked any prerequisites or credentials that would have impressed Cope from a strict scientific point of view. In response to Charles' request for provisioning funds so that he could mount an expedition into the Kansas chalk, Cope followed with a draft for $300, and a note which said, "I like the style of your letter...go to work."[7]

This small retainer was to link Sternberg to Cope for what he termed affectionately as "four long years." During this period Charles was to go from novice without any training to a self-taught and highly skilled technician. It was really thanks to Cope that the world can claim many of the world's finest fossils, for it was he who allowed Charles the opportunity and means to become an expert in his field.

Before Cope was to die in 1897 Charles was to actually spend 13 field seasons under Cope's command. These were not consecutive seasons, as were the just mentioned four years, but were seasons spanning the period 1876 to 1897.

One of the most interesting facets about this linking of fossil hunters is found not only with Charles' relationship with Cope, but also with the interconnections between a number of these early bone hunters. In many cases these linkings are elaborately convoluted; by teacher/student relationship, by marriage and by place of upbringing. It is noteworthy, for instance, that in 1874 (the previous season) Professor Mudge had led a party into the field that included among others: H.A. Brous and Samuel W. Williston. These men, as well as Mudge himself, were to play major side roles in the life of Sternberg, both on and off the fossil field.

Brous, for instance, was to initially work for O.C. Marsh, and then switched allegiance and went to work for Cope, under the supervision of Charles Sternberg. Williston, like Brous, had studied under Mudge at the Kansas State Agricultural College, and then after working for Marsh for a number of years had a falling out with him and left to pursue his own career. While Williston was with Marsh, Sternberg had worked for a period directly under Williston. Williston's brother Frank was also to prove traitor, beginning in the Marsh camp and eventually opting to work for Cope. Sternberg was to later have a fairly close (although turbulent) correspondence relationship with Williston and they often traded fossils and fossil information to their mutual benefit.

In addition to everything else about this odd relationship or linking of characters, it is remarkable that Mudge continued to violently dispute the theory of evolution right up to the time of his death, and yet he worked extensively for Marsh who was one of its prime proponents. Williston had apparently not been an evolutionist in his early years, but was drawn to accept its general tenets in later years. Williston credited Mudge, the antagonist, with having led him to this acceptance.[8] There is certainly great irony in that fact! Mudge's position on evolution is clearly defined in a series of articles published posthumously in the *Kansas City Review of Science and Industry*.

If it were not for the high integrity that these men displayed in their personal relationships, and the strength of character they showed under pressure, one might have thought that this bouncing around, and switching of positions and such, was a sign of plain fickleness. These men, however, always remained deadly serious about what they were doing, and it is this fact that probably attests to their lasting successes.

Having been finally given free rein to begin his expedition Charles lost no time but immediately bought a team of ponies and hired a young boy to drive them. They then headed for Buffalo Park, which he was to use as his temporary headquarters while he was exploring all of the chalk exposures in that vicinity. Buffalo Park could be described as being much like an oasis on the desert because of its clean water, and it was an ideal place to start and stop an expedition. He was to describe this first expedition in his autobiography as being one of "countless hardships and splendid results."[9] Although there is no solid indication as to why this area around Buffalo Park appealed to Charles' bone-hunting sense, it may well have been suggested to him through his army contacts as being a place of likely success. It should be said, however, that it had the "look" of good fossil territory to the practiced eye.

The hardships he referred to resulted in part from the fact that there were no roads in the area (not even vestiges of such) and they therefore had to blaze their own wagon trails. Some of the trails they were to blaze would become roads of consequence in later years. Besides this, they had to be constantly vigilant for marauding Indians. Western Kansas was then the home of the Kansa (Kaw), the Arapahoe, the Cheyenne, and the Osage tribes; and all of these tribes could be considered at best as unfriendly to any would-be trespassers. Sternberg thus took the precaution, in light of this fact, of using a brown duck coloring for their tent and wagon coverings so that they could more easily hide their campsite from inquisitive Indians.

Although there are conflicting and biased historical records as to just how troublesome the area Indians were at the time, it is clear that the Arapahoe and the Cheyenne were the major instigators of whatever trouble there was, and there is substantial evidence that they did actively kill and maim both settlers and each other in this vicinity. Kansas history is replete with examples of murders, rapes, burnings, plunderings, desecrations, and other sorts of inhumanities on the parts of both red men and white.

General Winfield S. Hancock's campaign in 1867 had done much to stir up the Indians. His infamous campaign had a reverse effect in that it had been designed to allay the possibility of Indian trouble and actually turned out to be the main cause of it. Around this time the current Governor, Samuel J. Crawford, stated that "the plains of Kansas [were] swarming with blood-thirsty Indians."[10] Crawford's attitude was not all that uncommon, perhaps even being the majority view.

Not only were the Indians considered dangerous at this juncture, but there was an extreme cultural gap between the Indians and the whites which greatly contributed to the fear and mistrust. Francis Parkman, in his famous overview of life in the plains states, *Oregon Trail*, defines many of the not-so-obvious differences between the two cultures. He was able to recognize these differences between the two cultures due to his having cultivated a rapport with the Indian that went far beyond the average man of his day. Parkman, although he too was endowed with the prejudices of his own day, did seem to have a genuine feeling for the Indian that few of his compatriots would have understood, and therefore we can approach him without the usual trepidation.

One of the innate differences between the cultures was astutely gleaned by Parkman, and it seems to accentuate how far apart the cultures really were, and thereby how utterly doomed the Indians were to ceaseless conflict with the white hordes:

> Yankee curiosity was nothing to theirs. They demanded our names, whence we came, whither we were going, and what was our business. The last query was particularly embarrassing; since travelling in that country, or indeed anywhere, from any other motive than gain, was an idea of which they took no cognizance.[11]

Taking this passage into consideration, one can better sympathize with Othniel C. Marsh's plight in trying to describe to Indians, with clarity, that it was only bones he wanted, since this was simply something totally beyond the Indians' comprehension. Their culture simply refused to allow them to understand anything that couldn't be measured in terms of gain. Considering their sparse and rough existence, where the plentitude of buffalo could mean the difference between life and death it is not hard to understand their pragmatic approach to life on the plains, and why their loathing of white settlers led to such vehement destruction.

Although Parkman thought Indians were "thorough savages" he also had an odd love for them. He nicely caps off the whole subject by saying that the Indians were doomed to be "abased by whiskey and overawed by military posts" and that the unfortunate bottom line was that the whole country would suffer for "its danger and its charm will have disappeared together."[12] This, in part, helps to explain why Charles was to move to Canada for a period.

It is clear from all this that the bone-hunter was then bound to be misunderstood both by the Indians and the settlers, (since the settlers couldn't understand such a leisure-man's occupation either), and would be targeted for Indian wrath because they were generally ill-guarded and ill-prepared for any kind of hostilities. Charles' dad's response to his son going to the fossil fields was typical of the day, for he could just not understand why one would take on a job that was unsure from day to day, where pay was always open to question, and that seemed to be in all aspects a job that only someone independently wealthy could maintain.

Charles described being in "constant danger from Indians," but made the decision to take only reasonable precaution (i.e. brown duck tenting, keeping a look-out, etc) and therefore spurned the carrying of his Spencer carbine with him everywhere he went. He felt the rifle was of little value since they were there to hunt fossils, and not Indians. In all likelihood Indian numbers would have made resistance futile anyway, and perhaps he recognized this also. On later expeditions, however, he was to carry a Sharps rifle, but its use was mainly for supplying camp game and for scaring off predators. He did, however, find it most useful during the Bannock Indian War in Oregon.

Since the Kansa, the Osage and a number of the other tribes were culturally aggressive and mobile, and their cultures were founded upon the principle of warfare being all-important, it is not really hard to understand why there might have been room for genuine concern. The plains warriors gained social prominence through their wartime achievements, and even though their overall power and position was waning in relation to the white settlers, there were still isolated but substantial instances of cruelty and mayhem. Early Kansas history in the western part of the state is strewn with stories of Indian atrocities at the very time Charles was traversing the same area.

Important to keep in mind in consideration of this Indian trouble was the major contribution that the swift killing off of the buffalo was making. Since the buffalo was the main life source of the plains Indians (providing food, clothing and shelter) this alarming reduction in the size of the herds was to have a direct correlation to the volume of Indian trouble in the region. It is a clear indication of the power that even such a few white settlers maintained when one considers how slow the Indians were to react. By the time the Indians really began to do something about their dwindling homeland it was already too late, for the herds were greatly decimated, and the battle already lost.

The buffalo count was estimated in 1865 to be an enormous 15,000,000 throughout the plains states area, and yet the buffalo was virtually exterminated in Kansas by the end of 1874, and buffalo hunters had to travel south for a livelihood. This wholesale slaughter came at such a rapid pace that the Indian never knew what hit him. Lt. Colonel Richard Irving Dodge in his ranging and lively view of the times, *Plains of the Great West*, estimated that in the years 1872-74 alone there were approximately 3,150,000 buffalo killed.[13]

While such devastation seems incredible to us, one only has to consider that the hunters were deliberately careless and abused all natural laws because they thought that the herds were so substantial they could never be destroyed. Buffalo Bill Cody was said to have personally killed a total of 4,280 buffalo in a short 18-month period, and Josiah Wright Mooar, a buffalo hunter by trade, admitted to killing over 21,500 buffalo himself (and it is likely that this was not the record for killing).[14]

As the buffalo became more scarce, and the Indian began to recognize his plight, their frustrations found their vent in additional and more terrifying exploits to try and stop the white invaders. Meanwhile Sternberg and his young assistant worked the chalk.

Professor E.D. Cope described some of his experiences with Indians in early expeditions in this manner:

> The expeditions have not been conducted without risks. My exploration in Western Kansas was made during a state of hostility of the Cheyenne Indians, and in a region where they were constantly committing murders and depredations. During my expedition of 1872 I was abandoned by some of my party, who robbed me of mules and provisions, and placed me in some peril. My expedition of 1873 was in the Cheyenne country and constant vigilance was necessary. The year following my visit the whites were driven from the region, or murdered, by the Indians of that tribe.[15]

This was the kind of country that these early bone-hunters had to spend their days and nights in. It is hard to imagine, from our perspective today, how someone could possibly concentrate on fossil finding or digging under that kind of adversity. You'd think one's allegiance to bone would have to take a back seat when one had to constantly worry about the possibility of receiving an arrow in the back, and yet we see little real evidence to substantiate such a condition either with Sternberg or with rival camp members. Since there remain those detractors who complain about early methods of fossil removal and preservation, I might suggest they consider this fact before they get too critical. How would they measure up under the same conditions?

Sternberg's experiences in that first of many expeditions were plentiful and varied and helped to give him a taste for the life, beyond the mere imaginings that had held him transfixed to this point. He describes narrow escapes from Indian bands, malarial bouts, incessant toil under the harshest of conditions, great heat and thirst, and great fossils. He indicates that all the discomfort and deprivation were worth it since he still maintained "the one great object of my life--[is] to secure from the crumbling strata of this old ocean bed the fossil remains of the fauna of Cretaceous Times."[16]

Sternberg had a special gift for description, and in his own poetic manner, he pictures for the reader the Kansas countryside wherein his first exploration tentatively explored:

> Both sides of my ravine are bordered with cream-colored, or yellow, chalk with blue below. Sometimes for hundreds of feet the rock is entirely denuded and cut into lateral ravines, ridges, and mounds, or beautifully sculptured into tower and obelisk. Sometimes it takes on the semblance of a ruined city, with walls of tottering masonry, and only a near approach can convince the eye that this is only another example of that mimicry in which nature so frequently indulges.[17]

It was on this first expedition that Sternberg was to discover the first of many-to-be discovered Kansan mosasaurs. The mosasaur was a particularly fierce beast of the water known for its sleek shape, flattened tail and razor sharp teeth, and Kansas was to prove to be the main area in the world where they were to be found. This particular species was to be named *Clidastes tortor* by Cope. Sternberg's initial response at its finding was typical, for he tells us that he shouted out "Thank God! Thank God."

Additionally, he was to find two other mosasaurs that first season in the chalk. These were to receive the name *Platecarpus* after their "flat wrist." These were taken out only after the most extreme hardship, as they were being removed during a torrential downpour of exceptionally cold rain and the camp existed for three days on handfuls of parched corn before the fossils could be extracted, for they feared leaving since the fossil material would have been damaged beyond repair through the brute strength of the one winter.

Further work that season produced a fine specimen of *Tylosaurus dyspelor* (a most formidable beast) and the great fish *Portheus molossus*, a species that was to later make his eldest son George famous when he was to find another fish within a specimen of this great fish. The *Portheus* was a major find in any expedition because of its sheer size. An average skeleton of this huge fish would be about 15 feet in length. He was also to find a small mosasaur named *Clidastes velox* that same season that was to later prove to be a prolific denizen of the chalk.

Charles also relates how it was on this trip that he developed a great lesion on the palm of his hand from cutting away at the Kansas chalk with a butcher knife. This lesion was devastating as it kept him off the field and sleeping very unsoundly for 10 full days. As with most occupations there was decided financial loss that accompanied the down period.

This mishap, and a general physical deterioration at the time, led him finally to ask Cope to send a qualified assistant. Cope magnanimously sent J.C. Isaac who had already worked for Cope for some time, and who was a trusted and valuable employee. As talented as Isaac was, however, he was of little help to Sternberg because he was totally preoccupied with the fear of being attacked by Indians. Isaac had just prior to the expedition lost five members of his previous party to a warring band of Indians who had killed and scalped the five. Isaac had just narrowly escaped the same fate, but could not forget the horror of it. Sternberg says that Isaac "saw an Indian behind every bush."[18] Isaac's fear proved to be somewhat infectious for Charles too began to fear things that he had never given a second thought before.

Charles wrote to Cope, including full particulars as to the circumstances, and was consequently ordered home from the field. In terms of success one would have to rate that first expedition very highly. Not only was he self-training in the art of bone collecting, having no one to instill him with the elements of fossil collecting and preservation, and yet he was to succeed when he had no help and very little equipment.

He did prove his worth to Cope clearly, making it obvious that Cope's choice to send him to the fields was the right one, and in the process he even found some magnificent material that first year. Assisting Sternberg in his efforts that novice year was his own great ingenuity. This trait, honed in the farmlands of the midwest where he'd become familiar with equipment and method, gave him a particular benefit in knowing how to best tackle a problem.

Following his return home Sternberg was to go to Omaha with J.C. Isaac where he would finally get his chance to meet Edward Drinker Cope, one of the three most prominent paleontologists in the country, face to face. He savored the experience, since he had already developed a fondness for Cope for having allowed him the opportunity to pursue his dreams, and to have done so sight unseen. This trust that Cope bestowed on him was not to go unanswered as we shall see, and for all that Cope gave him Charles was to return that and much more.

NOTES

1. Katherine Rogers, *The Sternberg Fossil Hunters; A Dinosaur Dynasty*, Mountain Press Pub Co., Missoula, 1991, Pg. 22 [According to Rogers a "younger brother" enrolled with CHS in 1868]

2. Samuel W. Williston, "Professor Benjamin F. Mudge," *The American Geologist*, Vol XXIII, #6, Jun 1899, Pgs. 343-344

3. Elizabeth N. Shor, *Fossils and Flies*, Univ of Oklahoma Press, Norman, 1971, Pg. 22

4. Julius Terrass Willard, *History of the Kansas State College of Agriculture and Applied Science*, Kansas State College Press, Manhattan, 1940, Pg. 37

5. *Kansas State Historical Collections*, Vol XIII, 1913-1914, "George Washington Martin," Kansas State Printing Press, Topeka, Pg. 8

6. Richard Swann Lull, *Ntl Memoirs Acad of Sciences*, Vol XVII, #5, Pg. 116

7. Charles H. Sternberg, *The Life of a Fossil Hunter*, Henry Holt & Co., NY, 1909, Pg. 33

8. Elizabeth N. Shor, Pg. 20

9. CHS, *Life,* Pg. 34

10. Homer E. Scolofsky, "Great Plains Studies," Part I, *Journal of the West*, Vol XV, #3, Jul 1976, Pg. 24

11. Francis Parkman Jr., *The Oregon Trail,* Caxton House, Inc., NY, 1910, Pg. 68

12. Ibid, Pg. 142

13. Richard Irving Dodge, *The Plains of the Great West*, Archer House, NY, 1959, Pg. 142

14. Mari Sandoz, *The Buffalo Hunters*, Hastings House Pub, NY, 1954, Pg. 88

15. Henry Fairfield Osborn, *Cope: Master Naturalist*, Princeton Univ Press, Princeton, NJ, 1931, Pg. 159

16. CHS, *Life,* Pg. 43

17. Ibid, Pg. 39

18. Ibid, Pg. 48

CHAPTER THREE
LIFE WITH COPE

J.C. Isaac and Sternberg met Cope, as requested, on Aug. 1, 1876 in Omaha, Nebraska at the train depot. This was Sternberg's first view of Cope and Cope of him, and both were a little dismayed at the other. Charles was filled with awe at the sight of the prominent paleontologist, and Cope was astonished at the limp which Sternberg sported (as a result of a childhood accident when in a freakish barn-fall he had dropped 20 feet to the ground). This injury at age 10 had left his leg permanently crippled. Cope went so far as to wonder aloud to Isaac whether Sternberg could ride in that condition. Isaac supposedly responded with vehemence saying, "I've seen him mount a pony bareback and cut out one of his mares from a herd of wild horses."[1] These were skills presumably learned while working on the Ellsworth family ranch.

From Omaha they took a somewhat unorthodox route to their next destination. Sternberg is not too clear as to why their travels appeared to be so circuitous, but one would have to suppose that the route at least in part was chosen because they were the more frequently traveled lanes with greater protection and more frequent departures on to the next point. They were also probably the cheapest routes, as both Sternberg and Cope were frugal men. Undoubtedly Ft. Benton was chosen because it was at the head of the upper section of the Missouri River, and hence a good spot from which to ship their return fossil load by steamer.

The route as described by Sternberg showed the various legs of the journey as follows: from Nebraska to Ogden, Utah, from there to Franklin, Idaho to Helena, Montana to Fort Benton, to Claggett at the mouth of the Judith River, and then on to the Valley of Dog Creek. The leg from Ogden was traveled by means of narrow gauge railway, which terminated in Franklin, and from whence the next 600 miles were traversed by coach. At Franklin, where there was no hotel, Cope and Sternberg found it necessary to sleep on the depot's platform before catching their early morning coach.[2] The first 600 mile section of the coach journey was very forbidding and Charles commented particularly upon the discomfort caused by the dust. The dry alkali desert of Idaho made their travel slow (Sternberg indicated a speed of around 10 miles an hour), but even more problematic was the fact that the coach kicked up great clouds of dust which left them all enveloped in a "death like appearance." Their discomfort was heightened by the fact that they were only able to stop three times during the course of a day for an hour each at the "dirty stage stations" for rest and food. Charles was not impressed with the food at the stations, and even less with the cost. Meals were one dollar, something he considered a king's ransom, and he further commented that the proprietors had "evidently settled here not for their health, but for dollars."[3]

To add to the excitement they found themselves thrust into placing their lives into the hands of a driver known as "Whiskey Jack" (hardly a name to inspire confidence). Although it was said that he was one of the line's best and fastest drivers, it must also be said that he was often so drunk that he had to be tied into the seat for fear that he would kill both himself and passengers.[4] Considering Cope and Sternberg's aversion to strong drink it is not surprising to hear of their complaints about the driver, but they were apparently in vain and they were forced to continue on rather than to change plans.

Charles further indicates their travels led them through Port Neuf Canon where the danger from road agents was very real, and that the balance of the journey from there was fairly uneventful. Although Sternberg did not elaborate much on the journey's beginning he was very specific in indicating that the final destination was in the "interminable labyrinths of the Bad Lands." Cope would comment further, however, in a letter of Aug. 14, 1876 to his wife and state that the "stage ride from Franklin here is an interruption to the ordinary course of affairs which is a good introduction to camp life."[5] It was clearly less of a diversion for Charles, and his writings clearly indicate his being bored with the tedious journey.

It has been said that the term Bad Lands (or Mauvaises Terres) gets its name originally from the Jesuits, who used the term to describe denuded terrain that was both water and wind-worn by the elements. Others have said the name comes from the Indians who called it "mako sica" (mako, meaning land and sica meaning bad). In either case the term is direct and to the point and has remained in vogue since its inception. Sternberg described the badlands in desolate terms, but there always seemed to be evident a real love for that kind of land, even at its bleakest. You could always tell he enjoyed being there, despite the dark language.

Before leaving Ft. Benton, Cope had purchased a work wagon, four work horses, three saddle ponies, and gear and provisioning at a cost of slightly over $800. He also managed to hire a cook, and a scout who was supposed to look out for rampaging Indians as well as to provide game for the camp's table. Hiring a crew at the time was not easy since area events tended to provide a built-in reluctance on the part of most hireable men. When in Helena the party had learned of the slaughter of Custer and his men at Little Big Horn, a major event that was of concern to all wise men. Custer's men had been annihilated just over a month prior to Cope and company entering the area. For that reason Charles was to describe the scene in Helena concisely when he stated "all was excitement."

Cope, however, was not to be dissuaded even in the face of such grizzly evidence, and so the expedition proceeded despite the fact they had no military escort or real protection against any armed enemy. This seems typical of the thoughtless action for which Cope was famous and of the type over which Charles never ceased to be amazed. In this instance, however, one must credit Cope with being right. Charles indicates that Cope "reasoned when the country was astonished by the news of the fearful slaughter of Custer and his brave 7th Cavalry, that all the able-bodied warriors would be south fighting the soldiers..." and hence he felt travel was as safe as it could be until the Indians were finally subdued. Charles goes on to add that Cope's "reasons were good and in proof we were so fortunate as to add forty new species of strange animal life to science."[6]

Cope, however, was to paint a decidedly different picture when writing to his wife, as he was to call the pervasive comments of the populace "cock and bull stories about the Sioux."[7] In what was a probable effort to assuage any fears his wife might have had as a result of the hysterical stories that found their way east he stated "everybody including Capt. Williams who is commandant of the fort says there is no danger in the region to which I am going."[8] One can only wonder what J.C. Isaac thought about all this since he was coming off such a negative Indian experience, but perhaps he too was even caught up in Cope's infection.

In any case, they traveled to their destination with some trepidation but without encountering any major obstacles. The Judith River area in Montana was to prove a veritable treasure trove for the expedition before the season was ended, and it holds a special position in the annals of American paleontology as a result.

The Judith River group, or more precisely the Ft. Pierre Group of the Cretaceous, was replete with dark shales that disintegrated quite easily, to the effect that climbing slopes became both difficult and dangerous. Sternberg indicates in many places a man would "sink a foot" into the loose dirt and that the loosened dirt was carried downward by the rains and contributed to the mud banks along the "Big Muddy." Maintaining one's balance under such conditions can be quite difficult, and the dangers of the terrain therefore make fatal falls or major bone-breaks a matter of constant concern. These same shales, however, also proved to be rich in the bones of marine mammals and the rapacious reptilian Mosasaurs. Before the season was complete they were to collect a boatload of fossils, but not before they were to encounter their share of mishaps and adventure.

The typical camp day is variously described by both Sternberg and Cope, and in both renditions a pleasant picture emerges. Mornings would begin early, with the camp arising at the first hint of light, and then all but the cook would be off into the field following a hearty breakfast. Cope himself was to sleep fully clothed so that he would be ready to jump up and begin work at sunrise without having to bother with dressing. Lunch would normally be at midday, barring great discoveries, and would usually consist of some combination of cold bacon, hardtack or crackers. They would not return to camp for lunch except in rare instances, but would carry the light repast with them for consumption in the field. Evening meals provided much more variety, as well as providing a special time for socializing or other relaxation outlets after a hard day.

The balance of the typical workday consisted of either searching out new fossil deposits, or in carefully working over older ones. The work itself was notably close and tedious whichever type it was. When searching this terrain one had to stay close to the ground and had to literally strain one's eyes looking for the merest scrap of bone, and when taking out a bone one had to be extremely careful as one wrong move could destroy the most valuable part of a cherished fossil. The shale's coloration also made it particularly difficult to easily locate fossil material since the bone tended to remain very similar in color to the matrix.

Lilian Brown, wife of Barnum Brown (known to him as "Pixie") had some of the pixie in her. In her light-hearted poke at her husband's profession, *I Married a Dinosaur*, she has fun in describing her non-traditionalist view of the fossil crouch:

> ...those early days at the dig, I was kept strictly in the spectator class, permitted only to watch while my husband showed off with

pick and shovel...'Observe closely,' he kept reminding me. But all I could observe was the seat of his pants and two elbows going like mad. Sometimes not even that for the dust. Head and neck were lost to sight.

This is the traditional pose of all self-respecting bone-diggers, and although it bears a superficial resemblance to the stance of a football center in the act of snapping the ball, more than likely it originated with the ostrich...In this posture a fossil-hunter is dead to the world. Come earthquake, fire, flood or wife...on he digs, oblivious.[9]

Charles Sternberg's description of himself in this same classic position is relatively similar except that he did admit to occasionally removing his shoes before lying down on the prepared "holy ground."

While we might laugh at Pixie Brown's novel view of the fossil collecting position, we have to admit that there is truth in her view, and it is that element of truth that makes it funny. Fossil hunters are indeed an unusual breed, and she caught the essence of the bone-hunter in this small gem. This was the true essence of the bone-hunter then, and it hasn't changed. Even with all the modern tools of the trade the work remains as close and tedious, and as ridiculous to watch as it always was, and still draws men with the same allure that it has since the science began.

Using tools such as awls and picks and knives the fossils would have to be painstakingly encircled until the fossil exposure was clearly discernible, and no bone's position unknown or in a position to be damaged. At this point various methods were used to keep the friable fossil and matrix from crumbling and ruining the whole effort.

The importance of equipment, serviceable equipment, can make or break an expedition, particularly expeditions into remote areas. Sternberg expeditions were no different in their dependence upon tools. In his fieldnotes of July 27, 1915 Charles was to comment:

Levi had 5 chisels sharpened by the blacksmith and I broke all within 10 minutes. This is the second time he has tempered them so soft the first blow of the hammer turns the point, the second breaks it off. I resolved to see what I could do in my old stove. I succeeded in tempering them so I could use them all day.[10]

Around five or six in the evening the workday would officially end (depending upon how far from camp one might have strayed) and the party would head back to camp. Dinner would have been ready since the cook had remained in camp all day, and was welcomed with ravenous appetites. Staples such as rice, beans, bacon, fish and coffee were the norm and were supplemented with occasional treats when the scout would come upon game. In addition to these variations Cope indicates they were able to further augment their diet with red berries (garambujos), dried apples, prunes, succotash, tomatoes (canned), potatoes, molasses, ham, pickles, cakes and pies. Pickles certainly must have been a staple for Cope records having written a letter home while straddling a five pound keg of pickles. Sternberg also mentions that the country was replete with mule-deer and antelope, as well as many mountain sheep, and in the high plains one would often encounter the grizzly after wild artichokes.

They also used to bring along butter as well, and in later years margarine. In order to protect the butter from turning into grease in the hot summer weather they would dig a hole in the ground and line it with burlap. The butter apparently kept much better in this primitive but effective ice-box.

Even out in this primitive countryside they were not without respite since Ft. Claggett was reasonably nearby. Ft. Claggett was a private trading post and essentials could be purchased there, but as one would expect, at a dear price. While there Cope managed to regale the Crow Indians who surrounded the fort with the awe and wonder of his false teeth. This is very reminiscent of Benjamin Mudge's use of the same ploy when he was enveloped by a band of hostile Indians. It seems clear that false teeth were a real wonder to the Indians.

Other utilitarian camp duties were sandwiched in between meals, and rising or going to bed. Additionally Cope made it a regular practice of reading aloud from the Bible most mornings. He seemed especially pleased with this particular expedition in this regard as their attitude was respectful:

> We have just had a reading in the Bible and prayer. I took Ephesians iv and a part of v, which are of a practical kind. Sternberg is a very religious character and Isaac is a good fellow and I am fortunate in having secured a very respectable man for a cook. They behave different from my Colorado party who scrubbed their teeth and chopped wood when I read. I feel comforted in the small service and hope that the effect may be known.[11]

Cope found the Ephesians chapter "practical" most likely because it describes the transition one undergoes in becoming a Christian. This is described in terms of "putting on the new self" and letting "the old pass away." To Cope this would have been a major New Testament chapter, and this would explain why he would be so upset with those who refused to listen attentively, or did not show proper respect while he was reading from its pages.

This fact, and the excellent capabilities his party displayed led Cope to compliment them highly:

> The boys prove themselves excellent collectors. J.C. Isaac is an invaluable fellow. So is Charles H. Sternberg at a bank of fossils...altogether my camp is the best I ever had.[12]

Coming from a man of Cope's qualifications in both collecting and descriptive fossil work this is a high compliment indeed! But then again we cannot forget that the quality of the work performed by this crew was great, and therefore deserving of such notice.

In keeping with their Christian upbringing, the camp would always close down from collecting activities on Sundays, and Charles' later camps kept up the custom as well. Sundays were the day when chisels were retempered for future use, boxes were made, fieldnotes were brought uptodate, correspondence was handled, baths taken and clothes were washed.

In confirmation of this fact, Katherine Rogers, a Sternberg family biographer, has noted that all of the camp pictures she has had opportunity to peruse that had Sunday dates attached show the Sternbergs in shirt and tie. They always observed the Sabbath no matter where they were.[13]

Both Cope and Sternberg indicated that the Dog Creek campsite was not a suitable site for fossils, as they were not able to find any complete specimens, only fragments. In fact, they found not one complete fossil skeleton in the area. Cope, however, gives the area slightly more credit than Sternberg for he said in another letter that they were to collect "a good many fossils" here, but adds that they "were generally alone, and badly broken." Because of this fact, Cope and his scout prospected in the Cow Island area nearby to see whether it would be a realistic alternative, and it proved to be, so they decided to move the entire outfit to that region for work. This involved a move of about 40 miles over prairie land, and their eventual campsite was to be about three miles from Cow Island Station, on the opposite bank of the Missouri.

Since the route to the Cow Island area required traversing very steep hillsides, the trip itself was not without its share of hardships. Sternberg credits J.C. Isaac, and the active use of hand axes, picks and shovels, for seeing them through the arduous journey. During the course of the move Isaac was threatening his team with calamity if they did not ford a particularly difficult ridge, when the wagon tipped, and before they came to rest the wagon, its rider, and the horses all completed full revolutions before the wagon arighted itself intact. Isaac was also unhurt, except for some obvious damage to his pride.

The area around Cow Island proved ideal for their purposes, for not only was it filled with good and whole fossil material, but it also provided river accessibility. Cow Island itself was a small but important station for the steamers that traveled the Big Muddy, and as such there would have been medical aid and other conveniences nearby. It also would have served the more immediate profit of being a spot from which they could actually ship back the fossil material once it had been collected.

Following this difficult transitional move, the hired guide, James Deer, and the cook of whom Cope had spoken of so highly, Austin Merrill, abandoned the expedition leaving only Cope, Sternberg and Isaac in the party. The despicable nature of this departure is accentuated by the fact that these two men had already been paid in full for the total expeditionary services they were to perform, and their leaving actually further jeopardized the lives of the remaining party members.

The scout Deer apparently decided to depart as a result of finding on his daily reconnoitering that Sitting Bull's camp was less than one day's ride, and the cook's reaction appears to have been equally cowardly. Sternberg, a very loyal man himself, cited the cowardly departures as "dishonorable conduct," and claimed that he and Isaac willingly submitted to the "double work...even if it were to mean working our fingers to the bone."[14]

Cope's rendition, taken from a letter of Oct. 3, 1876 to his wife, is different in the ordering of things. He says:

> ...I went to my camp and found that the boys had done good work, but that the cook had left; cause, fear of Indians. Then my worthy guide Jas. Deer announced his intentions of doing the same; cause, chance to make more money--but I don't believe that to be the real reason. I never had any unpleasantness with him.[15]

It is not clear how Deer could make more money elsewhere, particularly since they were encamped in the wilds and he had already been paid (which suggests the wages were satisfactory), so it seems more likely that Sternberg's account is the likely explanation of Deer's motives.

Sternberg was to say that the bones to be found in the area were not found easily on the surface, but had to be worked for. It was only possible to find the hidden bones "by noticing the color of the surface dust above the bones, which in all cases differed from that of the surrounding disintegrated rock." He also noted that they had to dig beyond the area penetrated by frost in order to obtain their treasure.[16]

It was soon after their move to Cow Island that the first dinosaur "with horns" ever discovered in America was found by the party. Cope was to find and name the great beast *Monoclonius crassus*, while Sternberg was to collect two relatives, *Monoclonius sphenocerus* and *Monoclonius recurvicornis*. These

Ceratopsian dinosaurs ("one-horned") had one nose spike, and had the characteristic frill common to the Ceratopsids, although the *Monoclonius* has a small frill compared to its more embellished brothers such as *Chasmosaurus*. Sternberg's finds were collected from the north side of the Judith a few miles from Cow Island. It was these great and little known beasts that were to make this expedition.

The importance of these new discoveries was substantial, and it is interesting (considering their worth) that Cope, who had a reputation for being both quick and accurate in publishing on extraordinary finds, did not publish on these horned dinosaurs until quite sometime later. In a letter to Charles Sternberg dated Dec. 21, 1887 Cope was to inquire:

> I am now going over the fossils you & Mr. Isaac collected on the N. side of the Missouri river in 1876. I send you a paper showing how far I have got along with the study.
>
> I want you to answer me some questions. 1st. Did you get the Monoclonius you marked at exactly the spot as where I dug up (with your help) the bones of the animal I so named? If not, how far off did you get them? I refer especially to the animal figured on Pl. XXIII figs. 2 & 2a. Where did you get it?
>
> There are 4 separate animals, all supposed to come from the place where I got Monoclonius crassus. Two of these I got. Both are M. crassus. Two you got, one larger (M. Sphenocerus) & one smaller (M. fissus) than mine. It is about these latter that I want information.
>
> Marsh has been duplicating this work in his newest shameless style. According to him nothing has been done in this field before! He made a good beginning by describing the horns of one of these fellows as a new species of bison! Answer soon. [17]

The reader will note that Cope calls Sternberg's find *M. fissus* while Sternberg calls it *M. Recurvicornis*. This was caused by the fast and furious changes that were occurring in the science, and many duplicate names were being uncovered. It was not uncommon later for two or three different species to be boiled down to one form. Such was apparently the case here.

One is also a little surprised at Cope's letter to Sternberg since some of the information he was inquiring about certainly should have been in Cope's fieldnotes. While the science of stratigraphy had not yet been elevated to modern standards as yet, it would seem that Cope would have had better fieldnotes in this case, since his questions to Charles were very basic. This is particularly hard to understand when one considers the value of those finds.

According to Cope they managed to amass 1,700 pounds of fossils on the expedition, and were able to get them down to the river so they could be shipped down river to Bismarck. The process of getting this awkward mass down to the river, however, proved to be a difficult task. Cope and Sternberg in a hair-raising adventure had to negotiate a steep incline in order to be there in time for the steamer's departure. The most difficult aspect of the descent was that it had to be done in the dark, and this was a treacherous area for travel even in bright sunlight.

In a letter to his wife Oct. 8, 1876, Cope himself, who instigated the night descent, is very matter-of-fact about the blind excursion:

> At last I made the descent with Charley [Sternberg] and the horses, and we took a delightful drink at the river. But we were six or more miles above our camp and the river at several points struck the bluffs so that we could not follow it, but had to climb the precipice again. At one point I made three attempts before I could get down again to the high bank and my horse came down some perilous places...[18]

Sternberg's version certainly has more zest and clearly highlights the uneasiness he felt in the matter. It is also evident that he would have waited until morning rather than to attempt such a foolhardy undertaking (for it certainly was that) if he were to have had the decision, yet even in the face of such a dangerous action he would remain loyal:

> Just as the sun was sinking behind the rockies he [Cope] came out of a narrow ravine with the head of a large mountain sheep on his back...we mounted and set off at full speed for home,

remembering the three men whom we had met on the prairie at noon, who had been lost for three days in the intricate passages of the Bad Lands. I did not like to think of trying to find the way there after night.

The Professor dashed over the prairie without once drawing rein, clearing bunches of cactus ten feet, sometimes, in diameter, at a single bound...I knew the uselessness of trying to combat his iron will, but I pleaded with him against the folly of attempting to thread in the darkness those black and treacherous defiles, where a single misstep meant certain death. I begged him to wait until daylight...He paid no attention to what I said, but dismounting, led his horse into the canyon. He had to cut a stick to shove in front of him, as his eyes could not penetrate the darkness a single inch ahead. I cut another to punch along his horse, which did not want to follow him...I have never known another man who would have attempted this journey. It was both foolhardy and useless, but we could say that we accomplished what no one else ever had in reaching Cow Island through the Bad Lands after dark.[19]

Sternberg's devotion to Cope is clearly evidenced in these lines, and that devotion was to deepen as time went on. This form of hero-worship was not unusual as we see much the same kind of relationship established between Samuel Williston and his mentor Benjamin Mudge. It seems the strength of teaching and fostering one in the fossil profession brings with it a special bond that is not easily broken. While Sternberg and Williston were perhaps hardly normal men in this respect, and therefore may have reacted more admirably than would have the John Bell Hatchers and those of that ilk, it still seems there is an unbreakable bond that generally develops when men are thrown together under such circumstances. I personally have no doubt that the profession by its very nature is instrumental in heightening that bonding, in much the same way that war does.

Cope's efforts to make sure that the fossils were ready for boarding upon the steamer *Josephine* was not without wisdom. Although the Upper Missouri was fairly heavily traversed by steamers by this date, they did not arrive every day and hence there would have been a delay; perhaps even a prolonged delay. The section of the upper river, particularly that above Cow Island (which was only

130 miles below Ft. Benton, the terminus of river traffic), was only traversable at high tide, and then only by those steamers with a design that allowed them to draw just a few feet of water when fully loaded. This obviously limited traffic on the upper river.

The *Josephine*, newly built but already a four-year veteran of the upper river run, was the perfect ship for this river. A smallish ship, running only 183 feet in length, it drew only 4 feet of water when loaded to the maximum. A sternwheeler with two boilers and twin capstans it was a powerful boat, and it was perfectly designed for the area above Cow Island especially, since that was a low water point that had once been the nemesis of all water traffic except for keeled boats. The *Josephine* was decked out with a large set of deer antlers over its wheelhouse, and like the *C.K. Peck*, it was run on wood-power. Apparently the boat would burn approximately one cord of wood per hour, and hence her crew played a very large role in helping to decimate the cottonwood population along the upper river.

The *Josephine* was additionally of interest because it was under the command of one of the two most famous river captains on the Missouri, Captain Grant Marsh. Marsh, who was known to be strong-willed and cantankerous, was exactly that with Cope. Cope indicated to Marsh that his four-horse wagon and outfit were just 3 miles below Cow Island at a steamboat snubbing-post, and would not the captain kindly stop there on his way back down river? Marsh balked, for who knows what reason, and made it clear to Cope that if he and his gear were not at the Cow Island station by 10 a.m. the following morning they would leave without him. It is a mark of Marsh's command that Cope argued no further but immediately made plans to get his wagon to Cow Island without delay. Although Grant Marsh was not related to O.C. Marsh, both his given name and his boisterous and commanding way must have made it very hard for Cope to accept the reprimand graciously.

The terrain along the river was not conducive to driving the wagon and team easily over the three-mile route since at varying points the bluffs abutted the river, and Cope was unable to hire any river transport to assist their effort except for an old scow for which he had to pay dearly. The wagon and its entire contents were transferred to the scow, and were with great effort finally dragged and manipulated by rope and sinew to the *Josephine's* berthing place at Cow Island. Sternberg notes that they only reached Cow Island at sundown, and that they were drenched and mottled with water and mud.

Being bone-tired from their exhausting ordeal they relished their return to "civilization." Cope must have been absent-minded it seems, for he had not brought any winter wear along on the trip and hence had to change to a "summer suit and linen duster." Considering the lateness of the season and their vicinity he certainly must have found himself uncomfortable, but as usual Cope was resilient.

Cope was to take the steamer *Josephine* down the Missouri towards Yankton, with his fossil trove aboard. Before arriving at Yankton, however, he was forced to transfer to the larger but less comfortable *C.K. Peck* as the *Josephine* was commandeered by troops under the command of General Hazen for use as a troop carrier, as the army wanted to more quickly move troops against the confluence of Indians that Sitting Bull had amassed. The change in vessels made little difference, however, as the fossils and the master fossil man arrived safely at Yankton and beyond. The *Josephine's* captain, Marsh, was probably delighted to have his steamer used as a troop carrier, since the army always paid well. The average pay around this time was in the area of $300 to $350 per day when an entire boat was leased.

Cope was particularly impressed with the *Josephine* since there was "no liquor on board, and the amount of swearing was small."[20] Undoubtedly Sternberg's sentiments would have been exactly the same had he been on the steamer with Cope, but passage was dear and his job was not yet completed. Isaac and Sternberg remained there at Cow Island until November 1, at which point the severity of the weather made further work impossible and they returned to Ft. Benton. From Ft. Benton Charles took the stage back the 600 miles under conditions similar to their trip to the region, except that the weather was much more inhospitable. Since they had to ride over the mountains they encountered temperatures as low as 20 degrees below zero. It appears that Isaac did not return with him, judging from Sternberg's description, but it is really not clear either way. It is clear, however, that Sternberg then took the Union Pacific railroad east towards Philadelphia. He had been gone a total of three months.

J.C. Isaac, Sternberg's companion of that summer, was to work the following winter (1877-78) as assistant to Jacob Boll in Texas. Boll was working for Professor Cope and so Isaac again made a ready and able assistant. Boll and Isaac spent their time that winter primarily in the areas around the Wichita and the Little Wichita Rivers.[21]

As a final footnote to the whole question of how concerned they should or should not have been about the Indians, one need only refer to Sternberg's comment that both he and Isaac were to reach Ft. Benton in safety, and that he was to learn later that Sitting Bull and his braves had attacked the small station there at Cow Island, killing all of the soldiers who had been left there.[22] From Sternberg's comment one would have to say that they had therefore gotten out none too soon from Cow Island, except that in this case he appears to be in error on a number of different points relating to the incident. It was not, for instance, Sitting Bull's Sioux band at all who was involved in this skirmish, but the Nez Perce under Chief Joseph, and this did not occur until almost a year after Sternberg had left the area (Sept. 23, 1877). At that time Sergeant Moelchert (presumably the same sergeant who had helped their party) with 12 soldiers and four citizens were serving as the total contingent at the station. The Indians apparently burned the supply depot which the detachment was guarding, and did kill a few soldiers and teamsters in the area, but they did not kill the entire group as Sternberg intimates.[23]

Charles was wrong on these aspects perhaps as a result of his memory having become clouded on the finer points due to the passage of time. Since the

Cow Island incident occurred in 1876, and Charles didn't publish his account of the ordeal until 1909, it is perhaps understandable how some of these things could have lost their clarity.

NOTES

1. Charles H. Sternberg, *The Life of a Fossil Hunter*, Henry Holt & Co., NY, 1909, Pg. 61

2. Dan Cushman, "Monsters of the Judith: Dinosaur Diggings of the West Provided Competitive Arena For Fossil Discovery," *Montana, Magazine of Western History*, Vol #12, #4, 1961, Pg. 27

3. Charles H. Sternberg, "Explorations in the Judith River Group, *Kansas City Review of Science & Industry*, Vol VII, #6, Oct 1883, Pg. 326

4. Dan Cushman, Pg. 27

5. Henry Fairfield Osborn, *Cope: Master Naturalist*, Princeton Univ Press, Princeton, NJ, 1931, Pg. 220

6. Charles H. Sternberg, "Explorations in the Judith River Group," Pg. 328

7. Henry Fairfield Osborn, Pg. 222

8. Ibid, Pg. 223

9. Lilian Brown, *I Married a Dinosaur*, Dodd Mead & Co., NY, 1953

10. Fieldnotes Jul 27, 1915 (Provided by Saskatchewan Museum-Tim Tokaryk)

11. Henry Fairfield Osborn, Pg. 223

12. Ibid, Pg. 227

13. Personal Letter from Katherine Rogers

14. CHS, *Life*, Pg. 83

15. Henry Fairfield Osborn, Pg. 227

16. Charles H. Sternberg, "Notes on the Fossil Vertebrates Collected on the Cope Expedition to the Judith River and Cow Island Beds, Montana in 1876," *Science* n.s., XL, #102, July 1914, Pg. 134

17. Henry Fairfield Osborn, Pg. 323

18. Ibid, Pg. 228

19. CHS, *Life*, Pgs. 89-91

20. Henry Fairfield Osborn, Pg. 230

21. Robert W. Hook (ed), *Permo-Carboniferous Vertebrate Paleontology, Lithostratigraphy, and Depositional Environments of North-Central Texas,* Field Trip Guidebook #2, SVP, Austin, 1989, Pgs 40-46 Entitled *An Overview of Vertebrate Collecting in the Permian System of North-Central Texas* by Kenneth W. Craddock & Robert W. Hook

22. CHS, *Life,* Pg. 98

23. Helen Addison Howard, *Saga of Chief Joseph,* The Caxton Printers, Caldwell, ID, 1941, Pgs. 309-310

CHAPTER FOUR
BADLANDS IN BALANCE (1876)

The importance of Cope's expedition into the Judith River Group cannot be overstated. Not only did his trek bring to light the first horned dinosaurs ever to be discovered in America, but it also signaled the beginning of a new age for the science of fossil collection in the territory. Other groups which followed would expand upon what had been found by this expedition of course, but the later groups would have the precious benefit of knowing what could probably be found, and what to expect in the area geology. Such a boon is beyond measure since it allows preparation and forethought and saves a great deal of time as it thwarts tentative action. We shall see later how this was proved out with Sternberg's first trip to the Texas Permian.

Although Hayden had earlier made a trip into the same general vicinity his expedition had not found terribly enlightening fossils, and they had to give up all but a few of those fossils when confronted by belligerent Indians. Hayden certainly had a much more cursory understanding of the geologic history of the area and had no need for anything more substantive at the time. One does not become an overnight sensation in the fossil business, and it takes more than a cursory glance at a fossil bank in order to properly gauge its merit and unlock its treasure.

As mentioned, the effects of the Hayden survey were not ameliorated when Hayden found himself driven from the territory by Blackfeet Indians, who spared his life, but forced him to leave the larger and better fossils he had found. He managed to slip a few fossil pieces of turtle shell and a few scattered teeth into his pocket, and these were therefore the sum total of his treasure trove. They also constituted the limited understanding men then had of the territory. The Cope expedition in contrast was not a "discovery" party, but a party bent upon searching out and extracting fossils. Such was their intention, and such was their success.

Henry Fairfield Osborn, head of the prestigious American Museum of Natural History, in his biography of Cope described the overall time period in this manner:

> The decade 1870 to 1879, as he [Cope] was passing from thirtieth to his thirty-ninth year, was the golden period in Cope's life: the period of his greatest achievement of his incomparable western explorations, of the acquirement of the greatest part of his fossil collections, of the foundation of his principal scientific discoveries, and of his continued domestic happiness and devotion to his family.[1]

Perhaps one could as easily expand upon what Osborn said to add that not only was it Cope's heyday, but it was that of paleontology in America also. The sheer volume of fossils discovered and extracted, and the sheer extent of virgin territory surveyed almost defies description. The feud between Marsh and Cope, including all the bit players in that drama, if viewed aside from all its negative aspects, most certainly helped to provide the impetus that the young science in America needed in order to begin to develop an American following. Up to this point, and certainly not before Agassiz's establishment of his European-styled foundation with its added twist of fund raising, geology in America had proved to be either an uneducated science, or a European-schooled science, with little visible innovation and designed along age-old lines.

American paleontology was to soar from this point onward, and Cope and

Marsh were to provide the impetus for that happening. Feuding aside, it must freely be admitted that they provided the success, whether either directly or indirectly. There are those who would say that such behavior is typically American, and that the western scene envisioned by the world was perfectly in keeping with this feud: madcap dalliance in the midst of carousing Indians and rampaging buffaloes. To this we can only reply that there has never been a science feud quite like this one, with the possible exception of the feud that occurred between Agassiz and Huxley (although one must admit that they were not native-born Americans). In the case of Marsh and Cope one can only finally say that it was a battle of wills, and that each man simply violently disliked the other which is ultimately what caused the feud. Their being American was strictly incidental.

W.E. Swinton, the great and famous dinosaur expert from England, places proper emphasis on the North American dinosaur fields in his book, *The Dinosaurs:*

> Of all the localities in which dinosaur remains have been found the most important are those of North America, not only in point of the numbers of specimens excavated, but also in the literature upon them and the number of leading paleontologists who have worked them.[2]

Beyond the broader and more obvious aspects of the badland's importance, one cannot overlook the advancement in technology that occurred at this time, particularly relating to the preservation and transportability of fossils. Prior to this juncture it was not uncommon for major fossil discoveries to be ruined, not due to any particular inefficiency on the part of the collector, but due to the fact that a method had not yet been devised whereby crumbling bone and matrix could be made to travel well. Since bone was to bring both prestige and money when the real war began it became important to solve this major problem. One only has to consider the remoteness of most badland areas to recognize how difficult it must have been to attempt to haul fragile fossil material for many miles over the roughest terrain. From the vantage point of our shock-absorbered cars it is hard to know just how abrasive these trips over ribbed and rutted roads could really be, but a short trip down an unpaved country road gives one at least an inkling of the difficulties one would have in protecting fossils from the jolting and jarring.

William Berryman Scott in his autobiography, *Some Memories of a Palaeontologist,* was to describe the then-current method for extracting bone, and the usual consequences:

> With the crude methods of extracting fossil bones which were then in use, we could not avoid breaking a large bone into several pieces in taking it out. Each fragment was carefully wrapped in cotton, or tissue-paper, and the parts of a single bone, if not too large, were wrapped in heavy paper and tied up in a single parcel. The parcels, in turn were packed in wooded boxes, with straw, or sawdust, for their long journey by wagon and rail to Princeton.[3]

Grass was to prove to be another material that was often used in the early days as packing. It was obviously cheap, available, and capable of providing fairly good cushioning. It had the advantage of not having to be hauled in from great distances since it could be found within short distances even in the remote badland areas. On the other hand it had the disadvantage of rotting and losing its heft so that fossils tightly packed in the material would eventually loosen up at some point and fossils could then be damaged. Charles himself used this method with some frequency, as well as using corn meal in a like manner, before he discovered his new method. Bran and excelsior were two other packing materials that Sternberg was to use at one time or another.

Sternberg claims in his autobiography that he was the one who established the first true innovation in technology for preservation, and that other methods by other collectors followed. Perhaps even he did not fully realize how major a breakthrough this was, and how long-lasting the general concept would be (as the same basic kind of procedure is still followed today, although somewhat modified). He explained the circumstances surrounding the innovative change in this manner:

> When we [Cope's expedition] uncovered these bones [*Monoclonius*] we found them very brittle, as they had been shattered by the uplift of the strata in which they were buried; and we were obliged to devise some means of holding them in place. The only thing we had in camp that could be made into a paste was rice, which we had brought along for food. We boiled quantities of it until it became thick, then, dipping into it flour bags and pieces of cotton cloth and burlap, we used them to strengthen the bones and hold them together.[4]

Considering Sternberg's reputation for honesty and loyalty, and the fact that he was not contradicted by Cope or other workmen aware of the circumstances, it would be hard to dispute his story.

The true sign of a great technician, an artisan if you will, the sort that manages to rise above the everyday doldrums of drudgery, can be best determined by looking at the innovations, and time and quality-savings that they bring to their work. Since large-scale dinosaur collecting was such a young science it left the door wide open for men like Sternberg to set the standards; to develop the rudimentary techniques that would be routinely followed by those who came after. Charles Sternberg was more than equal to the task, and if for no other reason than this his name should resound the halls of all major college earth sciences buildings, and yet we find few who really are aware of the evolution of this technique and to whom they owe homage.

It would be unfair to say that there remains no controversy in all this, for there are some dissenters, but there is clearly no question that Charles Sternberg himself firmly believed that he personally was the founder of this modern method. He follows with a wider explanation as to where this innovation was to lead:

> This was the beginning of a long line of experiments, which culminated in the recently adopted method of taking up large fossils by bandaging them with strips of cloth dipped in plaster of Paris, like the bandages in which a modern surgeon encases a broken limb.[5]

There is no doubt that Sternberg was convinced of his role in the matter, but it does appear strange that Cope remained totally mute on the subject.

Considering the importance of this breakthrough, and Cope's boisterous and combative personality, one would have thought he would have been beating the publicity drums; and yet he failed to comment at all. Perhaps the answer is simply that Cope was just more interested in the fossils themselves (in this case the horned dinosaur finds), and therefore gave the new method little thought. Considering this was Charles' "living," however, he could not have failed to recognize its full import, and he went on record by saying such.

Barnum Brown apparently believed that he was the first to use the plaster of Paris method (admittedly a spin-off), and at least implied so in his "mystery dinosaur" article of 1938. He began using this method in 1897, considerably after the first steps had been established by Sternberg in 1876. Some of Marsh's collectors, and most notably S.W. Williston, had used an interim method as early as 1880.[6]

Harold J. Cook, son of Captain James H. Cook (the owner of the famous fossil property called the 04 Ranch, but which later became known as the Agate Springs Ranch), later took ownership of his father's property and its famous fossil quarries. He was to become a prominent collector/paleontologist in his own right, as well as an acknowledged expert on mid-to-late Cenozoic mammals. He made the following related claim in his memoirs:

> After the bone has been allowed to dry, it is exposed more and more, further hardened, then coated with a thin paper such as Japanese rice paper, or Kleenex, with Shellac dribbled over it. Strips of burlap dipped in plaster of paris are then fitted with care down over the exposed fossil. This process was adapted from the medical profession, being first used by a doctor who was interested in fossils.[7]

It is not clear as to whom Cook was referring to as the doctor in the passage above, nor as to whether or not this information was based upon his personal observations or whether it was a case of hearsay from another party. Without knowing this it is hard to give much credence to his claim.

It is interesting, however, to reflect on the fact that Charles had some interest in the medical profession, both through his brother George, who was to become Surgeon-General of the United States, and through other contacts. Beyond that we already noted his letter discussion with Lesquereux that emphasized this interest in the medical field. It would appear then to be in keeping with what we know of Sternberg that this medically-related innovation (for that part certainly sounds plausible enough as a source) was indeed something which he originated. It is also noteworthy that Sternberg was to become good friends with Samuel W. Williston, and Williston was to become a doctor. Two other assistants, H.A. Brous and Russell T. Hill, were also to become medical men. There is certainly no question that an interest in fossil bones is often closely correlated with an interest in the medical profession, and "live" bones.

Edwin H. Colbert in his *Men and Dinosaurs* clearly gives the initial credit to Sternberg and Cope for coming up with the innovation, and he at least implies that the modern-Marsh method (or Brown method depending upon whom you believe) is an evolutionary byproduct:

> ...Cope and Sternberg devised an improved method for collecting fossil bones. Hitherto it had been the practice to dig fossils out of the ground in rather primitive fashion...no matter how carefully this technique of collecting was practiced, there was inevitably a great deal of damage to the brittle fossils. It occurred to Cope and Sternberg that if the bones could somehow be encased, even before they were completely freed from their rock matrix, a great deal of protection between field and laboratory would be afforded to the fossils...thus specimens could be shipped back to Philadelphia in much better shape than heretofore had been possible.[8]

Colbert, however, seems to give Cope too much due in all this, since a careful reading of Sternberg's *Life of a Fossil Hunter* seems to refute Cope being there at all:

> The species I discovered [*Monoclonius sphenocerus*] were collected on the north side of the river, three miles below Cow Island, *after the Professor had taken the last boat down the river*. When we uncovered...[9]

In other words, it appears that Sternberg is referring here to either the second or third horned dinosaur taken up, and not the first (the one he apparently helped Cope collect). It also seems to be saying that more than likely Cope had already left the area when the innovation occurred. Most likely it was Sternberg and Isaac who were the proponents, performing this novel work while Sternberg was heading south on the *Josephine*.

In one other spot Sternberg makes an obvious reference to this innovation when he stated, "Later I hope to tell of a method, originated by me, by which the most delicate fossil, even if preserved in very loose, friable rock, may be detached and transported safely."[10] Sternberg's words "by me," are certainly sufficient proof for us to assume Sternberg felt he singly was responsible for the change, and that Cope (if he were indeed even there) or Isaac, who most likely was there, were not relevant to the change.

If we truly consider all the evidence before us; Sternberg's known reputation for honesty and his unblemished participation in the Cope-Marsh feud, the silence of Cope and Isaac on the matter, and Sternberg's early claim to the innovated technology, can we really believe anything other than that he was indeed the originator, and that he is therefore also due the recognition that has so long eluded him?

It should be noted too, however, that even Sternberg himself probably did not totally recognize all the implications of his discovery right from the start. This seems to be hinted at in his failure to mention the innovation itself in his article of 1883, which appeared in the *Kansas City Review of Science and Industry* and was entitled "The Judith River Group."

After his inauguration of the rice-method Sternberg was to go on to other methods, since rice glue was merely the beginning and not the end product. Like most of the other men of his trade he was to progress to plaster and burlap (which he did at an early time as well), and he quickly learned the advantages of shellac, gum arabic, and the commercial cement ambroid and added them to his camp kits. Sternberg was never one to ignore things that worked better, and always looked for new ways.

In a late-life interview son Charlie was to comment upon their methodology during the years when they worked Wyoming beds. Interestingly enough he gives credit to both Brown and to his father, although it appears he is speaking about one of the spin-off routines and not the original rice glue method:

> What we did in a method developed by Barnum Brown or my father, we would work around the specimen and then wrap it in burlap dipped in fluid plaster putting in a stick to support it.[11]

Son Charlie was to highlight his own personal technique in a letter of June 22, 1932 to Harold S. "Corky" Jones. It surprisingly differed from that of his father. Charlie suggested that fossils extracted from friable shales should be wrapped in cotton batting, while hard rock fossils should be securely wrapped up in newspapers. He suggested both should then be boxed tightly, using hay or oat straw as filler.[12] It appears evident that since even father and son were to have their own methods that it was most likely kind of an industry norm for different camps and collectors to do things their own way, although most were a variation on the main theme.

As a footnote to this discussion of better technology for collection and extraction it should be noted that Sternberg had a passion for the impossible. If something proved to be "too difficult" it would annoy him to the point of distraction until he could find a way to solve the problem. Just such a situation was what occurred when he and George had tried to collect the massive *Inoceramus* shells that had always only been taken from their Fort Benton Group limestones in pieces. Sternberg was to remark that in the Kansas Niobrara "acres of the chalk" were filled with the fragments of these large shells. The sheer size of these fabulous shells (in some cases well over three feet in diameter) made them a prize of great worth. But that same size made them almost impossible to collect as it was in fact hard to even find one complete in situ. After much thought the father and son team discovered a way of preserving one whole. Charles was to describe the process this way:

> Although they [*Inoceramus*] strew the rocks of the Kansas chalk in

great numbers, they are always broken into small pieces, and these are scattered by the winds of heaven. It seems impossible to preserve them. But George and I learned the secret, and after finding a shell with lips or hinge exposed, we carefully removed the loose chalk above it, then put a frame of two by four lumber around it, in which we poured plaster. On hardening this stuck securely to the shattered shell holding the fragments in place. Then we dug beneath and turned over the panel, and in the shop removed the chalk, leaving one side of the shell exposed in the solid plaster.[13]

G. Dallas Hanna in his book *Methods in Paleontology*, that he was to co-author with Charles L. Camp, was to refer to this specific case of the *Inoceramus* preservation as being an excellent example of forethought in fieldwork and praised the Sternberg effort as something from which all students could learn.[14]

Sternberg had apparently discovered at least the rudiments of this style of fossil collecting at some time before 1903 because he was to comment in a letter to Dr. A. Smith Woodward of the British Museum that he had finally learned how to collect Kansas chalk fossils, and that the secret was "to make the plaster panels in the field before the specimen is taken up."[15]

Charles claimed the advantage lay in other areas as well, for it kept "fragile rock" and the fossil bones from shifting and breaking, and from the mixing and scattering which would perhaps destroy one's ability to reconstruct a given fossil once back in the lab.

In this day and age, where the emphasis is on doing things quickly and with the least amount of effort, and where jackhammers, dynamite and four-wheeled drive trucks have replaced the tools and equipment of Sternberg's day, it is hard for us to understand such devotion to a project. But devotion it was. Charles Sternberg's byword as it relates to his work is perhaps best summed up in a phrase from *Red Deer*, "But patience will always win, no matter what the obstacle."[16]

Another very clear instance where this quality of persistence showed itself was in 1911 when Charles had sent his son George again into the Kansas chalk.

George then discovered a marvelous specimen of *Portheus molossus* that when alive had probably weighed in at around 1,000 pounds. Unfortunately the skeleton was in a pitiful state of decomposition. The elements that had laid bare the specimen were now in the process of destroying it. This is certainly the dichotomy of fossil work, and explains why timing is everything. In any event Charles was not to be dissuaded. He covered the exposed and badly damaged section of the huge fish with a heavy coating of plaster of Paris, and then dug out and around the magnificent skeleton in order to extract it unharmed. Once the fossil had been removed to his workshop he promptly turned the fossil over and worked from the bottom up. The resulting specimen, at that time the most complete skeleton of this specimen ever found, was shipped overseas and now resides in the British Museum of Natural History, a prized tribute to perseverance and forethought.[17]

Charles was to call this find at the time his "triumph," and Dr. A. Smith Woodward was to write him in a complimentary vein causing Charles to respond, "I was glad to receive your letter...and thank you for the kindly words...they go further with me than any other living man, as your great work as a student of ancient fishes likely exceeds any living authority."[18]

In 1930 George was to claim that there were only 15 known mounted specimens of *Portheus molossus* in existence, and George had personally collected seven of them. This was indeed a prodigious feat when one considers the size of these fossils and how difficult they were therefore to extract and mount.

Charles was quick to realize after his work in the badlands that if a man was to excel in the art of fossil finding and collecting it was essential that he first and foremost be comfortable. He must not be "over hungry, or thirsty or sleepy" or the mind would "dwell upon...discomforts, and they will accomplish little."[19] He likened this to a hungry boy who keeps watching the waning sun as the measure of his quitting time, who is so wrought up in watching that he accomplishes nothing.[20] He therefore always took stock of camp needs from the point of view of comfort as well as practicality, and he made sure that wherever possible he added the extras that ultimately would contribute to the expedition's success.

This devotion to a project is indicative of the kind of effort that the Sternbergs, father and sons all, put into each and every expedition they were on. It is also the reason that they were so well thought of by all their rivals, and the reason why American and foreign museums would go to them in full confidence of receiving an honest and worthy day's work. Would we not all like to be so well thought of in terms of our life's work? Most certainly.

NOTES

1. Henry Fairfield Osborn, *Cope: Master Naturalist*, Princeton Univ Press, Princeton, NJ, 1931, Pg. 273

2. W.E. Swinton, *The Dinosaurs*, Wiley & Sons, NY, 1970, Pg. 289

3. William Berryman Scott, *Some Memories of a Palaeontologist,* Princeton Univ Press, Princeton, NJ, 1939, Pg. 172

4. Charles H. Sternberg, *The Life of a Fossil Hunter*, Henry Holt & Co., NY, 1909, Pg. 88

5. Ibid, Pg. 88

6. Roland T. Bird, *Bones for Barnum Brown,* Texas Christian Univ Press, Ft. Worth, 1985, Pg. 213

7. Harold J. Cook, *Tales of the 04 Ranch: Recollections of Harold J. Cook, 1887-1909*, Univ of Neb Press, Lincoln, 1968, Pg. 210

8. Edwin H. Colbert, *Men and Dinosaurs*, E.P. Dutton, NY, 1968, Pg. 83

9. CHS, *Life,* Pgs. 87-88 [italics mine]

10. Ibid, Pg. 42

11. Interview with Charles Mortram Sternberg by Free-lance Broadcaster Laurie LaMaguer, Courtesy of Tyrrell Museum, Dept. of Paleo.

12. Saskatchewan Museum Ltr, Jun 22, 1932 (To H.S. Jones)

13. Charles H. Sternberg, *Hunting Dinosaurs in the Bad Lands of the Red Deer River Alberta, Canada*, Privately Published, The World Co Press, Lawrence, 1917, Pgs. 25-26

14. Chas L. Camp & G. Dallas Hanna, *Methods in Paleontology*, Univ of Calif Press, Berkeley, 1937, Pg. 84

15. BMNH Ltr, Feb 7, 1903 (To A. Smith Woodward from Lawrence, KS), Courtesy of Natural History Archiver, London

16. Charles H. Sternberg, *Hunting Dinosaurs in the Bad Lands of the Red Deer River Alberta, Canada*, Pg. 31

17. Ibid, Pgs. 11-12

18. BMNH Ltr, Mar 2, 1912 (To A. Smith Woodward from Lawrence, KS), Courtesy of Natural History Archiver, London

19. CHS, *Life,* Pg. 38

20. Ibid, Pgs. 38-39

CHAPTER FIVE
MORE KANSAS (1877)

Following his triumphal year of 1876 Sternberg traveled to Pennsylvania to spend the winter as the guest of Professor Cope. He first went to the summer home at Haddonfield, and then from there to Cope's famous "museum-house" on Pine Street in Philadelphia.

Being at Haddonfield must have been a special treat for Charles, since it was near here that the first dinosaur in the new world was discovered. This fact, plus the mere proximity to his hero Cope would have made it a very special winter indeed. Although Charles does not specifically mention it, one suspects he could hardly have visited the area without taking leave to check out the hadrosaur site from which Joseph Leidy had collected this most famous of American dinosaurs for the Philadelphia Academy of Science.

While at Haddonfield, however, we do know that Charles was to make a visit to his aunt at nearby Morristown, New Jersey. It was this aunt who had agreed to safekeep Charles' boyhood fossil collection that had been garnered from the hills around Middleburgh, and the "fossil worms" and oddities which had been collected by his Uncle James from Ames, New York before being passed down to him. He was forced to give them to his aunt because his father Levi had refused to allow him to bring the items to Iowa since the freight was so high for such heavy items. Charles was most happy to again look upon these treasures of his youth.

While visiting at Haddonfield Charles boarded quite comfortably with the amiable Jacob Geismar, who was for many years Cope's one and only preparator. Charles spent his time that winter working with Geismar in the "commodious loft of a large, old-fashioned barn" which had been converted into a workshop. Although Charles does not reference anything in his writings about the specifics of his time with Geismar, it would seem fairly obvious that Geismar's expertise in preparation would have made him an invaluable teacher, and that Charles could not have failed but to benefit greatly from seeing the work done firsthand by a professional. Since this was a part of the industry that Sternberg could only guess at, never having received any formal training in preparation beyond what he devised on his own or read somewhere, and because he had to that point worked mainly on leaf impressions, this experience must have been invaluable. It is somewhat odd, in light of those facts, that Sternberg didn't specifically comment upon this further. Perhaps Geismar was not really amiable after all, or perhaps he was more of a worker than a teacher? We'll most likely never be able to answer that question.

If Sternberg had not already been sold completely on the merits of his boss he certainly was so by the time that he finished his Philadelphia stint that winter. He describes in most vivid terms the awestruck wonder that he experienced in being able to eat dinner each Sunday at Cope's residence. He particularly relished being able to partake of the Professor's learned conversation at these times, and actually likened this to a feast.

Cope had a knack for imparting his own personal enthusiasm about the fossil profession to those around him, particularly to those students and youngsters lucky enough to become part of his inner circle. This infectious behavior, that drew in many young men to the profession, is well described by Henry Fairfield Osborn who was a recipient of some of that enthusiasm:

> I often thought during my long, dull days of plodding research that

> a visit to Cope's study was like a draught from the Pierian Spring. It is impossible to describe the invigorating nature of his companionship...His half century of research, extending from his seventh to his fifty-seventh year, was ever buoyant in its influence; it lifted him far above disturbance by such commonplace matters as where he could get the next meal or pay the next bill.[1]

And we could perhaps easily add to that last statement saying "or climb down a mountain at night, that was impossible to come down without exposing oneself to the possibility of mortal jeopardy."

Sternberg's attachment to Cope was firmly cemented by the end of that winter, and he maintained a deep respect for Cope that went well beyond the normal boss/employee relationship. Even when Charles went to work for Marsh in 1884 he continued to stay aloof from the controversy (and from Marsh as much as possible). He maintained that same regard for Cope right up to the time of Cope's death in April 1897.

This does not, however, mean that he wasn't at times critical of Cope when he had reason to believe Cope was wrong. No one, Cope nor Marsh nor Williston, was perfect in Charles' eyes, and he took them all to task when he felt they were in error. A good example of this can be found in the proceedings for the Kansas Academy of Science for 1905 when he referenced:

> To me, after all these years, having collected many specimens of this large tortoise [*Protostega*], it has been a wonder how he [Cope] was able, from the bones he collected himself and restored with infinite care and patience, to make such a nondescript of the animal...One mistake leads to another. The professor thought the skeleton he discovered lay on its back...It is useless to try and understand how he could have made such mistakes when he had so much of the skeleton present.[2]

He then takes on all of the major players when he further states, "my discoveries have often proved that when men of science guess on the structure of an animal which they only know in part they usually guess wrong. Our works in paleontology are full of such errors."[3] This kind of flagrant disregard of intelligent reasoning continues today when outlandish claims are made on slim bone evidence.

It appears that not only was Charles critical of Cope scientifically at times, but they managed to have other differences as well. In contrast to Charles' glowing reports on Cope (which he continued up to his own death) we find he also had exasperations with Cope that led him to complain to Henry Fairfield Osborn. In an undated fragmentary letter housed at the American Museum Charles was to write that Cope "used to hurt me beyond human indurance (sic) and I fought him back."[4] He was to say this in reference to Cope's failure to respond properly to the clear evidence refuting Cope's erroneous assumptions about the "Pliocene Man" incident in the John Day (see chapter six).

Apparently, however, their differences at times went even deeper:

> In business also we could not agree and I had to actually threaten to sue him to get my money and for years he refused to open my letters. When he sold his collection of mammals to your museum I was the first man he employed...

He goes further in explanation:

> I only look at the miserable failings of that great mind, and stoop as low as some do now even to recount them. No. I always honored his grand intellect, and sorrowed for his sins and mistakes, and he recognized the absolute honesty of my life and motives, and the fact that I would not bow my knee and honor him for what he did not deserve. Nor would I knowingly uphold his errors. Now the honor he paid me in the letter I have shown you is a greater treasure to me than if he had left me an estate.[5]

This is indeed an interesting exchange of thoughts between two great men who both held a great love and respect for Cope. With certainty Charles loved Cope and often said so, but he too was aware of Cope's humanity, a humanity that often brought him up short.

The summer of 1877 found Charles again back in the Kansas chalk in an attempt to find more of the same in the Cretaceous. The discoveries of the first half-year were relatively uneventful, and hence Cope ordered Sternberg to the Upper Miocene beds in Nebraska. These beds had been labeled the Loup Fork beds by Hayden, who had taken the name from the Loup River which serves that area.

Sternberg's resourcefulness again showed itself when he took advice from an old buffalo hunter by the name of Abernathy, who steered him towards either an "elephant" or a "mastodon" skull near Abernathy's cabin in Decatur County. The skull was supposed to be in a sandstone ledge that ran above Abernathy's cabin, but upon arrival there Charles was skeptical. Charles claimed that the "rocks were too old for the elephant," and on viewing the fossil itself finally he found it to be the carapace of a great land turtle, *Testudo (Xerobates) orthopygia*. The carapace was situated in the rock in such a way as to make it appear to the novice Abernathy to be the "back of the head of a large beast."[6]

Such incidents as this were legion with these early fossil hunters, since the general populace was largely unaware of what a true fossil looked like, and certainly could not distinguish between a valuable fossil and a modern bone. There were many times in Sternberg's career where amateurs played both positive and negative roles as relating to fossil finds. In most of the cases, however, the find rarely ever turned out to be what it originally was claimed to be, or the bones were too few or too damaged to be of any value. There is more sophistication today in this respect since people are generally better educated in the sciences, and the media often highlights fossil finds, which helps to make procedure more evident to the novice. Destruction of major fossils by the uninitiated is less common today, although certainly not extinct.

Following this light excursion Sternberg went back to work with a passion. He explored the northwestern Kansas portion of the Loup Fork Group, which heretofore had been an unknown entity, and found it to be very rich in fossil treasure. This area was at the time almost totally devoid of human habitation, and in that sense was very primitive. Virgin countryside generally breeds danger, and such proved the case for poor Abernathy. Sternberg tells us that Abernathy was killed and scalped at his cabin door the following year by hostile Kiowas.[7]

The rocks of the Loup Fork Group as mentioned by Sternberg were somewhat unusual. He credited the formation with resembling old mortar, since it was made up of disintegrated gray sandstone, and indeed had the consistency of mortar. The material was difficult to walk on, but ideal as a building material for it was used as a form of mortar on sod-housing in the area.

It was in this formation that Sternberg was to make his greatest find, that of the Sternberg Quarry (later called the Marsh Quarry). This, however, was not to occur until 1882, at which point he was working for Marsh and not Cope. Although he was not yet to find the quarry he did find the skeleton of one of the quarry's denizens, *Aphelops fossiger (Teleoceras)* which would prove dominant in the Sternberg Quarry. *Aphelops fossiger* was an early form of rhinoceros.

The expedition of 1877 was to prove to be most successful, particularly in the fact of its uncovering a wide variety of fish specimens. Most of these finds were apparently collected in the vicinity of the south fork of the Solomon River, some 37 miles from Buffalo Park where they were headquartered. But they also collected a number of fishes as well on the Smoky Hill River. Charles was to note as early in the season as April 21 that they had already collected some 30 fish specimens, while finding only one saurian, a *Platecarpus*.

The weather in late April was troublesome with substantial downpours of rain. On April 22nd things were further complicated with a late snow. Charles was to describe their ordeal at the time as exasperating, and specifically recalled their being "cooked to the skin." The weather took its toll on the animals as well. Charles' horse, Buttons, was in terrible shape, was passing blood and almost died. After many deprivations caused by the lack of a useable horse he was finally able to replace Buttons with a horse called Rowdy. Unfortunately this horse was to suffer a similar fate because of the harsh weather at the time.

In early May Charles' assistant was to discover a fairly complete skeleton of *Hesperornis*, an atlas of *Clidastes* and a perfect skull of a turtle called *Toxochelys*. They had over 50 specimens of fish by that point. By the later part of May they had added a number of different kinds of skeletal remains of

Pterodactyl, although apparently no complete specimens of this rare fossil. By the end of May, however, they had amassed a valuable enough load in order to transport a full load back to Buffalo Park.

Charles was to note that their total seasonal fish finds at the end of June stood at 186 different fossils. This seems to be in accord with Samuel Williston's comments to Marsh that the party Williston headed for Marsh purposefully left all the fish remains for the Cope party. This was because they presumably placed less value on these finds. Considering the ultimate value of many of these finds this might well have been a case of the Marsh group's ploy backfiring. It certainly made Cope into an expert on the fish fossils of Kansas.

This part of the expedition ended with the removal of fossils to Buffalo Park, and the party then moved on to other areas for collecting. Charles noted that the balance of the season (June to August) was spent exploring Sappa Creek, Beaver Creek and other areas in northwestern Kansas. These Loup Fork exposures of Tertiary age brought further diversity to their camp bags that year, and included mastodon and rhinoceros remains.

Sternberg's assistants on this expedition for Cope were Russell T. Hill and Wilbur Brouse. Russell Hill, was obtained when Charles was in Philadelphia. Hill had been working for the Philadelphia Academy of Science in some capacity, and he seemed the perfect crew member because of the varied knowledge he brought to the field. Hill was hired as Charles' chief assistant, and Brouse as the cook, teamster and general assistant. Although Hill could have had an outstanding career as a fossil hunter it was not to be since he had aspirations to be a doctor. This is what he ultimately became after leaving the fossil profession. But this was not before he performed a valuable service for Sternberg and Cope. When Cope determined that he needed Sternberg's services elsewhere it was ideal because he could hand over the responsibilities to Hill with full confidence. In August of that year Sternberg received a letter from Cope directing him to immediately proceed to new fossil beds discovered in eastern Oregon, in the vicinity of Silver Lake, and he wanted Charles there with all speed.

Before Charles could turn the expedition over to Hill completely, however, they had to transmit their existing cache of fossils to Cope. The nearest shipping point for their large load was at Buffalo Park Station, which was some 75 slow miles away. This was a full two days journey away, and so the allspeed was subject to the limitations that Kansas was always to impose on those who chose to work there.

Charles was to travel to Fort Klamath in the southern section of Oregon, just below the famous Crater Lake, and from there he was to proceed east to Silver Lake to meet with the local postmaster, Mr. Duncan. Duncan was to be his guide to the final destination. As was usually the case it seems Cope was concerned about confidentiality and instructed Charles to use the utmost discretion in seeing that he was not followed, and to be on the safe side he should take a circuitous route to get there.

But before leaving for such a major expedition (where length of fieldwork was so unpredictable) Charles made sure to stop by the family ranch in Ellsworth County before leaving. His visit, however, was only an overnight one. While Cope would undoubtedly have been upset had he known that Charles had possibly compromised his route by going home first, Sternberg figured he needed to visit the family before getting under way. He was not as paranoid about security as was Cope. This was probably due in part to the fact that Sternberg knew most of his rivals personally and felt that he could elude them should the need occur. By the following day he was again at his departure point at Buffalo Park Station, and he was off and away with a blanket roll, fossil tools and equipment, and his luggage, westward by train.

NOTES

1. Henry Fairfield Osborn, *Cope: Master Naturalist*, Princeton Univ Press, Princeton, NJ, 1931, Pg. 584

2. Charles H. Sternberg, "Protostega gigas and Other Cretaceous Reptiles and Fishes From the Kansas Chalk," *Kansas Academy of Science Transactions*, 1905, Pg. 123

3. Ibid, Pg. 123

4. AMNH Ltr, n.d., to Henry Fairfield Osborn [written certainly after 1900 and perhaps as late as 1923], From the Archives of the Dept. of Vert. Paleo.

5. Ibid

6. Charles H. Sternberg, "The Loup Fork Miocene of Western Kansas," *Kansas Academy of Science Transactions,* 1905, Vol XX, Part I, Pg. 71

7. Charles H. Sternberg, *The Life of a Fossil Hunter*, Henry Holt & Co., NY, 1909, Pg. 120

CHAPTER SIX
OREGON DESERT AND THE JOHN DAY

Sternberg, as directed by Cope, headed for the eastern Oregon desert area around Silver Lake without any delay. In his only published poetical volume, *A Story of the Past or The Romance of Science,* Sternberg indicates that the change to Oregon was a pleasant and refreshing one for him:

> One day a welcome order's given
> That lifts my soul to gates of heaven,
> The Kansas Plains I leave behind
> And other fossil regions find,...[1]

Again, transportation being what it was at the time, a circuitous route had to be taken; west to Sacramento by train, then north to Redding, California by rail, and further on north by concord coach pulled by an eight-horse team. Sternberg was obviously taken with the terrain in this part of the west, and stated exuberantly in his autobiography that "it laid a spell upon us; we were dumb before the invisible presence of the Power that had reared this stupendous pinnacle."[2] This outburst was primarily in response to his seeing Mt. Shasta in all its splendor rising above the heavily treed areas of Northern California.

Silver Lake, the first destination, was where his guide George Duncan lived. Duncan was both a local rancher and the first postmaster there at Silver Lake. The lake itself at the time was said to be about 12 miles long and eight miles wide, and was fed by fresh water from neighboring Silver Creek. The lake itself was on the alkaline side, most likely due to the volcanic character of the low hills that surrounded the lake. Today, however, Silver Lake is regrettably only a dry remembrance.

Charles apparently encountered some difficulty in arriving at this first destination for he was to recount an incident involving the local Indians in an article in the *Rocky Mountain News*. Charles noted that he and a friend (since he appeared only to have a hired companion here, it is really not clear who this individual might have been) were looking for Silver Lake, and were following a "government" department map given them by Professor Cope. The map was wrong to the extent that it failed to tell them which way to go at a particular fork in the road. At that moment an Indian lad came along and offered to guide them to a nearby sheepherder's camp, and they allowed him to take them there. They were, however, dismayed to find themselves led instead to an Indian settlement. Charles indicates that five Indians with raised Winchesters demanded "White man lost. We tell right way for two dollars."[3] Not quite knowing what to do, he yet realized that in this potentially explosive situation that he must take decisive action, so he quickly decided on the following course:

> Well, I wasn't going to give them two dollars for anything. But pretty soon they were trying to get five dollars out of me. In those days I was a great smoker, so I put a wad of tobacco in my pipe and drew in a lot of smoke. Then I blew it, as hard as I could, right in their faces, put the spurs to my horse and we were off.[4]

Shortly after this incident they discovered that the Indians whom they had encountered the evening before had apparently stolen their larder of bacon and other essentials, and they lost their bread and coffee rations in their mad dash away from this encounter. This caused them to continue into the next day without any food. Fortunately they arrived at their destination soon after this and were able to locate both food and guide.

The boneyard that they had been so circuitously summoned to was reportedly about 40 to 50 miles from Silver Lake in the arid desert area of Eastern Oregon.

Charles was to name the immediate site where they finally began collecting "Fossil Lake," and he described it as a clay-bottomed and ancient lake now partially covered by sand. A local rancher in later years, Reub A. Long was to note that Fossil Lake was not always dry as it had water in it in 1908 when he was driving horses through that territory.[5]

Fossil Lake still bears the same name today, although it actually got its name at an earlier date than that noted here. Although Charles was of the opinion that it was he who named the lakebed, it turned out that the state's governor had done so before him. Their choice of names had been exactly the same. Governor John Whiteaker was said to have given the lake its name in June 1877 when he purportedly collected over 200 pounds of fossil bone there on a quick trip through the area.[6] This hasty expedition inspired Thomas Condon, the great paleontologist of Eastern Oregon fame, to make a more complete reconnaissance of the same area a month later with the governor's son, Charles.

After Condon returned from his trip he wrote directly to Edward Drinker Cope, who was then visiting at the Dalles in Northern Oregon. Condon advised Cope of the wide variety of fossils at Fossil Lake and sent him a few fossil specimens to augment his comments. It was this communication that prompted Cope to summon Charles Sternberg to the vicinity.

According to Cope the boneyard of Fossil Lake had been initially discovered by cattlemen looking for lost stock. He also contended that these same cowmen had carried off many of the best specimens when they left, to the detriment of science.[7] These specimens might well have been ground up for use as a fertilizer as was to prove the case with the Sternberg Quarry fossil bones, but more likely ended up on a junk heap somewhere or in an amateur's display case with no definition, and therefore no longer of value.

Sternberg's arrival at the site was in the early evening, and he describes his exhilaration at finally reaching the lakebed by saying that he could not "think of anything else" but to get out into the field to begin his search. Directing his crew men, Mr. Duncan the guide, and George Loosely, his assistant from Ft. Klamath, to begin setting up the camp and preparing dinner he immediately grabbed his collector's bag and tools and headed out in the waning light to begin his canvassing of the ancient lake.

That first evening, and in very short order, Charles was to make discoveries of what he thought to be of great value. He found some signs of what appeared to be early man in the area. These consisted of a number of artifacts fashioned from obsidian, mostly spear and arrow points. Since Duncan was only to stay the night before making his return to Silver Lake, having performed his guide duty, Charles was to have him hand-carry a cigar box full of the early implements back for mailing on to Professor Cope.

Along with the box Charles was to write a letter which was to eventually find its way into Cope's magazine, *The American Naturalist*. This was published in the year 1878, in a revised form under Cope's own name. Charles was to note in a later article in the *Popular Science News*[8] that the major section of this letter is to be found in the parts "enclosed in quotation marks." It turns out that the article is not in the *American Naturalist* for 1877, but for 1878, and adds nothing that Sternberg does not describe in much more depth elsewhere. It is, however, interesting because as far as I have been able to ascertain it is the earliest published work by Charles Sternberg, even though it technically bears Cope's name. It is additionally noteworthy because Charles was to later disavow his more immediate conclusions as to the ancient age of the obsidian artifacts upon further discoveries. Although he reported these new facts to Cope (i.e. the discovery of the remains of an Indian settlement and the obsidian tool worker's shop location) Cope continued to uphold his errored conclusions even when various authorities authenticated Charles' revised findings.

Once Charles really got into the lakebeds he found much to quicken his pulse. Not having been previously involved in any appreciable way with anything relating to early man before the obsidian finds especially pleased him, and it also gave him an opportunity to contribute to the knowledge of the area geology as well. A search of the beds showed that evidence of fossil vertebrate bone was also abundant, covering the surface almost everywhere. Scattered in among the bones were a plethora of obsidian arrowheads and some spearpoints that he was to gather up.

Unfortunately, everything in the way of fossil material and artifacts had been disturbed, which meant that they had to sift the sand for weeks in order to separate out the bones and artifacts from other debris. Out of the impressive array of material they were able to collect only one major piece, the partial skull of *Elephas primigenius* which they gathered directly from the clay-bottom matrix. All the other material was disheveled and not embedded in rock. There is no question that the value of material when found in this mixed condition is lessened, and the classification of individual bones becomes more difficult than with bones found articulated and *in situ*, yet there was value nevertheless. When one has such a wealth of material to work from, and when it almost literally jumps into your camp bag, it provides real impetus to the forager and there is delight in almost every morsel found. In that sense both Charles and Cope were pleased.

As to other material found in the desert basin there was a substantial amount. Charles notes finding bones from llamas, horses, elephants, sloths, birds, fish, dogs, otters, and a wide variety of other smaller animals. Cope himself was personally impressed enough so that he mounted his own expedition there two years later, and John C. Merriam of the University of California at Berkeley would also visit the site in order to check out the fossils and artifacts. This notoriety proved to be a real feather in Thomas Condon's cap as it did much to promote national acclaim both for him and for the value of the Eastern Oregon fossil fields.

Reub A. Long, previously mentioned, was to write that his older brother in 1902 was to be fallen on by his horse in a freak accident, and this broke his leg. The accident occurred at Fossil Lake. Luckily there was a party of students working there, and one, a medical student, was to tend to his injury. Long was to further note that the student team was to haul away their fossil trove in a full, four-horse wagonload. He thought the fossils ultimately went to the

Smithsonian.[9]

Charles acknowledges both in his autobiography and in other writings that the implements and artifacts found were not really of great antiquity after all. He based this observation on the fact that he discovered not too far distant from the lakebed an abandoned, modern Indian settlement. He estimated its age at being no more than 100 years old. In this clearly defined village site he was to find the remains of the work area for the toolmaker or makers who apparently fashioned the bulk of the obsidian tools that were scattered over the whole of the lakebeds. The chips from the toolmaker's efforts littered the shop site and were unmistakable signs of what had occurred there. It was this discovery that made Charles change his mind and admit that the artifacts found were not of ancient age as had been initially suggested.

Cope was to note that his own later trip there in 1879 was valuable as well and although his expedition did not collect as much material as did Sternberg they still found much of value. Cope noted finding three different species of llama for instance: *Holomeniscus hesternus* Leidy, *H. Vitakerianus* Cope and *Eschatius conidens*. He also noted a huge sloth named *Hylodon sodalis*, the mammoth *Elephas primigenius*, and a variety of other vertebrates.

Since Sternberg had more to accomplish in the area, and since he was a long distance from Kansas he decided to spend the winter of 1877-1878 there, wintering in the nearby Pine Creek area of Eastern Washington. It is not clear as to why this particular site was chosen, but it may well have had to do with its fossil potential. This method of wintering near a fossil site was not Sternberg's usual practice (particularly in later years) for he would regularly spend the winters in his laboratory at Lawrence, Kansas, preparing the specimens recovered during that summer's expedition. Since most of his collecting was done as an independent collector, at least after his establishment in the trade, he would be preparing fossils to sell during the off months, as well as writing his popularized accounts of life on the fossil fields. He also would spent a great deal of time handling a voluminous amount of correspondence from the world's greatest museums. This practice kept him totally busy year-round.

Charles indicated that earlier in November of 1877, while traveling down the Columbia River from the Dalles in Northern Oregon, he was to encounter a young army surgeon who enlightened him about a fossil deposit of some worth that had been uncovered on the property of a farmer by the name of Copeland. Copeland's land included a swamp section of "mud-springs," and it was in this section of his land that Copeland was to uncover a large and complete skull, and numerous other large bones. When Sternberg finally was to meet Copeland in March 1878 he was told by Copeland that nine relatively complete mammoth skeletons had been unearthed todate and had been donated to a local college. Before this meeting, however, Charles was to try his hand at working in mud-springs, although these were presumably not on Copeland's property.

Sternberg's memory about this incident is certainly questionable at best, for in 1903 he was to write to William Diller Matthew of the American Museum of Natural History regarding this period:

> I send you the proof of my article on the 'Swamps of Whitman Co.' Washington. Fortunately I found enough of my diary made at that time to give exact information as I learned from others and from my own experience. Unfortunately I can not remember names, and this Mr. Copeland may not be the one who procured the elephant sold to the showman.[10]

Although the weather is extremely wet in the northwestern section of the United States, and rain more common than dry weather, it is also equally mild in its temperature. The rains themselves are not like the cold rains so common in the Northern Midwest and hence a kind of work can be performed even in the midst of the off season. Sternberg records that they explored "swamps" in the vicinity of their camp and fought "against water to secure specimens."[11]

Charles was to say that their terrible labor in that winter provided little material, and much frustration. The mud-springs turned up no similar mammoth material, and the primary finds were only a few quality skull finds of the buffalo. Charles, however, felt that the peat swamps of this location would make a prime searching arena for archeologists, and that such a search would eventually uncover implements and tools of early man in great numbers. He therefore recommended that others consider searching in that vein, but had no further opportunities to work there himself.

There was apparently an interlude between the Pine Creek, Washington, wintering and the next trip to Oregon for Charles notes in his autobiography that "In the winter of 1878, while in San Francisco, I received orders from Prof. Cope to go to Oregon..."[12] Just exactly what Charles was doing in San Francisco is yet unclear, but there is at least reason to suspect that he was working there for Cope in some capacity.

When the spring rains subsided he started for Fort Walla Walla, near the present city of Walla Walla in southeastern Washington, along with two newly acquired assistants, Joe Huff and J.L. (Jake) Wortman. Wortman had been a model student of Thomas Condon, and had shown his talent early as a student of geology. Wortman was to later gain fame in his own right as he became a prominent paleontologist, but was now looking to learn the subtleties of the craft from his new tutor, Charles Sternberg.

Wortman was to quickly prove an exceptional talent, and became a mainstay at the American Museum under the tutelage of Henry Fairfield Osborn. Wortman was to also prove to be a maverick. Initially trained in medicine, he tired too soon of fossil work, quit the profession, and moved to Brownsville, Texas, in 1908. Here he opened a drugstore and spent the balance of his life in apparent quiet satisfaction. Wortman was perhaps the greatest of Sternberg's disciples, but certainly not the last as we shall see later.

Charles describes Wortman at the time in the positive terms one might expect, considering our knowledge of his later successes. He called Wortman the "intelligent young man from Oregon," and indicated that Wortman had been referred to him by his brother George, then a post-surgeon at Fort Walla and later to be Surgeon-General of the United States. His brother George obviously had a real eye for talent also, and always seemed to be in the background when Charles needed help or advice. Wortman assisted Sternberg for a considerable time. It appears he began working for Charles in April of 1878, and stayed with the party until sometime mid-to-late 1879. However, Charles was to also write that Wortman, in the six months prior to April of 1878, had "been my guest at my camp on Pine Creek."[13] In any event he certainly had ample opportunity to learn the craft of fossil collecting from Charles Sternberg, and Wortman would go on to teach many others before his career was ended. This is another clear-cut case of Sternberg's contributions to the field.

The impetus that brother George provided in Charles' livelihood was substantial. George was in a position as camp surgeon to be the recipient of many gratuitous fossil specimens and other artifacts that were transmitted to his custody simply because of his position as doctor. This is most likely because he would have been the closest thing to a scientist in the minds of the men who lived and worked there. George had already been the donor of a number of items to the United States National Museum (Smithsonian), and was to later hold a committee position for one of the Smithsonian's award foundations. Because of George's interest and his actual assistance, Charles' preparatory surveying of the territory would have been decidedly easier. Cope in his major work *Tertiary Vertebrata* (more commonly known as Cope's Bible) emphasized this alliance:

> The Tertiary formations explored in 1878 were the John Day, Loup Fork, and Equus beds of Oregon. These were examined by C.H. Sternberg, who received important aid from his brother, Doctor George M. Sternberg U.S.A.[14]

The expedition in May of 1878 traveled south through the Blue Mountain region to reach the John Day Basin near Canyon City, Oregon. The Blue Mountains receive their name from the color of their hue when seen from a great distance. Up close, however, there appears to be nothing that would lend itself to use of that name.

The terrain through which the party traveled, so radically different from Sternberg's usual haunts, was completely foreign to him. Being from the relatively non-volcanic midwest he was certainly not prepared for what he would find in Oregon. He recounts amazement, and perhaps intermingles this with a mild disgust, when he talks about being met with the "disfigured earth" of the gold seekers. The aftermath of the gold miner's lustful cravings in the hills was always a sore spot with Sternberg and he could not mention the subject without some evident dislike.

Gold had been discovered at Canyon City in 1862, and also at neighboring Prairie City (then called Dixie), and there had been a typical "gold fever" explosion of people into the territory that came with the discovery. The hills were still not healed from the abrasive workings (if in fact they ever heal) and were all too evident to Sternberg. Although the rush for gold had been shortlived over $26 million dollars in gold was eventually mined nearby.

When later exploring in New Mexico Sternberg was to really vent his true feelings about the gold craze, which he saw so many men involved with to their detriment:

> It was, therefore, with a feeling somewhat of contempt that I learned the Navajos had given me the name of 'Gold Hunter,' an occupation that seems foolishness to me, when I search for treasures more rare than gold or silver.[15]

In obvious description of gold and coal hunters, as well as describing those who ruined fossils by indiscriminate work, Charles would often use the phrase "vandal hand of man" in his writings. This clearly seems to be in juxtaposition to his equally frequent comments regarding God's work of creation. His frequent use of the phrase points up the vehemence with which he approached this subject. When he has been labeled as a naturalist it is not without some merit, and this strong concern he had for treating the earth with proper respect since it too is part of God's creation was obviously indicative of this naturalist's bent.

Sternberg's move to the northwest territory for work is even more vividly highlighted by his perception of the area rock:

> In this same locality [John Day] there is a bed of rock so light that it floats. I threw a large mass of it at some object in the water, and

was amazed to see it float off down the stream. It was the first time that I had ever seen a rock lighter than water.[16]

The effects of ash and mudflow volcanics were new to Charles since his experiences had been in the plains states, and he was young and untraveled when he lived on the East Coast. He had therefore no occasion to see pumice or diatomite before and was obviously amazed at its properties. Since he didn't call the rock by name (pumice) in his autobiography, which was published in 1909, it is reasonable to assume that he still did not know the rock's name at the time of the publishing, even though 30 years had passed. But then again we should in all fairness note that he normally worked in sedimentary rock, because it is generally in this rock that good fossil material is preserved.

The John Day Basin received its name curiously from a man who never came closer than 100 miles of the area. John Day, the man, was a mountain man of some prestige who migrated from Virginia to work for the Pacific Fur Company as a hunter in 1811. In 1812 he and a companion by the name of Ramsay Crooks were attacked, robbed and stripped naked by Indians at the mouth of the Mau Hau (Mah Mah) River, some 30 miles east of the Dalles. They managed to survive the ordeal when they met up with a party traveling to Astoria. For whatever strange reason people began to associate Day's name with the Mau Hau because of the incident, and started calling it John Day's River. The name was eventually changed to the John Day River. The towns of Dayville and John Day received their names from the river as did the label John Day formation. It seems quite a legacy for a man who never made it much beyond the Columbia River.

The John Day Basin is bordered in the north by the basaltic Columbia Plateau and in the south by the Strawberry-Aldrich mountains. The area provides a wide spectrum of different volcanic formations ranging from columnar basalt, to andesitic and rhyolitic lava flows, to ashen mudflows, to consolidated ash. There is very little that the vulcanologist cannot find here, and find in proliferation. There are a number of distinct formations that cover vast sections of the John Day; the more prominent of these being the Rattlesnake, the Mascall, the John Day, and the Clarno.

The dull greenish tinge of the John Day ash, generally representing a mudflow of ashen material, is a clear marker for the best fossil hunting, but the buff-whites and the reds are exceptional as well. These terraces of chalk-colored ash rise thousands of feet into the sky in some spots, and are water-worn to the point of providing one with a visual plethora of sculpted spires, cones and other novelties which only nature could create.

The denuded cliffs of the John Day are in marked contrast to the bottom-land along the river and to the sloping hills nearby which are sparsely cloaked with bunchgrass and juniper. The mainstay of area plumage, however, is a variety of Artemisia, or common sage brush. There is little else to catch the botanist's eye.

Sternberg was to add two new helpers to his crew here: Bill Day (i.e. Bill Day from Dayville in the John Day working in the John Day!) and the other a Mr. Warwick. They were appropriate additions to the expedition since both had substantial fossil collecting backgrounds from having worked prior for Professor O.C. Marsh in earlier hunts for fossil vertebrates.

The John Day has generally been interpreted as being Miocene in its geologic history, and its badlands are in spots brilliantly colored in orange, yellows and greens. Although the general coloration of the cliffs is a dirty white buff and a more dingy green, the marvelous intricacies of the weather-tooling more than offsets the lack of color. The bone beds are in a consolidated ash, and therefore represent the death beds of a land fauna rather than the sea-life beds Charles was used to working. This radical change of scenery and locale significantly helped to make Charles a more well-rounded collector.

This first effort for Cope in the John Day fields did not meet with grand success, even though a number of items were unearthed. It might, at least in a limited sense, be described as Sternberg's first marginal victory (in light of the fact that a season's success is often irrationally measured in terms of the number and quality of skull finds). The lack of skulls found was not hard to explain as the new crewman Bill Day was to state that when he had previously worked for the Marsh party they had been instructed to take only the skulls of specimens and to leave all else. Day told Sternberg that they "were only looking for heads, though we sometimes saved knucks and jints." [17]

Charles demanded, however, that his crew collect all vertebrate material, and not just the skulls. This "heads only" policy is just another example of Marsh's wanton disregard for the science and points up his equally greedy attitude and apparent need for acclaim.

Since the lower beds in particular had already been extremely well-cleansed of material by the Marsh party, Charles had no choice but to have his party canvass the higher and more dangerous cliffs for material. This naturally increased the chance of major injury, or even death if one should plummet from the heights they had to work. They worked these walls by cutting out niches for footholds, and when they managed to find bone protruding from the wall they would dig platforms from which they could stand to work. It was perilous, tedious work and the dangers and mishaps inevitable. It gives one an appreciation for the difficulties of their work, and the commitment they made to their craft.

It is hard for one who has not experienced the scaling of upright cliffs in pursuit of its treasures to fully comprehend the abject terror that can grip one when they first experience the feeling of losing one's contact with solid earth, and the utter hopelessness one feels before one is again safely on solid ground. This passage from Sternberg gives one a true sense of that struggle:

> I could tell of a hundred narrow escapes from death. One day I was standing on a couple of oblong concretions, about a foot in length, with a chasm, fifty feet deep and three or four feet wide, immediately in front of me. After I had searched carefully the surface of all the rocks in sight, I started to jump over to a narrow ledge on the other side of the gorge. Suddenly both concretions flew from under my feet, and I was plunging head downward into the gorge when by a violent struggle in mid-air I managed to throw my elbows on the ledge; and I hung there until I could find a foothold and pull myself out on to solid rock.[18]

Loye Miller, a student and member of the first field party to the area from the University of California at Berkeley in 1899 was to ratify Sternberg's comments on the scariness of scaling the John Day heights, and he too spoke from personal experience. In 1899 Miller was collecting in the famous Turtle Cove when he had some trouble:

> Got stuck near the top of the cliff, and came as near Eternity as is altogether wholesome. Almost lost my nerve, had to speak to myself peremptorily to keep from breaking down. When I reached a place where I could relax I went all to pieces, falling weak and nauseated as sick kitten. Hadn't the nerve to climb any more cliffs.[19]

The University of Kansas was to send an expedition into the John Day beds in 1905 under the leadership of C.E. McClung. McClung was to describe the inaccessibility of the best fossils, and their relative preservation condition thus:

> The individual exposures of the middle and upper John Day are from 100 to 500 feet high, and in the search for specimens it is necessary to crawl over the surface of these almost vertical cliffs and to look very carefully for bone fragments. Compared with many other collecting grounds in the Miocene there are here comparatively few specimens...It is a very difficult matter to remove the bones in good condition because of the lack of homogeneity in the matrix. On digging into the cliff after a specimen it will be found that for a part of the distance the bones will lie in the hard nodules, and will then for a while extend along in the seams that are filled with much softer material. For this reason it is impossible to remove any large part of a skeleton in one piece, and the individual bones have to be carefully removed and pasted up in cloth. The bones themselves, however, are in an excellent state of preservation and make beautiful specimens.[20]

Considering the magnitude of the problems encountered in the John Day, and the always declared scarcity of good and whole fossil material, it is of note that Sternberg's expedition was still to be labeled as a success. Beyond the more immediate success one can measure in terms of found fossils, one can add to this the more intangible knowledge-building that Charles was to gain from this experience. He had approached a field totally alien to his experience and had valiantly urged his team on to a successful conclusion. He had taken a carefully scoured territory and had found even more of value. By any measurement this would have to be considered a good season.

Of the sites that Charles worked in his stay there the most productive and famous was "the Cove," or "Turtle Cove." The site received its name, which it still bears today, from the famous Oregonian paleontologist Thomas Condon who discovered a fossil turtle there. Loye Miller was to call the Cove "the mecca of our pilgrimage" and the "grandest amphitheatre one could wish to see."[21] It is indeed an intriguing place and seems to mesmerize all who come there.

Miller was also very impressed with the surreal nature of the setting and described the cliffs as "...cut and furrowed into chasms and pinnacles bare as a tombstone. The first impressions I received was that of Dante's illustrations of the inferno. To heighten the impression some of the strata are of a dull dirty green color most repulsive in tone."[22]

Among the discoveries that the Sternberg camp came upon that summer were Oreodont, a new species of camel (named *Paratylopus sternbergi* by Cope), a three-toed horse, a carnivore named *Archaelurus debilis*, a dog named *Enhydrocyon stenocephalus* and the skull of the rhinoceros *Diceratherium nanum* Marsh. All of the expedition finds were eventually sent to the American Museum of Natural History in New York.

The John Day itself, however, has coughed up a great deal of varied fossil material, and much that Sternberg never saw evidence of during his time there. Mighty Titanotheres, a variety of small horses, seeds and nuts, turtles, and saber-toothed tigers are just a few of the other fossil remains that have been taken from its cliffs. Discoveries continue to be made in the John Day with great regularity even today.

But the story of that season doesn't end here, at least as far as excitement is concerned. Although we generally don't think of the Pacific coast as having had "Indian problems," we find that it had its share as well in our country's early history. Sternberg and his small crew (Bill Day and Jake Wortman) were to find themselves caught up in what has come to be called the Bannock Indian War.

In the latter part of Sternberg's season of 1878 a band of Snake River Piute Indians (numbering about 300) had jumped their reservation at Malheur Agency in Southeastern Oregon, and under the command of their War-Chief Egan had headed north into the area around where Sternberg and his crew were working. Chief Egan apparently had reluctantly compacted to join up with a band of Utamilla Indians who were quartered at Fox Prairie, under the command of Chief Homely. The Snakes had stolen a great number of horses and were driving them before them, and they were simultaneously being chased by troops under the command of General Oliver O. Howard.

The Bannock War of 1878 is attributable in large part to the reputed poor, corrupt administration of the Malheur Agency agent, Major W. V. Rinehart, who held back rations from the Snakes and purportedly treated them harshly. When the troublesome Bannocks (who had also been at Malheur) broke the peace in response to an incident whereby a white settler was killed by a warrior named Tambiago, one of the more prominent chiefs, Buffalo Horn used it as an opportunity to rally various tribes to declare war against the whites.

Although Buffalo Horn himself was killed in an early battle (some said by a Piute Indian rather than a white man) the Indian contingents did band together with the disgruntled Snake River Piutes, and headed north towards the Utamilla Reservation in the hopes of picking up additional forces. Although the chief of the Snakes, Chief Winnemucca, in his wisdom was against war he was virtually a prisoner of his own tribe and had to follow the dictates of some of his strong, but younger sub-chiefs. Sub-chief Oytes, definitely a war hawk, was instrumental in pushing for war and he managed to convince War-Chief Egan that they needed to join forces with the Bannocks. Chief Egan then became the primary leader of the combined force made up of Snakes and Bannocks, which numbered somewhere around 700.[23]

Charles indicated that he and his crew were only dimly aware of the presence of the Indian band, even though they were encamped only about six miles away, and so they were just mildly alarmed at the onset. Charles, however, became much more aware of the gravity of the situation when he was returning to the settlement area around Dayville for further supplies and he found that all of the settlers with the exception of two stubborn men had left their homes and belongings to seek the shelter and protection afforded them at Spanish Gulch, where a stockade was being built. His associate Bill Day was particularly alarmed at the thought of being confronted by Indians and refused to return with Charles to seek out Jake Wortman who was still working out fossils at their campsite. They therefore parted company while Charles returned solo to warn Wortman of the impending danger.

Wortman was found quietly working away at another discovery when Charles located him, and although Wortman was anxious to leave immediately once the particulars were made known to him, he agreed to stay with Charles so that they could cache their fossil trove so that it wouldn't fall into Indian hands and be destroyed. They therefore camouflaged their efforts and headed back to the relative protection of the settlement where they found that the Indian fears had not been without justification, since considerable mayhem had already been done. They then had to spend some time secreting the fossils they had been storing at the cabin of one of the stalwarts, a Mr. Mascall, as they feared the cabin might burn if the rampaging Indians came around.

The Snake and Bannock band had killed a number of settlers in the general vicinity, had destroyed a number of other houses and properties, and had slain a considerable number of animals, including estimates of around a few thousand sheep. Much of the damage seems to have been perpetuated in spite, or out of sheer maliciousness rather than for real purpose. This wanton and senseless destruction was hard for Charles to understand. The loss of settlers and soldiers in the total war only amounted to about 40, and no reliable figures are available for the loss of Indian life, although one expects that they were probably higher.

Sternberg notes that he was willing, and even attempted to recruit others, to join in assisting the troops help oust the Indians. He could not, however, get anyone to come with him. This action evidently was of much interest to him as he indicates his following the trail of the band at varying times. He apparently found it fascinating and curious.

The Indian war was soon over and ended somewhat anticlimactically. Snake Chief Egan was assassinated by acting Chief U-Ma-Pine of the Utamillas in an act of pure treachery. The Utamillas performed this ignoble act, however, in the interest of their own peoples. General Howard had captured some of the Utamillas and he was holding them hostage pending Homely's cooperation. The Utamilla, described as opportunistic by the Snakes, had played both sides against the other, but when a supposed $1,000 reward was offered for Chief Egan by General Howard that tipped the scales. The Utamillas pretended to desire joining with the Snakes in their quest against the whites, but instead turned on Egan, killed and beheaded him, and then cut his body to pieces.

With the death of Chief Egan the Indians lost their focus, and it only took a few minor skirmishes before the disjointed bands were finally subdued. The last of the group was taken prisoner by September of 1878.

Sternberg's season of 1879 was spent in Haystack Valley where he also spent the balance of 1878 after the Bannock War was over. One of his new assistants, Leander S. Davis, who had also worked for O.C. Marsh for a number of years, was to team up with Charles during this period. Charles had developed a real appreciation for Davis when he found him to be the only man he could find at the Spanish Gulch stockade who was willing to return with him to Dayville to check up on the two men who had stayed behind rather then leave their homesteads during the Indian uprising. Jake Wortman, as previously mentioned, was again with Charles on this expedition too.

Leander Davis was to also be a member of Merriam's University of California expedition, the one on which Loye Miller was a member. Miller was to find him a rewarding companion:

> I rode on the wagon with Davis finding him a very nice companion. He is a native of the State, one brother is a cow

puncher in Colorado. He served four years as a justice of peace.[24]

and:

> Davis had decided to go and is rustling like a good fellow. He is a quiet and reliable man who knows what he is talking about, well acquainted with fossil fields and knows much about the different forms and systems. An Episcopalian and man in whom one places his confidence.[25]

Davis was to continue personally to work the general vicinity from his home in Baker City, Oregon for a number of years after this venture. A few years before the University of California expedition in 1889 he was with William Berryman Scott's expedition. Scott's crew had made their permanent base at the "Cove," the same base camp from which Sternberg and company managed to elude the Bannock Indians.[26]

In 1974 the John Day Fossil Beds National Monument was established. That monument now encompasses over 14,000 acres of fossil cliffs and surrounding desert. It is obvious that Sternberg had a great part in the making of the monument since he played such a prominent part in proving to others the value of this land as a true fossil haven.

The headquarters of the National Park Service today is located at the Cant Ranch just north of Dayville. The headquarters are housed in a beautiful rambling ranch house of 1901 vintage, originally the home of a Bill and Lillian Mascall for whom the Mascall Formation was to receive its name. Mrs. Mascall is still living in Dayville.

The Mascall's were almost certainly related to the aforementioned "Mr. Mascall," referred to by Charles. He was the stubborn ferryman who refused to be budged from his property by the threat of the Indians, and who assisted Charles in storing his fossils at his cabin. He also helped Charles hide them once the Indian threat became real. This same gentleman also provided various hospitalities for the Merriam expedition of 1899 in which Loye Miller participated.

NOTES

1. Charles H. Sternberg, *A Story of the Past or The Romance of Science*, Sherman, French & Company, Boston, 1911, Pg. 74

2. Charles H. Sternberg, *The Life of a Fossil Hunter*, Henry Holt & Co., NY, 1909, Pg. 145

3. *Rocky Mountain News*, Jan 10, 1939

4. Ibid

5. E.R. Jackman & R.A. Long, *The Oregon Desert*, The Caxton Printers, Caldwell, ID, 1964, Pg. 90

6. Robert D. Clark, *The Odyssey of Thomas Condon*, Oregon Historical Society Press, 1989, Pg. 300

7. Edward Drinker Cope, *The American Naturalist*, Nov. 1889, Vol #23, Pg. 978

8. Charles H. Sternberg, *Popular Science News*, April 1898

9. E.R. Jackman & Reub A. Long, *The Oregon Desert*, Pgs. 131-132

10. AMNH Ltr, Mar 27, 1903 (To William Diller Matthew), From the Archives of the Dept. of Vert. Paleo.

11. CHS, *Life*, Pg. 170

12. Charles H. Sternberg, "Explorations in Northeastern Oregon," *Kansas City Review of Science & Industry*, Vol VII, 1884, Pg. 74

13. CHS, *Life*, Pg. 171

14. Edward Drinker Cope, *The Vertebrata of the Tertiary Formations of the West*, Pg. xxvi, 1884, quoted from Henry Fairfield Osborn, *Cope:Master Naturalist*, Princeton Univ Press, Princeton, NJ, 1931, Pg. 255

15. Charles H. Sternberg, *Hunting Dinosaurs in the Bad Lands of the Red Deer River Alberta, Canada*, (2nd ed), Privately Published, Jensen Printing Co., San Diego, 1932, Pg. 223

16. CHS, *Life*, Pg. 172

17. Ibid, Pg. 181

18. Ibid, Pgs. 182-183

19. Loye Miller (ed. J. Arnold Shotwell), "Journal of First Trip of University of California to John Day Beds of Eastern Oregon," Bul #19, Mus of Nat Hist, Univ of Oregon, Eugene, 1972, Pg. 14

20. C.E. McClung, "The University of Kansas Expedition into the John Day Region of Oregon," *Kansas Academy of Science Transactions*, 1905, Vol XX, Part I, Pg. 69

21. Loye Miller, Pg. 13

22. Ibid, Pg. 14

23. Brigham D. Madsen, *The Bannock of Idaho*, The Caxton Printers, Caldwell, ID, 1958, Pgs. 200-230

24. Loye Miller, Pg. 4

25. Ibid, Pg. 2

26. William Berryman Scott, *Some Memories of a Palaeontologist*, Princeton Univ Press, Princeton, NJ, 1939, Pg. 175

CHAPTER SEVEN
FOSSIL HUNTER ETIQUETTE

Though the general tendency of the various bone-hunting parties was certainly to make every effort to avoid rival bone-hunters, and to keep their new destinations and discoveries top-secret from other camps, it is also equally true that life in the bone camps did not always border on the hysteria and madness that generally has been played up in the press. Situations did, however, with some frequency become such that rival bone parties found themselves dragged into intrigue and cloak-and-dagger behavior that was more in keeping with movie melodrama than real life. If you asked the players they would have undoubtedly described their actions as being strictly cautious, and necessitated by their need to see that they would not arouse suspicions in other camps, suspicions that might lead to some of their rich deposits becoming places of battle.

Adrian J. Desmond in his book, *The Hot-Blooded Dinosaurs,* describes the then-prevailing atmosphere as "reading more like a fictional spy drama than fossil exploration" and that "secrecy and even deception were symptomatic of the paranoia that had overtaken the rival camps in their quest for great saurians."[1] It seems he was certainly right as is pointed up in Sternberg's autobiography and the writings of others from the time.

A modern counterpart of this type feuding between rivals, a rivalry that might better explain the fervor created in Sternberg's day, is perhaps hinted at in the scientific feud between Richard Leakey and Donald Johanson as regards their opposite theories relating to man's origins. This well-publicized disagreement took most noticeable form when the two met face to face a number of years ago on Walter Cronkite's television program, *Universe*. Leakey was particularly unsettled over the meeting since he claims that he was rushed into the confrontation without recognizing what he was getting into. He was caught unaware that Johanson had come prepared with pre-made props and an obvious agenda. Historians of the science could not watch the program without being struck by the similarity of circumstances between this rivalry, and that of Cope and Marsh in their heyday.

Writings from any of a number of different bone men emphasize this cat-and-mouse-type behavior, and Sternberg himself was not above this kind of furtiveness, although his may have been due more to the dictates of his early bosses rather than his own volition. In his autobiography he describes a chance encounter with a friend and rival that seems to speak both to the friendship aspect of the hunters as well as to the necessity of secrecy:

> At Monument Station, I was surprised to see Mr. S.W. Williston get aboard with all his outfit. Williston did not know at first that I was on the train, and when he entered my car, he was greatly astonished, thinking that I was on his trail. He tried to find out my destination, but failed. We slept together at Denver, then he took a train south, while I went north toward Cheyenne and the West.[2]

In an article for the *Kansas City Review of Science and Industry* Sternberg was to own up to his participation in the same kind of behavior, a behavior he conveniently doesn't mention in his autobiographical account. In describing the very same incident he says:

> A day or two after I was on the west bound train and there met Mr. S.W. Williston, one of Professor Marsh's collectors. He seemed to be on his way to a new locality, and refused to tell me of his destination. I thought he might have told the conductor, and

therefore on my first chance asked that gentlemen (sic) where Mr. Williston was going, and he told me at once that W. had learned of a rich locality where fossil bones of huge proportions were found, in Canon City, Colorado,---I immediately wrote to Professor Cope...[3]

It seems inexplicable that Williston would have compromised the secrecy of this major bone field by giving out this much detail to the conductor, but such was apparently the case. This certainly was not in keeping with Williston's normal practice of a very tight security.

We get a further glimpse of the same kind of furtive behavior on Williston's part when he states in one of his letters, "Sternberg & one assistant is down on the Smoky...I ascertained his plans but kept my own counsel."[4] It also perhaps says something about Williston and his friendship for Sternberg that he did "keep his own counsel" rather than passing that information on to O.C. Marsh. Knowing the tantrums that Marsh could throw if he felt betrayed, it was not without some risk perhaps that he made this decision. Although we have at this point no way of knowing, one might speculate that Williston was already feeling somewhat disenfranchised from Marsh and found this as a limited way of getting back at him.

Sometimes it was the necessities of life that brought the rival parties together; and most often that necessity was to prove to be water. Williston refers to this very fact saying, "one of Sternberg's brothers, for example was agent at Buffalo Park, where the clean water attracted the rival collecting parties."[5] This brother referred to was William, a younger brother of Charles' who was agent there for the Union Pacific Railroad. Buffalo Park had been known as Parks Fort, having been named for a railroad contractor by the name of Thomas Parks. This station, as with others, was motorized at one time for its location was to move. It started in Ellsworth and finally ended up in Gove County. It was a fortified construction camp, designed to provide whatever kind of reasonable protection it could for the workmen who were laying rails. This encampment eventually evolved into a station in the modern sense of the word. Buffalo Park, also known at a later point as Old Ogallah, is now known simply as Park.[6]

Charles was to describe the oasis that was Buffalo Park in this manner:

> Here at Buffalo I had my headquarters for many years. A great windmill and well of pure water, a hundred and twenty feet deep, made it a Mecca for us fossil hunters after two weeks of strong alkali water.[7]

A crude drawing of the station by Samuel W. Williston gives one a good sense of its functionality. Elizabeth Shor has had the drawing reproduced in her book on Williston, *Fossils and Flies*. It shows that the station consisted mainly of a large water tower (to refuel both humans and railroad engines), and a few surrounding outbuildings. Life at Buffalo Park was simple at best, but it had pure water and that made it into an oasis.

Charles' further description of Buffalo Park (Park P.O.) when it was in Gove County was that it consisted of a depot and section house. It was both his base of supplies and his shipping point. The fact that his brother was in command here probably did much to allay any fears he might have had in leaving off fossils here and having their privacy protected. He depended heavily upon Buffalo Park to the extent that he always tried to stay within a day or twos ride from it because it afforded his only real link to the rest of the world when working in this desolate area.

The former town site of Park is located about a mile west of Wakeeney. There exists now only the foundation of the crumbling walls. The site in its founding days was strewn with the mouldering bones of myriad buffalo carcasses, and it was this visual fact that lent the "Buffalo" to its name.

The death of Buffalo Park, as it occurred back then, was not unlike that of the small town of Coyote. The death of this Kansas railroad town is recounted within the pages of *Harper's Magazine:*

> Coyote soon disappeared. The temporary terminus moved forward to Sheridan. If the noise of house-building, the blow of the hammer and the tear of the saw, are sweet music to the workman's ear, however jarring to that of the neighborhood, no such plea can be put forth for the sounds which proclaim a prairie building's

removal in situations where each man is his own carpenter. A liberal application of nails has done the duty elsewhere assigned to tenons, and the consequent breaking of boards and voice of the axe are discord most wonderful...In one short week not a house but that of the railroad section men remained.[8]

One perhaps cannot fully appreciate how important clean water was to these bone men until you have opportunity to read some of their accounts. When we simply turn on the spigot in our homes and out comes wonderfully clean water it is hard for us to understand their situation. Although Kansas water may not have been quite so bad, Sternberg described Texas water from the valley of the Big Wichita as it came down from the hills in this manner: "it soon becomes as thick as cream with the fine red clay, and to think of depending upon such water for drinking and cooking purposes is revolting..."[9]

Closely allied to this view of water is Dr. James W. Gidley's jesting comment about the prevailing opinion of Kansas water. Gidley was the Curator of Mammalian Fossils at the National Museum: "The cattlemen of the region insist that the only way to determine whether a given pool contains water or mud is to thrust a stick in it, if the stick falls over it's water; if it stands up, it's mud."[10]

It was also Charles' claim that the lack of good drinking water was the greatest single ill he encountered in the Kansas chalk. Judging from other fossil hunters' responses it would seem that this comment was almost universal. But then again, who would expect life's luxuries in the badlands of the world? It should be added too that most of the midwestern states suffered from this same plight, and in that sense at least Kansas was hardly unique.

Charles likened the drawing power of clean, refreshing water to "Mecca" as we saw before, and this illustration seems to hold true for two main reasons. Not only because of the water itself, but also because the areas around water always provided some respite from the elements since people always migrated to them. When at Buffalo Park, or its counterparts, the rival parties felt themselves to be on safe or neutral ground and they therefore could rest in a comfort unknown in the field: "At this well Professor Mudge's party and my own used to meet in peace after our fierce rivalry in the field as collectors for our respective paleontologists, Marsh and Cope."[11]

There were destructive aspects to the rivalry also, and the secrecy and distrust only seemed to heighten and proliferate when one would have thought it was at its worst. Although Cope's collectors were not always known for their angelic behavior, and most likely did things that would not stand up well under close scrutiny, there never was any direct evidence of fossil destructions. Charles, for instance, would never have permitted this on any of his parties. Marsh's collectors on the other hand, clearly carried out such destructions under either the implicit orders of Marsh or with his tacit approval. Marsh, who was apparently under the impression that fossils left would fall into Cope's hand, condoned the actions as we see from this quote from a Williston letter to Cope; a letter written after he had left Marsh's employ:

> Professor Marsh did once indirectly request me to destroy Kansas fossils rather than let them fall into your hands. It is necessary for me to say that I only despised him for it...[12]

Other Marsh collectors were guilty of the same. W.H. Reed, a former section foreman at Como, had partnered with W.E. Carlin, a former station agent at Como and it was the two of them who began the famous work at Como Bluffs. Together they began work at the bluffs for Marsh, but split company after a violent argument. Carlin was to later defect to Cope's side, and an intense rivalry between the former partners ensued:

> Carlin was the station master, and would crate up fossils for Cope in the comfort of the station room, while Reed had to work out on the exposed freight platform. When Reed finished a quarry, he smashed the remaining bones so that Carlin could not get them.[13]

As Marsh's men were all prominent in their profession and would often express outrage and sorrow over their performing such activity one can only suppose that it was Marsh himself who suggested and condoned these actions. One suspects that the only reason these men chose to destroy fossils rather than to quit was because they felt their employment would have been jeopardized should they not have done so.

Cordiality between fossil hunters, however, can exist and did at varying times. Barnum Brown's camp coexisted in general peace with the Sternberg family in the Red Deer, Belly River Beds in Canada. Although admittedly there was an eventual falling out when Peter Kaisen of the American Museum party felt that the Sternbergs were encroaching on spoken-for territory, but this was quite a different type matter. Although one might say that times had changed by then, since Marsh and Cope were both long dead, yet we do know that old habits die hard and many of the men who played parts in the Marsh-Cope feud were still living and practicing their craft. Even though the potential was there for just such a feud, it did not happen.

But cordiality came about during the heyday too at times. Williston shared some congeniality with Sternberg's camp near Buffalo Park: "We bought a pony of Mr. Chas. Sternberg who had been collecting that season in the region..."[14]

The relationship established between Sternberg and Williston always was a tenuous one at best, most likely made so by the dictates of their bosses. After the "Marsh Days" they had intermittent periods of good and bad in their dealings with one another. In 1904 Charles was to write to Henry Fairfield Osborn with an indication that Williston had actually offered him a job. He wrote:

> Now strange indeed, when I think I have no friends the eclipse passes and I find they are so many that they cover two continents. And more wonderful than all the man, who for over 20 years I have longed for as a friend has written me that he is trying to arrange things in such a way as to offer me the position of assistant to him for the rest of my life. I refer to Dr. Williston, and give you this information, as, when we met for the first and only time I told you my heart burns. Wonderful. Whether he wins or not he has bound me to him with fetters of love I hope only death can sever.[15]

Sternberg never did go to work for Williston as was suggested here, and in fact they became quite violent with one another (at least on paper) in a later dispute over collecting rights at the Craddock Quarry in Texas. Perhaps the best one could label them at any point would be merely as friendly rivals.

When William Berryman Scott and Henry Fairfield Osborn were just graduating from Princeton they made a special point of going by to see Cope. Osborn, after the death of Cope, looked fondly back upon this first visit:

> At this time only rumors of the warfare between the two great men had reached us; we did not realize that one feature of this warfare was the concealment of all information about localities and collectors. We were, accordingly, surprised and chagrined that, while Cope was courteous and agreeable, he neither gave us any information nor displayed any interest in our scientific ambitions. This conservative and reticent attitude was especially notable because it was in such great contrast with the thoroughly genial and open-hearted attitude he late showed.[16]

In the days after all the fighting and rivalry were over most of the hunters who were still alive worked together in one way or another. Letters between the ex-rivals were common, the trading of fossil piece for fossil piece to enhance particular museum collections was regular, and the men of the field began to trust others in a much more significant way. Once Cope and Marsh were dead and buried the rivalry was no longer fed and as such the industry could move forward, unimpeded by the bickering and grumblings that were to prove so unfruitful.

If there is anything to learn from this contact between rivals it is that man will occasionally rise beyond what he is given credit for, but that just as frequently he will sink lower than one might expect. Those who contribute to the risings are the ones ultimately worth learning from. Charles Sternberg was certainly one of these.

NOTES

1. Adrian J. Desmond, *The Hot-Blooded Dinosaurs*, The Dial Press, James Wade, NY, 1976, Pg. 109

2. Charles H. Sternberg, *The Life of a Fossil Hunter*, Henry Holt & Co., NY, 1909, Pg. 144

3. Charles H. Sternberg, *Kansas City Review of Science & Industry*, 1884, Vol VII, Pg. 597

4. Elizabeth N. Shor, *Fossils and Flies,* Univ of Oklahoma Press, Norman, 1971, Pg. 78

5. Ibid, Pg. 82

6. John Rydjord, *Kansas Place-Names*, Univ of Oklahoma Press, Norman, 1972, Pg. 454

7. CHS, *Life*, Pg. 34

8. *Harper's Magazine*, Vol LI, Nov 1875, "Air Towns and Their Inhabitants," Pg. 830

9. CHS, *Life*, Pg. 212

10. James W. Gidley, *Warm-Blooded Vertebrates*, Part II, Vol #9, Smithsonian Scientific Series, "Experiences with Fossil Mammals," 1934, Pg. 179

11. CHS, *Life*, Pg. 34

12. Elizabeth N. Shor, Pg. 118

13. Url Lanham, *The Bone-Hunters*, Columbia Univ Press, NY, 1973, Pg. 178

14. Elizabeth N. Shor, Pg. 61

15. AMNH Ltr, Feb 22, 1904 (To Henry Fairfield Osborn), From the Archives of the Dept. of Vert. Paleo.

16. Henry Fairfield Osborn, *Cope: Master Naturalist*, Princeton Univ Press, Princeton, NJ, 1931, Pg. 577

CHAPTER EIGHT
LAWRENCE, KANSAS

Although Charles enjoyed traveling, and through circumstances managed to find himself residing in a number of diverse places and climates, the closest thing to home was always to remain Lawrence, Kansas. Before his long life ended he was to spend fairly substantial periods of time in San Diego, California, and in Toronto and Ottawa, Canada, but Lawrence was always where his heart was to be bound. The hearts of those who knew him, or that know of his work and admire him, continue to hold an estimate of him that cannot be separated from Lawrence, Kansas. This is in much the same way that the Kennedys cannot be thought of outside Massachusetts, or the Leakeys outside of Olduvai Gorge. One cannot separate the Sternbergs from Kansas, whether they continue to reside there or not.

Charles was to move to Kansas at the behest of his older brother George, who was then post surgeon at Ft. Harker, some 200 miles west of Lawrence. George wanted the boys; he, Charles and Edward, Frederick, Albert, William, Francis and Theodore all together at the ranch he owned in the region around Ellsworth. Presumably he also expected their help in making the ranch a success. It was this initial move that was to first acquaint Charles with Kansas fossils, and that was to tie him to that state in much the same way that he was bound to Cope after receiving his initial letter and draught.

The locality around Ellsworth to which Charles and his brother Edward moved was semi-rural. It was a region fed by the existence of Ft. Harker. At this time Ft. Harker, already bore its new name and location. Ft. Harker had originally been known as Ft. Ellsworth and it held that name for three years before being changed to Harker. The young fort, in its secondary location, had been established in August 1864 in an attempt to protect the settlers from pro-slavery marauders and Indian attacks. Its new name had been bestowed in honor of Brig. Gen. Charles G. Harker, a valiant man who had been killed just prior to the forts founding at the Battle of Kenesaw Mountain. Ft. Harker was to play a strategic role in early Kansas history, and it allowed the surrounding countryside to grow and prosper, something that could not have occurred without the protection it afforded.

Ft. Harker was to remain in active military use for only a short period of time, and it was abandoned in early 1872, but not before it played an active and important role in the locale by serving as the main distribution base for supplying men and materials to the westwardmost forts. Ft. Harker's greatest contribution during this period of conflict, beyond its more pedestrian purpose, was the result of its strategic location relative to the Indian War of 1867. A large band of Cheyenne, with a smattering of Sioux and Arapahoe, attacked settlers in the vicinity north of the post and the campaigns made against this renegade band were to be centered from Ft. Harker because of its proximity to this enemy.

Ft. Harker in its heyday was a bustling and highly operational fort. It was designed to be functional, and it proved to be just that. The fort was also to contribute oddly when it was deserviced too, for in that process it left behind men of quality who remained after their tours of service were complete and settled the area around the fort. As hardy men, men used to the hardships of a rural and hard existence, they found the countryside much to their liking. The opportunities of a growing area, and wonderful farmland made it an area of promise for these enterprising veterans. Among the men who stayed in the district were some notable ones: David B. Long, Major Henry Inman, Theodore Sternberg, Ellsworth Sheriff C.B. Whitney, and others. Among some of the famous men who passed through its gates, but did not stay were the scouts Wild Bill Hickok and Buffalo Bill Cody.

Ft. Harker also played a significant role in the westward spread of the intercontinental railroad. Troops from Ft. Harker, and supplies routed through Ft. Harker, were literally the life blood of the railroad workers as they worked their way westward from Ellsworth. The Union Pacific, Eastern Division, passed within yards of the fort, and the railroad would have been doomed to sure failure had not its umbrella of protection been in place.

Ellsworth, a mere five miles from the fort and the most noteworthy town in the district, was an exceptionally rough town in those days, and was generally characterized as being lawless. It certainly had a reputation that was well-earned. It didn't require a Ned Buntline to embellish on happenings there, for everyday happenings were rough enough on their own. As the surprised and unlikely recipient of the cattle drive business from Texas that passed over the Chisholm Trail, at a time when Abilene was beginning to wane as the trail center, Ellsworth mushroomed. It soon had taken on all of the unsavory and sordid aspects of life that one connects with the rowdy Texas cowboy of that era (i.e. red-light districts, saloons, ruffians, etc.). The worst of the factions that quartered the town were to be found in the river-bottom district known as Nauchville. This hard-core red-light district was filled with disreputable saloons and other gambling joints, and even was to sport a race track. Sternberg was to describe Ellsworth as a dangerous place in this vivid assessment:

> I might tell also of the ruffians who at one time held Ellsworth City in a grip of iron, and how, until they killed each other off or moved further west with the railroad, the dead-cart used to pass down the street every morning to pick up the bodies of those who had been killed in the saloons the night before, and thrown out on the pavement to be hauled away.[1]

Lest anyone should think that this appraisal of Sternbergs was only that of a peripheral passer-by, one need only be aware of Charles' talk to a Boy Scout troop, reported in the *Lawrence Journal World* of Jan. 26, 1917:

> In 1868 General Hancock had a force at old Fort Ellsworth. The Indians were again roving about. It was pay day for the troops and Mr. Sternberg sold a big load of watermelons in the camp. He was induced to make a trip to Ellsworth with a detail of men after a drunken soldier who was detained in the town. It was full of gambling houses and saloons, and Mr. Sternberg and his detail were welcomed with a fusillade of revolver shots which the inhabitants fired in honor of their coming.[2]

An early resident of Kansas, Peter Robidoux, managed to capture the real flavor of Ellsworth life at the time. His recollections highlight the primitive and dangerous nature of the town as well:

> So I ventured across the street to a big saloon with a big sign over the door, 'U.S. SALOON.' It was a big one, about 125 feet deep. I took a chair in a corner near the front where I could watch everything. It was getting interesting. Soldiers from Old Fort Harker were coming and going. The dames and gamblers were there. Yes, and there were Indian scouts, teamsters, bullwhackers, and citizens of all sorts promenading the streets, as well as the dance hall. The orchestra was playing melodious tunes and the ball was on. Drinking, gambling and dancing were in full blast...Every now and then groups of long-haired men wearing high-heeled boots, and spurs, red underwear, cartridge belts full of cartridges, scabbard at side with pair of six shooters, and bowie knife would come; call for drinks, and as they went out, bang, bang, bang would ring out from their guns.[3]

In this passage, as in most descriptions left us from people having visited there, all were quick to label Ellsworth as rough and dangerous. Since both residents and visitors were all armed personal danger was enhanced. It was not a place for a peaceable man, and a man didn't dare venture there without recognizing the inherent dangers to be found in a lawless town, and taking appropriate measures to protect himself.

The Sternberg family was one of the first of the pioneering families in this vicinity, and in the Ellsworth County histories they always have a prominent place. Because of the family's being so active in all phases of life in the area (see

Chapter 17) it made them well-known to all. Levi took an active part because of his role as churchman, as well as for his official role with the Stock Grower's Association, and his sons each gained some stature due to their hard industry at both farm and professional works, as well as their entrepreneurial endeavors.

After Charles had made his break from the family homestead and had begun his career as a fossil hunter he needed a base of operations from which to work, and Lawrence seemed ideal. First of all, Lawrence was one of the largest cities in the young state (Manhattan, Wichita, Dodge and Topeka being others) and therefore it had all of the advantages that one could expect from a growing western city in terms of education, opportunity and other cultural elements. Secondly, it was ideally suited to fossil work since it was close to any number of the main Kansan fossil fields, and it was on a straight line with the main river arteries and the fledgling railroad lines. Thirdly, it was not near as rural as the area around Ellsworth had been, since it was on the eastern border of the state. It therefore was more civilized and conducive to the raising of a young family in a semblance of security.

Additionally, Lawrence's religious climate was such that it made it a pleasant place for a man to raise a family. Everett Dick in his fine book, *The Sod-House Frontier 1854-1890*, describes most of Kansas as being filled with the "spirit of worldliness and wild speculation," and he goes on to explain that Lawrence was decidedly different in that respect:

> Lawrence, Kansas was an exception to the rule. On Sunday a hush came over the city and the sound of the church bell reverberated over the prairie. Hymns of praise swept through the Kaw Valley. On that day it was a little New England town set down in the rude environment of the frontier.[4]

This "puritanical spirit" (as it has been called) was a zeal that did not find better voice than that in Lawrence, and perhaps was not better evidenced at any one time than it was in the town's celebration of the Fourth of July in 1855. While other towns around celebrated with dances and uproarious behavior Lawrence quietly celebrated in its own way with cake and ice cream.[5] This was a perfect destination for the likes of a gentle spirit like Sternberg. It was partially this religious fervor that was to make Lawrence such a hated place in the eyes of the border ruffians, although by the time Charles had moved to Lawrence the effects of the massacre had begun already to wane.

The growth in Lawrence during these early years, like that of Kansas and the other Plains states, was phenomenal. Lawrence had the advantage over some of the other young cities due to its geographic position; it was the first stop on the Oregon Trail. Those pioneers that left from Independence would have to pass directly through Lawrence, and those leaving from St. Joe or taking the Mormon Trail would pass just north of Lawrence. In 1890 Lawrence was to be already a town of over 12,000, thereby making it a major metropolitan area, since the population had mushroomed from the 3,000 it had claimed as inhabitants in the early 1860's.

The Homestead Act of 1862 was a major contributor to this growth. Between 1862 and 1880 around 150,000 homesteads were established in the belt formed by Kansas, Minnesota, Nebraska and the Dakota Territory. These men and their families placed their roots deeply because of the land, and didn't move around from place to place as with other segments of the population.[6]

Some of the additional major factors contributing to this early and quick growth in Kansas were the abundance of Texas cattle (said to number over 5 million at the end of the Civil War), high beef prices, the marching extermination of the buffalo, control of the western Indians, leniency and proliferation of open land laws, and the expansion of the railroad west. This burgeoning of the population seems to have been a mixed blessing to area residents, for on one hand it seems to have brought about a calming of hostilities as law moved westward, but on the other hand it brought with it the scurrilous elements that followed the initial pioneers. The end result, however, was a steady progression of colonization that would end in the gradual domestication of the area.

These were the more obvious advantages to Sternberg making Lawrence his base of operations for his fossil business, and so he moved to Lawrence shortly after his marriage. When he first came to Lawrence, Anna continued to live with her parents and Charles would occasionally stay there too when he happened to be home from the field. Shortly, however, he was to maintain a residence in the city as well as maintaining a little 20-acre farm about four miles outside of Lawrence to the southeast.[7] He had out of necessity mounted a tent on this property and it was from this crude housing that he was to begin his fossil preparation business. In each of the different Lawrence city directories (similar to our modern cross-street directories) he was always advertised as "paleontologist."

In the early 1880's Charles was to maintain a secondary business. Always enterprising, Sternberg never overlooked any opportunity to make a little extra money. This trait was certainly in keeping with the whole of the Sternbergs in America for many showed their inventiveness. In the year 1884, for instance, Charles' stationery proudly heralded him as "Charles H. Sternberg, Manufacturer of Sternberg's Silicious Soaps."

It was in Lawrence also that he was to later establish his permanent laboratory from which most of the preparation work he and his sons would perform during the cold Kansas winters would take place. This brick edifice, an ex-school building, was located at 617 Vermont street in Lawrence, and Charles was to rent the ground floor quarters from a Mr. John T. Constant, a local carpenter. These facilities were ideally located since they were just off the main business area on Massachusetts Street, as well as being only just around the corner from the famous Eldridge Hotel (owned by the Eldridge brothers who had also previously owned the ill-fated Free State Hotel) that had played such a prominent role in the Quantrill Raid. Besides being ideally situated the quarters were spacious, and therefore commodious enough to accommodate the legions of fossils that were to eventually pass through its doors. His sons all learned the art of fossil preparation in this building, as well as how to mount fossil exhibits (by trial and error mountings) in this room.

Views of the inside of the laboratory show it to have been well-kept; perhaps even meticulous in its layout. It was well-equipped with various mechanical apparatus that would assist hoisting heavy loads, drilling holes in metal and fossils, fitting mounts, etc. It was truly a workplace of an artisan and not a backstreet operation. The Sternbergs were to be as exacting about their preparation work and their mounts as they were about their collection activities, and it showed in the business itself.

In December of 1903 Charles was to purchase an expensive elevator that allowed him to take photographs on a platform from above a fossil. It was a good investment since the fossils would often be too unwieldy to move about, and in order to market the pieces he had to show them to their best advantage. There is no question that his photographic equipment often made the difference when selling his fossils, and even today various museums still carry prints in their archive files of varying vertebrates that he had offered them for sale.

George F. Sternberg was to later write that the family team was often able to collect enough material in a season to charter a return trip by freight car.[8] Apparently in these auspicious circumstances they would load the fossils in at one end of the freight car and the horses in at the other. Obviously a boxcar load of fossil material made the laboratory an important factor in the lives of the Sternberg family. The building was to serve them well, and only left their control because of an act of God: the demolition of the building by a Kansas cyclone, or as they are more commonly known to us today, a tornado.

Lawrence even at this early date had already a jam-packed history. Considering its short history as a town (it had only been formed in August of 1854), and the fact that one of the most memorable events in United States history had taken place just previous to Charles' moving to that city, it was a city thought well enough of to fight for. That memorable event was of course the infamous "Quantrill's Raid," which even today can't fail but to evoke emotion in the hearts of many.

As background to the raid one has to know that one of the main factors leading to the Civil War directly involved the drama over the pending statehood of Kansas. Kansas, which had been a territory until 1861, had been a divisive issue among both area and out-of-area peoples because of the slavery question. Abraham Lincoln was newly president at this time, had not yet had opportunity to address the question of slavery, and therefore the issue was undirected and ripe for contention. The climate might be compared to that of the abortion issue in our own time. The state of Missouri was the hot-bed of pro-slavery sentiment, and its neighbor Kansas was to enter the Union as a free-state, opposed to slavery and all that it stood for. This major decision, whether to be slave or free, had been brewing as an issue since 1854 when the Kansas-Nebraska Act had been enacted, but had now been really brought to a head. Because of the proximity of Missouri to Kansas, and of Lawrence to the Missouri border (it only being about 30 miles from the dividing line of the Missouri River), Lawrence was almost doomed to suffer a major catastrophe, and it certainly did.

The explosive nature of things can be better understood by recognizing that Missouri at this time was home to about 50,000 slaves. The volume of the trade therefore made it a substantial financial investment for that state, and Missourians viewed it as contributing to their financial ruin when they saw Kansas being enacted as a free-state.

The city of Lawrence had been named after Amos A. Lawrence of Boston, who had avowed to do all in his power to keep Kansas free, and he was willing to back up that statement with cash and other kinds of support. He had formed the powerful New England Emigrant Aid Company which acted as a counterbalance to Missourian ballot box stuffing, and which had been the channel through which guns and other supplies had been funneled to beleaguered Lawrence. He had also been an early supporter of both Charles Robinson, and of John Brown, who were both men singled out for special hatred by the border ruffians. Lawrence was a gregarious man, and a man who could draw out the best from another man, and his personal magnetism did much to make Kansas a free state. He was also to serve as Treasurer at Harvard College in Cambridge for a period, and was instrumental in assisting Louis Agassiz in establishing his famous museum, the Museum of Comparative Zoology. One cannot read the personal letters written by Amos Lawrence without recognizing his zest for life, and his genuine Christian character. In many ways his nature was similar to that of Charles Sternberg, since they were both gentle men, but yet men of power and intellect.

On Friday Aug. 21, 1863, two scant years after statehood was granted Kansas, Lawrence was to be the scene of what prompted the appellation that is still often used today, "bloody Kansas." Pro-slavery forces under the command of William Clarke, now having assumed the name of Quantrill, rode across the Missouri border under the cover of darkness and made their way towards Lawrence. This was made easy by the fact that Lawrence was only a few miles across the border and therefore easily traversable by horse in the course of an evening.

It is ironic that William Clarke Quantrill (1837-1865), who was a former school teacher, had previously lived in Lawrence in 1859 and yet was to become her destroyer. He had, however, been forced to flee Lawrence when he was indicted by a grand jury on a charge of horse stealing. He wisely had fled before he could be jailed for this offense. It is perhaps this early incident that added fuel to the flame for Quantrill seemed to have developed an especial dislike for the city. The former Lawrence resident, who had used the pseudonym of Charlie Hart while living in Lawrence, would certainly be remembered for what he and his men were to do that evening. It is interesting, but certainly not clear, that Clarke might have used this first of assumed names because he was even then anticipating the havoc he was to later cause.

There is no question that Quantrill was a vile and dangerous man, and that many of the men who rode with him were of the lowest sort (since there were many outlaws, or soon-to-be outlaws, under his command). Among these desperadoes were such infamous men as Frank and Jesse James, and Cole Younger. The deeds of these men leave no further explanation necessary. Quantrill himself, however, is typically vilified as an evil man, and one espousing an evil cause, but that is as patently unfair as labeling Nixon all bad because of Watergate.

In retrospect we view slavery as an unbelievable horror and cannot comprehend how any "right-thinking" individual could ever be a part of that whole ideology, but one cannot on the other hand overlook the might with which tradition and upbringing can rule. There is a wide disparity between being a cruel slave owner and a concerned and perhaps even loving slave owner. If one reads, for instance, of the southern plantation upbringing of the geologist Joseph LeConte, one has to recognize that he felt genuine sympathy for his slaves and a true concern for their welfare. Even being sympathetic, however, he still continued to remain a slave owner right up to the time of the Civil War when the issue was forced. In considering Quantrill's personal circumstances one cannot overlook too the atrocities that were perpetrated against the "border ruffians" by the "free-state men," because these undoubtedly fueled Quantrill's already negative feelings.

Renegade Kansans, for instance, under the command of James Montgomery and Charles Jennison had raided border towns in Missouri, pillaging and murdering as well. Yet these men have never been vilified in the manner that Quantrill has. Although these raids by the "red legs" were not officially sanctioned they did have a great popular sympathy in Kansas. These raids, plus an unfortunate and ill-timed accident whereby a number of female relatives of the Quantrill men who had been captured and imprisoned in a three-story house, and were subsequently killed when a freak and furious storm collapsed the house, caused bitter resentment on the side of the Missourians. It was unfortunate, but predictable that Lawrence would bear the wrath of the ruffians eventually, and it was to prove to be the case. On that fateful evening the town that only numbered about 3,000 in population was to lose about 100 of its homes (as well as about 75 other buildings) and 150 of its best men.

One of the prime focal points of the border ruffian wrath was the famous and fiery Senator James Lane. Alice Nichols, in her book *Bleeding Kansas* refers to

James Lane in not so complimentary terms as being the "chameleon politico." She accuses him of having less than humanitarian ideals in his actions involving blacks, and she cites in particular the heavily underplayed fact of Lanes' "Black Law."[9] This law was to state that no black, be they slave or free, was to be allowed residence in the state of Kansas. This is hardly the "free-state" Kansas that we have come to believe was being founded, but was really a "non-slave" Kansas. The distinction is important (at least as far as viewing Lane himself), and Lane seems to have been the prime-mover. Whether or not the average voter understood things in this same light is unclear.

Quantrill's Raiders, as they were to be later called, were on an organized hunt that fateful evening for this same Senator James Lane, as he was a staunch anti-slavery supporter with a taste for the flamboyant. Lane, one of the two state senators in Kansas, was a charismatic, rabble-rousing firebrand. Lane himself had participated in the role of military leader on a number of military excursions against the ruffians and was the most vocal of the Kansans. Because of his very vocal role and his public stature he was the central target of the Missourians' anger. Even with all the early warning systems and the Lawrencites preparations, the raiders managed to make it to Lawrence that evening unmolested.

Charles Mortram Sternberg was to recount a story involving his mother, who at age 10 was a resident of Lawrence and was there during the raid. Her father, Charles Reynolds, was away during that period of time and her mother took charge of the situation and managed to placate some of the ruffians who confronted them by feeding them fresh bread and milk. The house was spared since there were then no men present. Charles noted that an old man from the house next door climbed over their fence and asked to be hidden, since to be discovered meant certain death. Not having any immediate place to hide him they dressed him up as an old woman, and gave him their young baby to hold. At this point one of the ruffians decided to examine the "old woman" more closely, but one of the Reynolds' girls said "don't touch him, he's our old auntie." Luckily the ruffian didn't reflect on the "he's" in her response, heeded her admonition, and the old man's life was therefore spared.[10]

Although the main individuals that Quantrill's men had set out to destroy were to escape the net that was intended primarily for them, the city of Lawrence was not to fare so well. Quantrill and his 450 Bushwhackers (as they were then also known), literally burned and pillaged in "Attila the Hun" fashion. The main business section along Massachusetts street did not escape retribution as almost every building along its length was destroyed.

Charles' future father-in-law, the Rev. Charles Reynolds, was to lose a number of church members that fateful day, and his church. Trinity Episcopalian, was nearly destroyed by fire as a result of the raid. Reverend Reynolds was himself unharmed since he was at the time serving as Chaplain of Ft. Riley, and was therefore away from Lawrence. Shortly after this incident, but probably not because of it, he was to tender his resignation from Trinity Church.

The main focus for the abusing of Lawrence that evening turned out to be the hated Eldridge Hotel. It was here that a number of individuals who had failed to escape had been sequestered during the brunt of the raid. The Eldridge Hotel (formerly the Free State Hotel), was owned by the equally hated Eldridge brothers, and they had suffered substantial damages in the earlier sacking of Lawrence. This was the same hotel that was around the corner from Charles' business.

Only the Alamo tragedy in Texas was to weigh more prominent in the minds of Americans of that date. They would have rated this as the second greatest disaster to occur in the United States. In the taking of the Alamo there were only 37 more people killed than the incident in Lawrence, and the ramifications of the Lawrence incident were almost as devastating.

Quantrill's forces did not go unpunished, and in fact justice proved very swift. Purportedly about 100 or so of the bushwhackers were killed that very night as they were chased back into Missouri. William Clarke Quantrill was only to manage escape until 1865 at which point he was also killed in an attack by federal forces upon his haven in Kentucky.

Because of the breadth of interest in the Old West it must be mentioned that Kansas was additionally famous for the many gunhands and noted lawmen that criss-crossed the state. Buffalo Bill Cody, Wild Bill Hickok, Bat Masterson, Wyatt Earp and his brothers, Doc Holliday, Calamity Jane, Clay Allison, Luke Short and the Dalton brothers are the names of just a few of the prominent ones. Noteworthy also is the fact that it was often either difficult or impossible to tell the "bad guys" from "the law" as the actions of such notables as Wyatt Earp and Bat Masterson might testify.

It was to this kind of environment that young Charles was to move, and this was enlivened by the fact that the Civil War was only recently ended (1865) and the strong Ulysses S. Grant was now president. But it was evident that most people were ready for things to settle a bit after that disastrous confrontation and bloodbath, and things began to improve. So Kansas finally settled down to a life of relative peace, and so therefore did Charles.

Area peace was further enhanced by the fact that the end of the Civil War relieved Union troops so that they could concentrate on the Indian situation. Lawrence therefore had only to worry about its internal strife, and its rebuilding in the wake of the Quantrill Raid. Rebuilding, in this climate, was to prove to be quick in part because disasters often prove to be character builders and people work together in ways they normally wouldn't. Lawrence builders had the ability to build quickly, and had the adaptability to use available and varied building materials. There is record of over 50 dwellings being built in the space of about 60 days in 1855, and of the 50 there were a number of large stone structures. Everett Dick indicates that housing was varied and consisted of "dugouts, sod houses, log cabins, shake structures, and other odd dwelling places."[11] After the war, construction was to prove even more frantic since there were so many new settlers that were to pour into the state.

One other major factor impacting Lawrence's population explosion resulted from the centering of intellect that occurred with the establishment of the University of Kansas, for it seems the proximity of an educational institution is always a draw for persons interested in higher education. Such centers seem to draw like a magnet the kind of men and women who will domesticate an area and cause it to be a place of firm commerce. Annie Williston, wife of Samuel W. Williston, was to aptly describe the Lawrence climate when she said: "This is a one-horse town & the University is the horse."[12]

Charles was to remain a stalwart citizen of Lawrence right up to the time he made his move to Canada. He was to raise his family in this town, worship in its churches, and make many and lasting friends here. It was also here that he was to join the Kansas Academy of Science, an organization that was to do much to professionalize him as it put him in contact with many more men in his same profession or like industries. Interestingly the Academy had been co-founded by his old nemesis Professor Benjamin F. Mudge, in the year 1868. Charles himself was to be voted a lifelong member in the year 1896. The early years of the Academy were years of growth too and paralleled his. Charles was to be a frequent contributor to its meetings, its committees and its annual reports. In the years 1905 through 1912 he was in fact to be the most prolific writer of geological papers for their annual transactions volumes.

Charles' love for Kansas was obvious. And while he may not have exactly immortalized Kansas with his verse, he did often state his love for the Sunflower State in many of his poems. A good example is this from his work *A Story of the Past or the Romance of Science*, under the title of "Where the Sunflowers Grow:"

> We'll admit her bergs [Wyoming] are glorious,
> Clad in everlasting snow,
> But we'd rather be in Kansas,
> Where the sunflowers grow.
>
> Yes! here's where the wool is growing
> A million tons or so,
> But our thoughts are oft in Kansas,
> Where the sunflowers grow...[13]

Charles' eldest son George was also to particularly love Kansas, enough so that he was to come back there to live after many years of independent collecting elsewhere. He was to later become Curator for the then-named Ft. Hays Museum located on the campus of the now-named Kansas State College. This position allowed him the luxury of being able to remain a Kansas citizen.

An amusing counterpoint to the poem just quoted is a parody found among the memorabilia left by George F. Sternberg. This printed text was simply signed "L.M." and does not give one any further clue to its authorship. This poem is entitled "Gone Back to Kansas:"

>So the Sternberg tribe has vanished
>To the roll call they say, no,
>And will ne'er come back to Kansas,
>Where the sunflowers grow.
>
>As they reach the Northland happy,
>With their spirits all aglow,
>They thank God they're far from Kansas,
>--With apologies to Prof. Sternberg.
>A.L. and A.E.
>* * * * * * *
>They can have their Northland happy,
>With its frozen plains of snow
>We feel thankful we're in Kansas
>Where the Sunflowers grow.
>L.M.

This poetic work has all the earmarks of a community-penning, for fun in sending off George and his son Charles to Canada at a later date. George apparently enjoyed a good laugh, since among his things were a number of cartoons and other "fun" things.

In the college's annual publication called *Aerend,* George was to recall the halcyon days of collecting with his dad in an article entitled "Thrills in Fossil Hunting." Interestingly enough he used the same phrase in describing this desire to be back in Kansas: "As time went on we began to realize that finding dinosaur specimens was not as easy as we had expected. We wished that we were back in Kansas *where the sunflowers grew.*[14]

George and his dad were to both find permanent niches in the hearts of the

Kansans who knew them, and are included today in all comprehensive histories of Kansas of any worth because they indeed played a major role in the formation of the state as it related to the unearthing of its ancient history. Quite honestly they ought to be mentioned as well in the nations' earth science books with the same frequency, although they are often overlooked in this regard.

NOTES

1. Charles H. Sternberg, *The Life of a Fossil Hunter*, Henry Holt & Co., NY, 1909, Pg. 14

2. *Lawrence Journal World*, Jan 26, 1917

3. W.F. Thompson, "Peter Robidoux; A Real Kansas Pioneer," *Kansas State Historical Collections*, Vol XVII, 1926-1928, Pg. 286

4. Everett Dick, *The Sod-House Frontier 1850-1890*, Univ of Neb Press, Lincoln, 1979, Pg. 66

5. Ibid, Pg. 75

6. Alexander B. Adams, *Sitting Bull*, G.P. Putnam & Sons, NY, 1973, Pg. 285

7. CHS, *Life*, Pg. 25

8. George F. Sternberg, "Thrills In Fossil Hunting," *Aerend*, Vol #1, #3, Kansas State Teacher's College, Hays, Pg.147

9. Alice Nichols, *Bleeding Kansas*, Oxford Univ Press, NY, 1954, Pg. 42

10. Interview with Charles Mortram Sternberg by Free-lance Broadcaster Laurie LaMaguer, Courtesy of Tyrrell Museum, Dept. of Paleo.

11. Everett Dick, Pg. 45

12. Elizabeth N. Shor, Pg. 132

13. Charles H. Sternberg, *A Story of the Past or The Romance of Science*, Sherman French & Company, Boston, 1911, Pg. 39

14. George F. Sternberg, *Aerend*, Pg. 142

CHAPTER NINE
TEXAS PERMIAN & RED BEDS

Charles was ultimately to spend a great deal of time in Texas, since he was to make lengthy trips there in 1882, 1895, 1896, 1897, 1901, 1902, 1917, 1918 and in 1919. These trips met with mixed review, since their success levels were so different, and the fossils so scarce by Charles' standards. Yet Charles' Permian fossils are worthy of special note because of their age, and important because of that unknown quality a "new" fossil field brings with it.

Texas in the early 1880's was a fairly new field, and the red beds of Texas had just begun to be explored. This made it especially appealing to Charles who relished the challenge of an unchartered course, and such always seemed to provide him with new impetus to success in the field. His unbridled enthusiasm for new fields is emphasized in many of his writings. This area, although not replete at the time with arrow-shooting Indians as was Kansas when he began, was still similar in terms of its being out-of-the-way and basically beyond civilization's perimeter.

Early effort at geological exploration in Texas was primarily in the upper western portion of the state, and almost all was above the Balcones Escarpment which split the state on a line southward from Waco to San Antonio where the line headed west to the Rio Grande. What limited exploration that occurred was primarily due to the government's incentive. Pre-Civil War efforts were often sponsored by the federal government as part of its program to assist those in areas experiencing difficulties from Indians, unmanageable terrain and so forth. The state government participated because it wanted to promote settlement. It was not until the early 1880's that any appreciable growth was shown in the whole of the area.[1]

Jacob Boll in 1876 was the first to record discovering vertebrate Permian fossils in North America. He was a naturalist working on behalf of Agassiz's Museum of Comparative Zoology in Cambridge. Edward Drinker Cope was himself to be in the Permian as early as 1877, where he met Boll and hired him to collect vertebrate fossils on his behalf. Cope was at this time working for the Hayden Survey.

William Fletcher Cummins had collected fossils for Dr. Charles A. White, the "preacher who had collected vertebrate fossils in N. Central Txs. for E.D. Cope."[2] In 1889, in a turnaround, White was to assist Cummins through the auspices of John Wesley Powell. Theirs was an effort to investigate the Osage Plains Permian deposits. Cummins, who was to play a part indirectly in Charles' early Texas adventures, had succeeded Boll in collecting for E.D. Cope, working for him between the years 1880 and 1883. Boll in turn was assisted by J.C. Isaac, who had formerly assisted Charles in 1876.

Charles had not been in Texas prior to 1882, and as such had no real idea of what to expect except for what he might have been briefed on by his employer. Texas was considerably different from Kansas, and therefore he must have been relatively unprepared for what he would find. Like with Kansas he was aware that Texas had water problems, and he was equally not surprised to find that Texas was inhospitable to the badlands adventurer. He was to learn that Texas fossils were also very different, much harder to find, and were much older in their geologic history. He was evidently not fully prepared at this eventuality since he had always found quick success before. Charles had been used to collecting in the Recent and the Cretaceous, and therefore was unprepared for the scarcity of fossils in the Permian, particularly whole salvageable fossils. There is a general rule in Paleontology that the older the deposits the more scarce the fossils and Charles was to quickly learn this to be true. But even though Charles was to find it difficult to ferret out the unblemished Permian fossil it did provide real impetus for him to be successful. He was delighted at the opportunity to expand his fossil knowledge and cover virgin territory where he felt he certainly could make another indelible mark in the annals of paleontological history.

The Permian, youngest of the divisions of the Paleozoic Era, is dated to about 270,000,000 years ago by current scientific estimates. This period was so named by Murchison after a Russian province of the past called Perm. In the United States it is best represented in the vicinities of Texas and Kansas. Typically referred to as the "Red Beds," because of the iron oxidized composition of its rocks, it forms a very distinctive land mass that is clearly visible for miles during air travel over the southern Midwest. It was to this new and strange territory that Charles was to travel in search of the equally strange and elusive Permian fossils.

His first trip in 1882 was under the auspices of Agassiz's museum, the Museum of Comparative Zoology at Harvard. It can rightly be called disastrous overall because it was apparently poorly planned from the beginning. The museum was especially interested in finding out just what could be found in these little-known beds, and they held high expectations since Cope had at least a little success there in the past. They felt that Charles was clearly the man to head up such an expedition because he had a reputation already for finding great fossils even in areas where others had already searched. Undoubtedly they found him to be easy on their budgets as well since he worked mainly alone, worked inexpensively and could be trusted to provide full value for their dollar.

Since Texas has a relatively dry climate in this part of the state, and because the summers were scorching, they began the expedition at a time of the year when he would normally be pent up in his workshop preparing fossils. It was just before Christmas in 1882 that he made his way south to Dallas, with only the name and address of his contact from which to start. To his consternation upon his arrival in Dallas, however, he was to find the local post office could not identify the person listed as his contact, an A.R. Roessler by name. He also was to find that the address he was given was non-existent. Having had no previous experience in Texas and not knowing anyone, nor having any clue as to where the Permian beds were that he was to explore, he was in a fine predicament. He explained that he was relying "absolutely" upon Roessler for exact information, and that lacking that contact he felt as a "new-born child" since he was bereft of any personal knowledge that would help him locate workable beds.

The man for whom he was looking it seems was the Austrian born Anton R. Roessler. Roessler had worked for Benjamin F. Shumard's survey as a draftsman and had been a geologist for the U.S. General Land Office before his employment in the private sector with the Texas Land and Copper Company.[3] Roessler would have been an excellent guide, should Sternberg had been able to locate him, for in 1872-73 he was to publish geological reports for the *Texas Almanac*, and in 1873 he was to publish on Permian fossils from the Osage Plains in a German periodical.[4] He was mainly noted for his excellent mapping of the Permian of Texas by county. It was certainly regrettable that Roessler could not be found for it caused Charles to waste much time. Just where Roessler was at the time, and why the mixup in names and addresses occurred is also not clear.

Due to the kindness of the postmaster he was referred to a Professor W.F. Cummins[5], who it so happened had worked for Cope as an assistant the previous year, and he immediately proceeded to Cummins' residence only to learn he was away. Mrs. Cummins, however, had been with her husband on that expedition and was kind enough to give Sternberg exact enough information as to how to proceed. This is an amazing thing in itself when one considers how confidential and mistrustful these fossil hunters had been taught to be, and to be sure Mrs. Cummins would have been under the same strictures. She evidently liked and trusted Charles, or he never would have been given the necessary information. Or perhaps she was aware or was made aware of Charles' prior affiliation with

Cope, and this may have persuaded her to trust him.

It was a shame that Charles had not actually encountered Cummins himself, however, as Cummins was well acquainted with the area and could have provided detailed information of which his wife had no idea. Cummins himself had already managed to correlate the Pecos Valley Red Beds with the Red Beds of Oklahoma and northwest Texas because of his extensive work in the locale. This correlation was to be later confirmed by Charles N. Gould, a disciple of Williston who was to also work the area. Cummins was the dominant figure in paleontology during the early period in this part of Texas, and his writings were some of the first real descriptions of what Texas geology was like.

Charles was to later write, however, that Cummins himself refused to disclose any detailed information relating to the Permian regions that Cummins had worked before finally leaving the field. Sternberg was much displeased at Cummins' reaction to his overtures, since he felt Cummins was simply being arbitrary, particularly so since Cummins was no longer active in field work.[6]

Even with Mrs. Cummins' kind information, however, Charles was still to have great difficulty, for he was to place too much reliance on what he expected the local populace could or would tell him. It was to prove to be one of those instances where the people in the area were no help at all, because they apparently weren't well versed enough in the land in order to help. It might be possible too that some considered the Kansan a "foreigner" and therefore remained tight-lipped about what they considered to be a Texan's business. This insular attitude can be found in remote parts of Kansas even today, so such is certainly not inconceivable.

Charles was to elaborate further on this whole question of help from the populace in a letter to Henry Fairfield Osborn:

> I have had an experience of 6 years, of the people of that region few are worth their salt to me. I know, it is impossible at any price to hire a man and team who will do satisfactory work. They always hold the threat openly or expressed over me 'if you don't let me run things I will drop you in the brakes 20 or 30 miles from

a house and let you run things to suit yourself'. Of course there are many good men there, but if they are worth anything they already have a home & employment.[7]

He was always worried too about teaching the young men too much as he felt there was a real threat from young fellows finding out the "mysteries of my vocation" and leaving to compete with him. This is in odd juxtaposition to the openness he had displayed elsewhere in wanting to teach others the trade. It seems perhaps that advancing age perhaps made him somewhat mistrustful of youthful followers, and undoubtedly being in an area where he was treated as an outsider could have jaded him as well.

He was to further note that bringing his own team to Texas was necessitated by the fact that farmers in the area would generally be in a better financial position than he, and they could therefore afford to pay more for the use of a local team and outfits. As such he had to either accept the poorest of helpers and animals, or pay such high rates that he was doomed to financial burden unless he brought his own team with him.

Arriving in Seymour, Texas, which is in mid-northern Texas in Baylor County, and southwest of Wichita Falls, he finally was able to begin his work. He notes that he "wasted months of careful exploration over barren beds" since he had misrelied upon the cooperation of the townspeople, and it was some time before he was able to discover the fossil horizon for which he was searching. And even once he discovered the horizon he was frequently frustrated because the Permian is so very stingy in what it coughs up to even the ablest collector.

These first futile weeks were spent in the barren Clear Fork exposures. It was not until much later that Sternberg was to find significant sites in the Lower Clear Fork area of Coffee Creek. At these sites he was to make important discoveries that included specimens of *Seymouria* and *Diadectes*. Charles was to also make very important type specimen discoveries of *Dissorophus multicinctus* and *Labidosaurus hamatus*.[8] But these successes were preceded by great trials.

The difficulty caused by the poor trip planning is hard for us to understand today since in the same set of circumstances we would just pick up a telephone and call whoever (wherever) in order to find out the information needed or we would go to the nearest library or governmental agency for data. Why Sternberg didn't telegraph his original source or proceed in some other way is hard to say, but he probably just kept feeling that the next horizon would be the one for which he was searching, and it was too much trouble to get back to civilization for such an undertaking. Besides, telegraphing was an expensive operation for one on a limited budget, as Charles always was. In all probability he would have lost at least a day or two in actual search time had he even had to make a trip to the nearest telegraph office. It should be mentioned too that poor planning was certainly not indicative of Sternberg's usual expeditions. He was normally very thorough in laying out all of his trips.

Traveling southeast to Gordon by rail he managed to hire a novice assistant by the name of George Hamman, the son of the local hotel keeper, who came with a wagon. Hamman, although inexperienced, was both available and willing. After an eight-day wagon ride they were back to Seymour to begin their search, armed with the most precise information yet obtained from the local people for finding the fossils, which was "they're over in the brakes." Since his search had been narrowed, however, they now began to meet with success.

Charles, with his usual capacity for describing terrain with beauty, viewed the scene as "miniature Bad Lands, with rounded knobs, deep canyons, bluffs and ravines." The prevalent color in the area was "Indian red, but beds of white gypsum and of greenish sandstone relieved the sameness."[9] Like we have seen before he was filled with awe at the marvelous nature of the uninhabitable badlands; he saw beauty that few would find in that stark terrain.

Charles was to comment upon the "acres of hot concretions" he was forced to traverse while in the badlands near Seymour. The concretions were brilliantly colored and this further contributed to his discomfort because this caused them to reflect an intense heat. Charles was to complain that his body was often twisted and contorted after traveling over even small distances on these concretion beds, and the pain was often unbearable.

Because he was so uncomfortable while working these particular badlands he made it a routine of going back to Seymour on Saturdays. He would then stay over that evening so as to avoid the discomfort of having to sleep out. This weekly rest and regimen apparently allowed him enough rest to be able to bear up for the following week.

Although the land was inhospitable to man, it did have its share of wildlife. Charles indicates that rattle snakes abounded, and wild turkeys flourished (in the thousands by his estimate). He also noted an abundance of antelope, wild cats, coyote and even a lynx. These animals generally stayed on the level prairie when possible because food was more plentiful, and water more readily found. Water was noticeably a problem to both man and animal in the badlands.

Air flights over this whole region today give one a good sense of the roughness of the terrain as well as the great number of miles covered in red beds. Bailey Willis was to note that the length and breadth of the area labeled as "red beds" covered much more than just Texas, and are found in such neighboring states as New Mexico, Wyoming, Kansas, Utah and Colorado.

For the whole of that winter of 1882-83 he searched the countless hillocks and ravines of stark rock for their hidden treasures, and broke open a myriad of the concretions which were prolific at the site, and yet for all this effort the work was essentially fruitless. This never-ending monotony finally got to Hamman, who purportedly started a quarrel with Charles so that he might have excuse to leave. This he did in much the same manner that Cope's cook and scout had done in earlier days. Like them, he had also been paid for services that he was never to render.

Sternberg, even though out in the wilds some 30 miles from civilization managed to find a replacement, an Irishman by the name of Pat Whelan. He described Whelan as both good and honest. Owing to an unseasonable and horrific storm, Whelan's team of horses were lost not long after his having joined the camp and the team had to be replaced, and Charles almost froze to death while waiting for him to return from a trip for reprovisioning since Charles was ill-prepared for life in these harsh conditions, at least as far as proper tenting was concerned. It is another indication of Charles' lack of preparation for this first Texas trip.

Although Whelan was conscientious and served Charles well, he had farming duties that were to force him to cut his duties short. After only a short period he had to leave and return to his farm. It is to Whelan's credit that he would have preferred staying, and only left because of other duties that had to be met.

Discouragement heaped upon discouragement to the point that Sternberg was very close to leaving the field and trying other sites, when success finally arrived. After 40 days of no value whatsoever he was to find a new horizon that immediately showed promise. And then on the second day in these beds he finally met with his first real find of the season, fragmentary bones of *Eryops*, that great blunt-nosed salamander so familiar to us in museums today. A week or so later he was to find his first *Dimetrodon*, an animal of which he was not even aware at the time existed. This "ladder-backed" specimen was a reptile with long spined vertebrae that rounded the air much as a sail does on a sailboat. Similar to its carnivorous relative *Edaphosaurus* it was a reptile that lived near deltas, and most likely lived mainly on a diet of the sluggish *Diadectes*. *Dimetrodon* was slightly larger than its cousin and generally attained a length of about 10 to 11 feet or so. There was enough bone in this find, even though the animal was naturally small, to provide about 75 pounds of bone and matrix.

These small successes were a real stimulant to Charles, particularly after he had wasted so much time with fruitless searching, and so he was avid in wanting to continue his search. Unfortunately his Irish helper had to leave and take his team with him. Again, however, he had sent word for an assistant and a Mr. Wright soon appeared on the scene. The remote nature of the territory is very evident when we note that Charles indicated that this assistant had spent a full day and a half searching the Big Wichita brakes for him and his camp, even though he began with a general idea of Charles' whereabouts.

In March of 1883 Charles was to decide that he could not exist without a team of horses, since Wright did not come with a team, and he would have to rely on the letter of introduction he had received from the Secretary of War, Robert T. Lincoln (son of Abraham Lincoln), at the instigation of Alexander Agassiz of the Museum of Comparative Zoology. This letter was to prove instrumental in providing food for Sternberg and his camp as he traveled to Ft. Sill, Oklahoma, in search of a wagon and an escort, as well as providing the men and equipment that allowed him to complete his expedition without having to abandon priceless fossil material.

Thanks to the letter he was able to return to the field with an army escort consisting of a Corporal Bromfield and three army privates, a wagon drawn by a six-mule team, and a teamster to provide the drayage. The army also graciously provided rationing to last 50 days, and all this on the strength of Agassiz' request to Lincoln. It was the first time that Charles had ever benefited from the army's services in relation to fossil hunting duties, while he had seen Marsh's ability to use that to great advantage in past years. He felt a real sense of gratification as they drove back to the Texas campsite.

The rest of the season went relatively without incident, except that Sternberg's tent caught fire and burned to the ground. Thanks to the diligence of the army men, and Charles' concern for his treasures, they were able to save the fossils which were in the tent at the time. After finishing up his collecting in the region in late April they set out with their fossil load to the nearest railhead, which was at Decatur, east of Seymour. Decatur was at the time the western terminus of the Fort Worth and Denver Railroad, and they finally left off their precious cargo there on May 4, and from there returned to Fort Sill.

It was not again until 1895 that Sternberg was to return to Texas, and then he only did so since it was at Cope's behest. It had been 16 years since he had last worked for his friend and mentor, and now Cope wanted him to make further sallies into the Big Wichita area to search for new species. In this second expedition Charles was assisted by a Frank Galyean, a man who lived within the general vicinity. In this year they worked above Seymour at an area around Coffee Creek, and at first, as was so typical, they met with little success. This was a pattern for Texas it seems, for fossils were much harder to come by in the Red Beds, at least those of any real value.

Galyean, much as had the old trapper Abernathy, had indicated that an immense and complete skeleton of some past denizen had been discovered by him, and so Sternberg followed him to investigate. The fossil turned out to be ordinary, weathered and broken, and of no value to science and so was left to the elements. Charles indicates that his sorrow was great, but it was soon turned to joy when Galyean found a complete skull (unnamed in his autobiography) that proved to be mainly of value because it opened up a new vista for him. The area in which it was found looked decidedly unfossiliferous, and had been purposely bypassed by Charles since it seemed unpromising, and such was also the case with Boll, Cummins and other collectors over a number of prior years. It just looked unpleasing to the fossil hunter's eye and hence no one bothered to check the area, concentrating on the many more likely spots nearby.

From this skull they followed the trail to what he describes as one of two amphitheaters, which were to be the finest sites he was ever to discover in the Texas Permian. A number of skulls were quickly found in the concretions of red clay, with the rock being tinged with green and other colors. In a short time he indicates having filled his collector's sack with a veritable treasure load of fossil skulls, of wide disparity. He remarked that his sack weighed in at over 75 pounds, and contained skulls of the minutest variety (under 1/2 inch) to over eight inches in length, and all new to science at the time. He was so enthused with this load that he felt real sorrow at admonishing the good Galyean for offering to take the load from him. The sheer magnitude of the find was overwhelming to him, and he didn't want to lose one minute's worth of joy at its finding, so he could not bear the thought of relinquishing his treasures under any circumstances.

From this wonderful site he was to eventually uncover 45 skulls, that were either complete or nearly so, and additionally another 47 partial skulls (some of which were almost two feet in length!). The entire collection from this site when completed came to a surprising 183 specimens, and this from an area that gave up fossils with great alacrity.

The success of 1895 led to a repeat performance in early 1896, except in a different area of the Red Beds. This expedition was into the area around Bushy Creek, just 10 miles north of Seymour, and it broke the pattern since it proved a success from the very start. On just the third day Sternberg was overjoyed to find a good specimen of what came to be called the "ladder-spined" *Naosaurus,* which provided his greatest professional challenge for collecting. *Naosaurus* was thought to be an animal similar in makeup to *Dimetrodon* or *Edaphosaurus.*

This particular specimen had a number of perfectly preserved but broken spines, and he estimated the fragments from this skeleton as being in the thousands. Since the spines themselves could exceed three feet in length it would have been impossible for a preparator to match all of the tiny fragments from similar body parts together in order to properly bring the specimen to fruition. Recognizing this fact Charles carefully wrapped each spine and its related fragments separately, labeling all of the parts so that a proper and quicker restoration could be done. This skeletal portion of this animal was discovered on Hog Creek in Baylor County, Texas, and was designated as AMNH #4015. It includes all the vertebrae of the animal, most of the ribs, as well as the clavicles and the cleithrum. He was later to make two other discoveries of the same animal, one near Coffee Creek in Baylor County in 1895 and the other in the same general vicinity but in 1902.

In a letter to William Diller Matthew in 1907 Charles was to indicate that he had found some notes from the fieldwork of 1882 which related to the skull find of the "*Dimetrodon* or *Naosaurus.*" He apparently had discussed some skepticism relating to the skull of this creature with Matthew for he was to state "you may remember I did not think the skull you have attached to Naosaurus (sic) skeleton is correct, and I suppose the reason was on account of my remembrance of the fragmentary one I found in 82."[10] His notes indicated that the skull he found was of a large reptile with "long recurved serrated teeth." He found another broken skull, apparently of the same animal, on west Coffee Creek that had similar dentation.

It was with great pride that Charles acknowledged how pleased he was (since he hadn't quite realized how complete this animal was, being he had taken it out in bits and pieces) when he learned that it was to be the only mounted specimen of this animal to be found in the world. The elation, however, did prove ephemeral.

The animal that Sternberg speaks of, *Naosaurus*, has since been ousted from its spot in history. Professor Osborn of the American Museum was to have this mount constructed based on Sternberg's finds, as well as finds by Cummins, Ball [Boll?] and two other collectors. It was therefore a composite mount, which although based on good evidence, seems to have been premature. It is certainly worth noting that E.C. Case, Osborn, Matthew and Williston all seem to have been convinced of the accuracy of the mount when initially done since they each played a part in it. The University of Chicago was also duped by this composite animal as well for Samuel W. Williston was to comment in a letter to Handel T. Martin that "their" specimen was almost finished and ready for public viewing.[11] What was Sternberg's highest moment in the Permian came to an abrupt end soon after.

The main portion of this composite, AMNH #4015, has since been undone from the composite and is still on display in the American Museum today, except that it now bears the name of *Edaphosaurus pogonias*. It remains an impressive fossil regardless of what it might be called.

The intrinsic value of this find was in its being an index fossil for the Permian. E.C. Case was to sum up the value of *Naosaurus* (or *Edaphosaurus*) when he made it clear that the strata commonly known as the Wichita was Permian. He based this primarily on the strength of this particular fossil species:

> The evidence for the Permian character of the beds rests then, on the presence of a single genus, Naosaurus, common to the Permian

of North America and Europe, and on the community of many very primitive characters and numerous more specialized ones, which, however, reach down into the Carboniferous below or up into the Triassic above.[12]

This discovery, like so many in the Texas beds, was great while it lasted, but was followed by a real drought over the next couple of months. The ephemeral nature of fossil hunting is evident to any that have tried their hand at it, and Texas is one of the worst areas for obtaining quick success because of the age of its beds and the destructive power of its volatile weather. To make matters worse Charles was plagued with a variety of ailments, most of which seemed to fall under the general heading of "ague." Although this term is out of vogue in today's vocabulary it seems to best be described as a modern flu that finds itself manifest in fever and chills. He claimed the causative agent to be the infernal red clay that constantly stuck to his boots, and that seemed to always be in a state of wetness.

Charles takes Cope to task for the ill-success of work after the discovery of *Naosaurus*, and blames it on Cope's mistaken ideas about a fossil stratum that supposedly separated the Permian and the Triassic, and which would upon discovery yield fauna as yet uncovered. Since Charles had spent a great deal of time in his initial surveying of the region he protested strongly to Cope, only to be finally overruled by his "more powerful will." He commented that the result of this interplay between he and Cope was that he had to spend a whole month of backbreaking work at sites around the head of Crooked Creek and other nearby creek valleys, well away from the productive fields southeast of these sites. Charles was very discouraged at having to waste valuable time to no avail. It was simply not enough for him to be satisfied with mere pay, he needed the fossil incentive as well. It is eminently clear that he felt he had a dual purpose in the field, and that he had two masters: the museum-payer and science itself. Both claimed his equal preference after his allegiance to God.

In a letter from Cope in 1896 (that is shown in Charles' autobiography *Life of a Fossil Hunter*) Cope takes especial pains to try and mollify Sternberg's injured feelings over this lost time, and offers real balm to salve Charles' wounds. Cope was to write "the serious worker in science holds a high position among men, no matter what the great h---- may say about him. They simply do not know & their opinion is not worth considering."[13]

This is exactly what Charles wanted to hear, and once again he was filled with the wonder of Cope. There was still no question that Cope held greatest sway with him in all things fossil. He was so impressed, in fact, that he indicates he was encouraged enough to remain an extra month. Cope apparently did promise that he would never again send Charles upon an assignment where there was work to be done that went against Charles' better judgment.

By staying on he indicates that he found more good material, but not in the beds Cope had him work before. He found another salamander that Cope was to call *Diplocaulus magnicornis*. The head of this very unusual animal when looked at from the top down looks very similar to a boomerang with eyes. The enlarged head seems out of place on the small and flattened body, and the tips of the triangulated head come to points on the end. The animal was amphibious and most likely spent the majority of its life in water. To call it a grotesque caricature of an animal would not be far from the mark. It was to prove to be the most prolific species to enhance his collecting bag that season.

The skulls of these gruesome creatures have the typical sculptured bones found in amphibians, although the pits and grooves seem to have a consistency of pattern that are lacking in many other amphibians.[14]

It is appropriate enough that Cope was to label these unusual creatures as "mud-heads." He gave them this name because of what they became after death. The raw skulls were invariably covered with a "thin coating of silicified mud" which Charles indicates was extremely difficult to remove when preparing the specimen. Since this animal was deemed amphibious it seems in perfect keeping with its lifestyle.

In the spring of 1897 Charles was to once more head to Texas to work for

Cope. The age of the specimens from the Permian intrigued Cope, and he wanted to know all the more. Cope was so interested that he wanted Charles to always send the better specimens by express so that he might get them more quickly. Charles had by mid-April already made a long and arduous trip of over 100 miles around the Little Wichita, and was encamped back at Indian Creek when he was aroused from a heavy sleep by a livery-man bearing the news from Mrs. Cope that her husband had passed away on the 12th of April.

Charles was appreciably moved by the passing of his friend and was to comment "I had lost friends before, and had known what it was to bury my own dead, even my firstborn son, but I had never sorrowed more deeply than I did now over the news that in the very prime of life, in the noonday of his glorious intellectual achievements, as he was bending all his energies to the study and description of the wonderful fauna of the Texas Permian, the greatest naturalist in America had passed away with his work undone."[15] One cannot but feel the severity of sentiment with which this information hit him, and understand the love which he had for his first employer.

Charles would next see the Red Beds in 1901, only this time in a more organized way since he had learned well how difficult Texas could prove if one did not have his own team of horses. Because he knew it could be the deciding factor as to whether good fossils were recovered or not he made sure that at least this factor would not work against him. His son George, at 18, was now old enough to accompany him and so they loaded the team aboard a freightcar and rode with the car to its destination in Texas.

During this trip Charles was to be assisted by a man named J.S. Chesnut who acted as both surveyor and guide. Chesnut was particularly valuable in that he knew the whole of the countryside where they were to work.[16] This made him quite different from the other assistants he'd previously worked with in Texas.

This season they were to begin at Willow Creek, and they started in the heat of the month of June. This region swelters normally even during the early summer months, but this summer was exceptionally warm even by their standards. Charles was to claim that the temperature in the shade would often rise above 113 degrees. This heat naturally played an important role in determining how their expedition would go, and it did negatively impact their movement. It both depleted their water supplies quicker and added to the difficulties of their working upon the very hot soil, but impacted them most by exacting its terrible toll upon the fossil hunters themselves. They quite simply could not perform as well under these conditions as they would have under normal circumstances.

This season was under the direction of Dr. von Zittel of the Paleontological Museum of Munchen, although distance necessitated that he correspond with Charles rather than taking an active hand at collecting himself. He did, however, send his assistant Dr. Broili (now quite famous in his own right) to spend a few weeks with the Sternbergs. While there Broili managed to take a number of photographs of the formation as well as taking considerable notes. Broili was perhaps the first true scientist to spend any appreciable time in the Permian.

On this trip Sternberg was to find a real oddity; he was to find casts of 10 different skulls. The casts had no bone left to enhance the finds, but were of such excellent quality that Charles was to claim he got several that were actually "better specimens than the type." The value of skull casts is considerable since they provide definitive information on brain size and shape.

He and George were also to find three separate bone beds from which they managed to extract a wide variety of minute animal forms. The length of a number of these finds ranged from a quarter of an inch to just over an inch. In the beds he managed to collect over 20 perfect skulls of minute dimension, in addition to a large supply of more poorly preserved skulls. He commented that the miniature bones of these creatures stood out as white as snow against the reds of the formation and hence were relatively easy to find. So white were they, in fact, they seemed to him to be as of animals having only recently died.

At the tailend of the season they got a firsthand lesson in how water reacts in a desert environment. George had discovered a rich invertebrate locality (most likely ammonites and belemites judging from his scant description), and after a short downpour of rain Charles had crossed the creek to work at this location. Because George apparently recognized what was to come he yelled at his dad to come back across. Charles indicates he only just managed to scamper back across the creek, minus his tools which had to be left in his haste, before the "boiling flood of water covered the rocks in the bed of the creek over which I had just crossed dry shod, rapidly rose to eight feet, and threatened to submerge my camp."[17] Those working for any time at all in desert environs will quickly attest to just such a danger, and the quickness with which it comes on.

During his stays in the Texas Permian Charles was to often travel to Seymour as it was both close and hospitable. When in Seymour he generally stayed at the Maclain Hotel because it was comfortable and cheap. Rather than eat at the hotel, however, he would generally take his meals at the home of a friend of his named Patterson.[18]

His supplies he got through the grocers, Taylor and Mitchell, and he also had frequent business with the liveryman, William Adkins and the druggist, Steve Lee. His banking was done at the National Bank in Seymour. Because Charles always feared being robbed (perhaps because of his early Kansas encounter) he would have the various museums who hired him deposit directly there and then he would then handle his bill-paying by check.

The season of 1901 was to prove to be a fiscal flop. Charles was very disappointed in this fact because he always was in a bind for funds, and a bad season had major repercussions on the whole of his life and his ability to work. It also greatly affected his family who depended on him as well.

The 1901 season was the first of five seasons where he worked for the German museum. Although Charles' dealings with von Zittel were primarily by mail he was to develop a real affection for the man. This affection was to lead him to write an obituary upon von Zittel's death which was published in the *American Geologist*. Charles was to refer to von Zittel as the "truest lover of the ancient life of the earth" that he ever knew. This kind of language is unmistakenly similar to the enhancing language used to describe both Cope and Lesquereux before him. Charles obviously had more than one mentor.

In May 1902 Charles was to work again for the American Museum of Natural History for a period of about two months. Due to his not being able to get the special transportation rates through the railroad as he had the previous year, he was forced to leave his outfit and team at home and hire directly at Seymour. This greatly hampered his ability to collect, and impacted his opportunity to come out ahead financially. He was to complain bitterly about this fact to Henry Fairfield Osborn. Charles was to note that he managed to finally hire a man and a team at a total cost of about $70 a month, and that he had high hopes the man would prove worthy.[19]

His high hopes for the man he had hired, however, soon waned. Although he would describe the man as being "energetic," he would also describe him as being "densely ignorant." This therefore gave Charles "no society." Since Charles always thrived on society this was to prove a major deterrent to his work there. If this, however, were not bad enough, he was additionally inconvenienced by the fact that the man was "a miserable cook" and Charles indicated that "my stomach revolts against the food that he prepares."[20] Because of this he was to write to Osborn with an urgent entreaty that he be sent an able assistant with some knowledge of the Permian.

Charles was to admit to even greater frustration when he explained to Osborn that he personally felt very inadequate in recognizing the value of specimens uncovered in the Permian. In Kansas he could "in imagination restore it [a specimen] to life and put myself back to the time when full of life and vigor they fought for existance (sic). Here I find thousands of mingled bones of many forms, and only in comparatively few instances are enough of the skeleton present to enable me to get much of an idea of how it appeared in life. I have not then, the use of my imagination to help make the discomfort and labor endurable..."[21]

Sometime between mid-June and mid-July Charles left the field and returned to Kansas. He did write to Osborn that he had been hampered by matters financially to the point that he could not even care for day-to-day matters at home, and that his wife Anna had been forced to "endure many privations for lack of funds."[22] Whatever the stress was must have resolved itself to some extent because Charles was back in action soon after. He was, for instance, in the Niobrara Cretaceous of Logan County in October of that same year. One can safely say that the Sternbergs were always just a poor season away from starvation, so whatever the amelioration, it was short-lived.

Charles was to contemplate another Permian trip in 1910, for he was to write to Dr. A. Smith Woodward of the British Museum: "Now I have thought of another chance to get Permian fossils, they are so rare I dare not run the risk of an expedition there unless I have a guarantee that I am to have pay for the honest labor I put in work..."[23] Since Charles was forced by the very nature of his independency to be always concerned about wages, he was forced to abandon his thoughts of another expedition at that time since he could not obtain a sponsor under those terms. Since Permian fossils were so illusory it was not feasible to sponsor trips there without some guarantee of success. A man on a shoestring could not obviously provide such a guarantee. Unfortunately many of the world's greatest museums had to use that same logic due to their budget strictures, and as such, work in the Permian was at times greatly hampered.

As it turns out, Charles was to twice more go into Texas. In 1917-18 he would take his last hand at the Permian, at an age when many men have already retired from anything active. At 67, however, he was still spry enough to get around even the Texas badlands with some agility, although his recuperation times were undoubtedly longer.

In 1917 Charles was to collect at the now famous Craddock Quarry that had been discovered by one of Samuel W. Williston's assistants. The Quarry was named for the owner of the quarry who allowed Charles to collect there both in 1917 and 1918.

The work here was not easy, but was frenzied and demanding due to the heavy overburden which had to be removed before the close work could begin.

Charles estimated this overburden at about 20 feet in thickness, and indicated that it was a sluggish form of earth and clay that proved quite difficult to move. In order to proceed Sternberg had to hire a man with a scraper and a team of horses.

From this quarry Charles was able to collect in the first year a very fine mountable skeleton of one of the more difficult animals to collect whole, *Dimetrodon gigas*. This was sold to the United States National Museum, and again provided impetus to Charles since he was in the spare Permian. However, the success of 1917 didn't seem to help in the year 1918 as that year (at least in the Permian) proved very elusive in good fossil material.

Charles' working of the Craddock Bone Bed locality was to put him at odds with his old friend and foe, Samuel W. Williston. Williston was displeased to learn that the Sternbergs were working the site, for he felt that his "discovery" of the site gave him preference to the fossils there and he had not been asked for permission to collect there. Charles was to disagree most vehemently with him. Apparently Williston had communicated his displeasure over the whole matter directly to Charles for Sternberg was to respond by letter as follows:

> Yours of the 1st instant received. If I remember right, when you pumped my man in 1876, and went into a locality I discovered and gutted it of specimens and left there, it seems to me you set a very different set of ethics now. As far as I could learn, last year, you had not been in the Craddock Bone bed when I went to it last year for years. I believed, however, when you learned of the rich finds I made last year, you would attempt to work it again. I sent Levi ahead to take posession (sic) first. But you informed me in the vacinity (sic) of Snow Hall that you were working in Texas, and a wire from Levi also told me you were in the Craddock Quarry. He was obliged to sacrifice time and money until Mr. Miller holed himself in. We have been at work there ever since at an expense of over a thousand dollars. I will say for your benefit that we secured several fine skeletons of several species, some of them articulated. We have done ten times the work you have ever done.[24]

Williston in his letter must have threatened some kind of retaliatory action for Charles was to add this further comment:

> Now if you want to start at your age a newspaper controversity (sic) go ahead, I warn you however, that you smooth down whatever is personal, you have to say about me. If you attempt to injure my character, and lay yourself liable, you will certainly feel the full force of the law, that will protect me.[25]

This falling out with Williston was to contribute to the negative feelings that H.T. Martin already had for Sternberg. Martin was incensed over the whole episode, and he was to write George F. Sternberg noting his displeasure.

Interestingly enough, considering the very volatile language used by both Sternberg and Williston in this instance, they continued afterwards to have a relationship. Somewhere along the line they either made up, or just simply agreed to tolerate one another, perhaps due to business need.

Most likely to forestall any efforts by Williston to impugn his name at the American Museum due to this incident Charles was to write directly to William Diller Matthew:

> I have just received a remarkable letter from Dr. S.W. Williston of which I give herewith a copy fearing that he has already written in similar report to you...in regard to the deposit, as you know I have explored every nook and corner of these beds for many years before the Dr. went there. He has collected many specimens of the same ground, and it chanced that Miller ran across what he calls the Craddock bone-bed. I made a large collection from there several years ago, for von Huene. Doing more work there than he has ever done before. Last year I went there again and as far as I could learn, he had not been there for four or five years. I sent Levi down to the locality last March, and found Mr. Miller in posession (sic). We left him alone until he went in again...I found two horizons above the one Dr. W. got material from. All the specimens nearly described in my list are above his bed. The way Mr. Miller worked there, it is doubtful if he would ever have found it. He told Levi his whole object in going there was to get something to make a show, as he could always get bones there. If that is scientific ethics, I have never recognized them, because it would simply give one institution, as in this case a great bone-bed to exploit, or leave alone. I had, as I said full authority from the owner, nor did I go into it while W was there so my skirts are clear.[26]

Failing at the Craddock Quarry in 1918 Charles headed homeward to Kansas while George and Levi made a sidetrip to the Rock Creek Horse Quarry in Tulia, Texas. This quarry is about 40 or 50 miles south of Amarillo in northern Texas. They were disappointed, however, to find the quarry's condition unworkable at the time and so headed back to Kansas to join their father.

Charles in the meantime was in Logan County at Butte Creek, and in their brief time away he had discovered a *Tylosaurus* skeleton of some value. Its value was extra due to its containing the skeleton of a smaller animal, the plesiosaur *Clidastes*, that consisted mainly of "half-digested bones." The *Tylosaurus* was some 29 feet in length, and its meal a much smaller eight and one half feet.

They finished the season on a further upnote since they were able to collect two complete skeletons of *Platecarpus coryphaeus*, each of which was about 17 feet in length.

Charles' last trip to Texas, although not to the Permian, was in 1919. The beginning of the season of 1919 was begun in Kansas, and the season proved to be both a busy and prosperous one. Sternberg began work on Hackberry Creek in his frequently traversed hunting grounds around Quinter, Kansas, while much snow was still on the ground. The snow, however, did not prove a problem as Charles was able to uncover an 18 foot plate of the most beautiful *Uintacrinus socialis*. This crinoid grouping, a marvelous plate by any standards, was sold to the Albany State Museum in Albany, New York.

Further Kansas work in Gove County uncovered an almost complete *Pteranodon* skeleton *in situ* and another *Clidastes tortor* skeleton some 12 feet in length. From this site they moved to the Smoky Hill River in an area near Utica where Charles collected solo. It is a measure of his determination and skill that he was able to collect an 18 and one half foot *Platecarpus* skeleton that was very nearly complete, by himself. He claims to have taken up two massive sections of this great fossil, sections that weighed about 500 pounds apiece. In taking them up he had to turn them over for bandaging, which he also managed to do by himself. He presumably performed this task through the use of levers and pulleys.

In mid-August he was to travel southward into Texas, arriving in Bristow County, Texas on August 27. Their camp was made at Gidley's famous horse quarry, which was Pleistocene in origin. Ten specimens of *Equus scotti* had already been taken from the site, and Charles was to secure the eleventh.

Charles was to hire a local farmer, a Mr. Stephenson, and his three sons, who together with three teams, plows, and scrapers tackled a massive section of clay and sand bank on the southern perimeter of the quarry. When all of the work had nearly been completed Charles found that they still had no fossil horse to show for their efforts, even though the floor had been laid bare in a number of spots. Mr. Stephenson, however, was to surprise Charles when he arrived for lunch carrying with him a freshly extracted horse bone. This single bone led to the 11th *Equus scotti* skeleton.[27] This work at the Gidley Quarry was to last one full month, and was indeed a profitable one. It also proved to be the last work Charles Sternberg was to ever do in Texas.

NOTES

1. Walter Keene Ferguson, *Geology & Politics in Frontier Texas 1845-1909*, Univ of Texas Press, Austin & London, 1969, Pg. 50

2. Ibid, Pg. 92

3. Ibid, Pg. 41

4. Ibid, Pg. 61

5. Sternberg incorrectly identifies him as W.A. Cummins

6. AMNH Ltr, Jun 15, 1902 (To Henry Fairfield Osborn from Seymour, TX), From the Archives of the Dept. of Vert. Paleo.

7. AMNH Ltr, Apr 15, 1902 (To Henry Fairfield Osborn from Lawrence, KS), From the Archives of the Dept. of Vert. Paleo.

8. Kenneth W. Craddock & Robert W. Hook, "An Overview of Vertebrate Collecting in the Permian System of North-Central Texas," Field Trip Guidebook #2, 49th Annual Meeting of the Soc of Vertebrate Pal., Austin, TX, 1989, Pg.

41

9. Charles H. Sternberg, *The Life of a Fossil Hunter*, Henry Holt & Co., NY, 1909, Pg. 209

10. AMNH Ltr, May 16, 1907 (To William Diller Matthew from Lawrence, KS), From the Archives of the Dept. of Vert. Paleo.

11. Univ of Kansas Ltr, Apr. 30, 1918 (To Handel T. Martin from Samuel W. Williston while at the Univ of Chicago)[Courtesy of John Chorn]

12. Bailey Willis, *Index to the Stratigraphy of North America*, Dept of Interior, U.S. Geological Survey, prof. paper #71, 1912, Pg. 477

13. CHS, *Life*, Pg. 238

14. See *The Life of a Fossil Hunter* for an illustration, Pg. 240

15. CHS, *Life*, Pgs. 241-242

16. AMNH Ltr, Jul 22, 1902 (To Henry Fairfield Osborn from Lawrence, KS), From the Archives of the Dept. of Vert. Paleo.

17. Charles H. Sternberg, "The Permian Life of Texas," *Kansas Academy of Science Transactions*, #18, 1903, Pg. 98

18. AMNH Ltr, May 06, 1902 (To Henry Fairfield Osborn from Seymour, TX), From the Archives of the Dept. of Vert. Paleo.

19. Ibid

20. AMNH Ltr, Jun 8, 1902 (To Henry Fairfield Osborn from Seymour, TX), From the Archives of the Dept. of Vert. Paleo.

21. Ibid

22. AMNH Ltr, Jun 15, 1902 (To Henry Fairfield Osborn from Seymour, TX), From the Archives of the Dept. of Vert. Paleo.

23. BMNH Ltr, May 9, 1910 (To A. Smith Woodward from Lawrence, KS), Courtesy Natural History Archivist, London

24. Univ of Kansas Ltr, Jul 3, 1918 (To Samuel W. Williston from Lawrence, KS) [Courtesy of John Chorn]

25. Ibid

26. AMNH Ltr, Jul 3, 1918 (To William Diller Matthew from Lawrence, KS), From the Archives of the Dept. of Vert. Paleo.

27. Charles H. Sternberg, "Field Work in Kansas & Texas," *Kansas Academy of Science Transactions*, 1922, #30, Pg. 341

CHAPTER TEN
THE BONE HUNTER AS A TYPE

The turn of the century bone hunter was a curious mixture of both hardened workman with calloused hands, and meticulous scientific inquirer. He has been labeled almost anything from lunatic, to bug-catcher, to genius. Considering the task, and the territory within which they worked, one expects all brawn and little mind, and is surprised to find an unusual combination that pairs strength and innovation. One cannot, for instance, but be struck with the similarity between the opposite extremes of the bone-hunter as just mentioned, and the same dichotomy of traits as found in Teddy Roosevelt. Roosevelt, as both President and Rough Rider, had that curious combination of wit and vim, and it is not really surprising to find him President during part of this era (1901 to 1908), since he in fact typified the times. He also provides for us a good analogy that allows us to arrive at an understanding of a profession so diverse in its chemical makeup. This roughness, as typified in Roosevelt, is as out of place with our normal conception of the presidency of the Unites States, as the raw and roughened Charles Bronsons and Clint Eastwoods might seem today in their stark contrast to past leading men.

Charles Sternberg was to describe himself as an explorer in addition to naturalist, paleontologist and geologist. His definition of an explorer was:

> The explorer always suffers a great deal from exposure. He is away from all comforts of civilization; absent from home and friends. He works with unfailing enthusiasm, and is happy if success crowns his efforts...In the interests of science there is no place on land or sea that he will not explore...He is quick to read the mighty volume the Creator has written in solid rock.[1]

The early bone hunter had, out of necessity, to wear many hats. He had to variously serve as organizer, as administrator, as teamster, as boxmaker, as shipper, as technician, as explorer, as Indian scout, as doctor, as stratigrapher, as salesman, as fund raiser, and as journalist. He had to be all these things, and at times much more. He had to have the native intelligence required to decipher the clues of the most unpredictable terrain, to both keep himself alive in harsh environs, and so that he could follow the trail to the elusive and hidden skeletons of the badlands. This proclivity for bone-finding had to be in-built, for while it is true one can learn the tricks of the trade, yet one cannot "learn" to find bone. One can only look at the clues and take away some of the guesswork. This, considering the lack of documentation of the young science, made it both intriguing and difficult for the fledgling hunters since they were then often forced to learn almost entirely on their own. The bone hunter also had to have the physical stamina that would allow him to be able to withstand the sheer brute force of the elements in the badlands, as the deprivations created by the land were substantial, and the unpredictable nature of the weather legendary.

We tend to cultivate an erroneous and glorified picture of the bone hunter, one that focuses too heavily on the success column, and one that ignores the true facts. In this industry the debits certainly outweigh the credits, even today. We overlook the fact that for every fossil found there are dozens of unprofitable hours spent in stooped and concentrated investigation of ground, under the harshest of desert or badland conditions where not even the merest scrap of bone turns up. William Diller Matthew was to capture the heart of the industry when he was to write, "I collected a little sunburn, killed a rattlesnake and admired the flowers."[2] From the comforts of our cities and houses of today it is easy to forget that the badlands get their name for good reason, and they are not the tourist spots of the world for the comforts they provide. Their beauty is hard-won, even when approached from the luxury of a recreational vehicle.

The bone hunter had to be exceptionally fit, for the physical rigors of the work were substantial, and sometimes overwhelming. Only the hardiest of individuals could survive. The term "raw-boned" perhaps best describes these men. The bone hunters were constantly being worn down by the relentless elements, in the same real sense as the fossils were being eroded from the land. A man's eyes suffered greatly from the careful and close work, the eyes and lungs were debilitated by the airborne rock and dust particles, the knees suffered constant abrasion as well as being constant reminders of what arthritis can do at an early age, and every major bone was subject to the whims of the terrain since they provided numerous opportunities for a man to fall, stumble and break. Such was the daily scourge of the bone hunter then, and things have not changed greatly since that date.

In a letter from the field W.D. Matthew of the American Museum was to further comment upon the harsh results of the collector's strenuous work:

> Last night my hands were so sore that I could not hold a pen, so I missed my usual contribution. Today they are not so bad, so I'll try to write a few lines. Yesterday I did a lot of heavy work with pick and shovel, trying to find all there was of my prospects. Not much success.[3]

This kind of pain and discomfort was not all that unusual. Sternberg describes a similar incident that occurred when he and his son George had taken an expedition into the Dog Creek area of Montana, and Sternberg was to suffer the effects of camplife. Judging from journal entries of other early fossil hunters one would have to judge this as commonplace:

> I know that when I got to camp at night, and had sat down at our camp table, to eat...my feet would swell so badly that I would often be obliged to crawl on my hands and knees to my tent and cot. There, stretched at full length, with lamp above me, I read until bed time, never thinking of getting on my feet until the next

day, when I again went through the same experience.[4]

Yet for all the wearing down and beatings these men took they were right back at it the next day, and sometimes they even seemed the better for it. Being renewed in the "refiner's fire" made them into men of real substance.

Sternberg was to share one of his camp secrets when he noted his trick for staving off stiff finger joints and cracked and bruised hands. Using what they had available was a necessity and so he found that cooking Crisco served wonderfully as a salve when used in front of the open camp fire. Another example, similar in nature, was his choice of a soothing balm to help ward off the tiny but ferocious black gnats that attacked he and his camp fellows when they were in Canada. Although they undoubtedly didn't enhance their appearance by doing so, they found great relief in covering all their exposed skin with bacon grease! It is a mark of his thriftiness and his adaptability that he made do with camp materials at hand, although one can easily imagine what men who met them on the trail must have thought to see them all greased up.

Perhaps it borders on redundancy to mention that Sternberg was one of the most durable fossil hunters that ever lived since his name is synonymous with longevity. There is not likely to ever be one to compare with him in this respect because of the changes that have occurred in the trade. Sternberg began working on the fossil fields of Kansas in 1870, and was still doing fieldwork as late as 1924. This is indeed a stupendous feat for a man who was deaf in one ear, lame in one leg, slight of build and diminutive in stature. Charles was considerably shorter than his son George who measured in at only five feet five inches in height.

With obvious respect for Sternberg, and perhaps even awe in light of his own failing field capabilities, William Diller Matthew wrote from Clarendon, Texas in 1924 to his wife Kittie:

> ...I don't know how many years I'll be good for this western field work; perhaps I can judge better at the end of this season. Granger [Walter Granger] talks as though he and I were pretty near the end

of our string; yet I have known men, like old Sternberg, to be still sticking to it at seventy. I don't think I am as good at climbing around badlands all day as I was, but that may be temporary effects of my tonsillitis.[5]

At the time Matthew wrote this tribute to Sternberg he himself was 53 years old, and already a veteran of 30 years in the field. Walter Granger, to whom he attributes the initial statement, was 51 years old, and the seasoned veteran of the famous 1921 Mongolian Gobi Desert Expedition that successfully uncovered the first recognized dinosaur eggs known to the world. Although both these men were extensive field workers they couldn't begin to touch Sternberg's time on the field, not even when matched year for year. Matthew was only to live to be 59 years old, while Granger was to die at 68 of a massive heart attack. As hardy as these two fine men were, and rugged from years of strenuous work on the fossil fields of the world, it is certainly ironic that slight and fragile Charles Sternberg was to outlive them both, on the field and off.

If one were to picture Sternberg in his later camp years one would envision him as a lean and hardened man with a slight paunch, who had thinnish hair and a grayish grizzled moustache. We would note quickly that he was a man with piercing eyes, but yet gentle in their nature. He would be wearing denim overalls that were covering up his long cotton underwear (which he wore in all weather in order to absorb moisture), while a floppy and weather-beaten hat with a broad brim would round off his campwear. His gait would be noticeably troubled due to an old injury and his hearing would be sub par from an early childhood ailment. The total picture, however, would be of a man comfortable in his element, and dressed for the job.

When one considers that Sternberg had a crippled leg from an accident in early childhood it seems beyond belief that he could have accomplished so much in terms of physical exertion and stamina. Since the deprivations of camplife (lack of water, unbalanced diet, long hours, etc.) must have taken their toll over the years it seems utterly impossible to comprehend his being in his 70's when he finally quit the field.

As a final footnote to this point one cannot but remember another very large looming fact: the average life expectancy of a person born in the early 1880's was into the mid-40's. Sternberg was to keep booming right into his 90's! It wasn't a sedentary life either, but one fraught with all sorts of physical challenges and life-threatening sequences. Perhaps it was this thrilling and stimulating lifestyle that added to his lifespan. Hard physical work beats the sedentary life any day, and as proof you quickly note that you don't find living any old non-stimulated senior citizens. Blood flows best when it is warmed from exertion. Benjamin Franklin was right years ago when he stated that "the used key is always bright."

Although there are many other things that can be said about the fossil hunter as a type, it would certainly be unfair to not at least comment upon their social acceptance. For the more elite of the group (generally those born of money and well educated) there was probably only the merest hints of disdain from others for their avocation, but for the everyday plodder, the workers and the field technicians there was no such insulation. They had to bear sometimes the full brunt of the behavior that their profession evinced in the populace. Captain James Cook, the famous western personage, compatriot of both Professor O.C. Marsh and the famous Chief Red Cloud, and owner of the equally famous Agate Quarry ranch site, perhaps best summarizes this early sentiment when he describes the unkind attitude his peers took to the early fossil hunters:

> The early fossil hunters surely had to endure hardships in the days when West meant West. I can well remember how most of my western friends regarded the early fossil hunters and naturalists who came to do collecting. They were usually spoken of as bone- or bug-hunting idiots. For anyone to go chasing about over the West hunting for petrified bones, or even bugs, was conclusive evidence of his lack of good horse sense, especially in sections of the West where the Indians were still wild enough to want to stick their arrows into anything wearing a white skin.[6]

The California geologist Olaf P. Jenkins, in an early century encounter with a cowboy (when Olaf was yet a camp aide to a party of geologists), took much amusement at the cowboy's assessment of their work: "Finally he [cowboy] said he didn't think they knew themselves what they were doing and that I had a much better job because I knew exactly what I was doing."[7]

Even Harvard professors weren't exempt from the ridicule. Louis Agassiz recounted a story about a group of them who were bug collecting in the White Mountains. The men would jump from their coach at period intervals it seems to collect specimens, and then they would cache their trophies by pinning them to their hats and coats. The coachman, unfamiliar with this kind of behavior, was to later explain this peculiar behavior to another party by stating:

> Last Thursday I had the queerest lot of passengers you ever saw; they were men grown and dressed like gentlemen; but they kept jumping out of the coach, and like little children ran about the field chasing butterflies and bugs, which they stuck all over their clothes. Their keeper told me they was *naturals;* and judging by their conduct, I should say they was.[8]

While this may seem somewhat outlandish, considering the number of odd characters that populated the Plains states at the time, one must recognize that the scientist was really new at this time both in America and in the West. Such behavior as Cook describes was new to the white inhabitants also, and not just to the local Indians. It is not hard to see how this kind of attitude could be taken against the early bone hunters. Once the scientist was established as an integral part of the western scene social mores were to change to some extent, and a more lofty position awarded to the bone man. People such as Agassiz and Cope and Marsh, although somewhat odd in their personal habits perhaps, helped to stimulate the populace into thinking about Paleontology in a more positive light. They paved the way for the modern scientist in the field by making the profession respectable. Once the dinosaur wars actually began the reporters that hounded these bone men helped in their odd way to also build up respectability.

The profession of bone hunter was clearly a tough one, and it would be easy to understate the case when describing its rigors. Sometimes even those in the profession found it hard to account for why they themselves remained in the profession. Even though they personally may have enjoyed traipsing around in the wild badlands, being chewed alive by various and sundry kinds of insects, being chased around by all sorts of wild animals, and subjecting themselves to every describable kind of menacing weather, they still more likely than not could not bring themselves to recommend the profession to others. Charles Gould, a pioneering geologist himself, in his autobiography was to credit his mentor Samuel Williston with making the following remarks to him about the geologist's life:

> It's a dog's life, Gould, and there is nothing in it. A geologist never makes any money, he works hard all his days, he is called a fool and a crank by nine-tenths of the people he meets, and he lives and usually dies unappreciated. I would advise you to try something easy, like law or medicine or selling shoes, or stick to school teaching. There is more eating in those jobs than there is in geology.[9]

Of course Gould didn't listen to Williston in this regard, and love-hated his profession all the way to its apex for him which turned out to be his becoming the first head of the State Geological Survey in Oklahoma. No real geologist, or geologist at heart, could possibly be surprised at Gould's response or at Williston's advice.

Charles Mortram Sternberg, not normally known for the fluidity of his prose, was to rather elegantly describe what kept him in the profession in a letter to the amateur collector H.S. "Corky" Jones:

> It is not only you westerners that live for 'next year,' I fear that many of us would get discouraged if it were not for 'next year.'
> I think most of our joys are from looking back at what we have accomplished or looking forward to what we hope to accomplish.[10]

Another major trait of the early bone hunters involved their exceptional knowledge of the land, and their abilities to recollect sites where they had found particular fossils, or had discovered minute traces of a particular strata in an area where it had been previously unknown, and so on. Since this was such a fledgling science it lent itself well to the sort of unschooled and raw talent that many of them possessed. Just as George Sternberg was to recall in later years how he most assuredly could go right back to the spot where he had many years before discovered his first fossil, it is as easy to say that the same could be said for most of these early hunters. They had that same keen sense of direction and the explorer's knack for searching out these places and then keeping them in their mental catalogue. Most modern bone hunters, with all their hardware, don't have to rely on their memories as much, and hence have lost something in the process. This raw talent seems to have been lost along with their close communion with the land. With the deaths of Charles Sternberg and Barnum Brown the science definitely changed, and field work has not been the same since.

As the occupation gained in stature, mainly as a result of Cope and Marsh's western dinosaur finds, the professionalism of the itinerant bone men had to keep pace. They had to learn to cope with the ever-present forces of the media, because the public was always trying to ferret out the latest sensational find, and it did so through the writings of the keen-eyed newspapermen who would follow after the bone men in pursuit of another scoop, and they had to then be able to describe intelligently what it was they were unearthing. Although the media was certainly a plague at times it did help in the end.

A fairly typical example of such a media blitz, for instance, is cited by George Sternberg who wrote that while he was working in the Canadian Red Deer that there were over 400 people who visited their remote camp in the space of four short months, as well as two separate movie crews.[11] This was certainly in contrast to earlier days when such visits were a real rarity and generally purely by chance.

The feud that developed between Marsh and Cope certainly added both fuel and flavor to the fire as far as the media was concerned, and made the bone hunters take on a stature that had not been theirs previously. A modern-day analogy might be found in the realm of the college basketball star. It used to be that all the player had to do was play ball well, but now they are expected to be personal spokesmen for the college and the game, and have to be given lessons in handling the media along with their game plans.

The general sense of the reporting can be seen in this excerpt from the *New York Times*, which tells of a sister papers efforts to obtain a story:

> A *Bee* reporter, hearing that Mr. S.W. Williston, the assistant of Prof. Marsh, of Yale College, who has been digging up fossils for several years in the mountains, would remain in Omaha during Sunday, made a call upon the gentleman.[12]

Williston's remarks are such that he describes the inaccuracies that had plagued him, and that might be used today in the same context in terms of reporting in this day:

> 'The press,' said he, 'has made a great many blunders and misstatements concerning these discoveries, and while I often hesitate on that account to be interviewed, I shall gladly comply, hoping you will be somewhat careful in presenting the facts not to distort them.'[13]

The interview that was to follow certainly had an air of authority to it that might even be described as a form of pomposity. This attention that the scientists were now receiving was something new to them, and sometimes they did not know exactly how to react. It must be said also, that it is not uncommon for the reporter to make an interview seem loftier by rearranging the comments or by changing their emphasis to change the attitude displayed.

By around 1900 the bone hunter had finally crested the wave, but it had taken great ado and commotion for him to do so. He had the dinosaur to thank. As a result the interest in the past has never been greater than it is today, and even the youngest of boys can generally give one a better lesson in nomenclature on dinosaurs than can his parents. It is this interest that has spurred a reawakening of interest in the men who initially scoured the countryside in search of yet another giant. We have much to thank these early adventurers for, but it is seventy-five to one hundred years too late for most of them.

One wonders how pleased they would be to look down today upon the throngs of natural history museum-goers in the main foyer admiring a dinosaur extracted by sheer brute force from the cliffs of some distant badland, and then telescope into the museum office to see the resident paleontologist poring over his books and writing his papers, never having spent more than a few short hours on the field in comparison. His whole life having been devoted to after-the-fact summarization of fossils extracted mechanically by others. I think I know the answer.

NOTES

1. Charles H. Sternberg, "The Triassic Beds of Texas," *Kansas City Review of Science & Industry*, Vol VII, Dec 1883, #8, Pg. 456

2. Charles L. Camp (ed.), "The Letters of William Diller Matthew," *Journal of the West*, Vol. VIII, #2, Apr 1969, Pg. 284 (Ltr of Jun 5, 1908 from Agate, NE)

3. Ibid, Pg. 273 (Ltr of Jul 1, 1906 from Porcupine Creek)

4. Charles H. Sternberg, *Hunting Dinosaurs in the Bad Lands of the Red Deer River Alberta, Canada*, Privately pub, World Company Press, Lawrence, KS, 1917, Pg. 121

5. Charles L. Camp (ed.), "The Letters of William Diller Matthew," *Journal of the West*, Part II, Vol. VIII, #3, Jul 1969, Pg. 471

6. James H. Cook, *Fifty Years on the Old Frontier*, Yale Univ Press, New Haven, 1923, Pgs. 280-281

7. Olaf P. Jenkins, *Early Days Memoirs*, Bellena Press, Ramona, CA, 1975, Pg. 59

8. *The Observer*: The Outdoor World, Vol VI, 1895, #3, "The Agassiz Association," Pg. 43

9. Charles N. Gould, *Covered Wagon Geologist*, Univ of Oklahoma Press, Norman, Pg. 48

10. Saskatchewan Museum Ltr, Nov 22, 1939, Courtesy of Tim Tokaryk

11. *Calgary Daily Herald*, Oct 8, 1921

12. *New York Times*, Jul 17, 1878

13. Ibid

CHAPTER ELEVEN
COPE-MARSH FEUD

Certainly one of the foremost, if not the world's foremost paleontologists of the past, was Othniel C. Marsh. On almost anyone's list he would certainly be rated within the top five American paleontologists of all times. Although Charles Sternberg's associations with Marsh himself were not all that extensive, it is yet amazing how often their paths crossed in some manner or other. Perhaps not always personally, but through associates, they managed to have a great deal of contact; some favorable, but most not.

Sternberg's first association with Marsh occurred when Charles was a student at Kansas State Agricultural College in Manhattan, Kansas. Around this time he was taking one of the few college courses he was to ever take, and he had just missed the opportunity of studying under Professor Benjamin F. Mudge who had previously taught the geology courses. Mudge had been eminently qualified to teach paleontology courses as he had already taken a number of field expeditions into the Kansas chalk. It was only because of his having been dismissed from the college over a political matter that Charles missed the opportunity to have him as a teacher.

Mudge's first expedition had been in 1870, and he had more recently led expeditions into the chalk, expeditions which had shown a great deal of success.

Mudge had worked for O.C. Marsh during these expeditions and hence if Charles had made it onto the expedition he probably would have been a Marsh rather than Cope disciple. There appears little question that Charles' thoughts of college would have turned on Mudge as teacher, and with Mudge gone the real attraction of the college was missing.

In the year 1875 Mudge was to form yet another expedition to enter the chalk, and Charles as a student of the science was very interested in being a part of that team. Although the reasons remain unclear as to why the expedition was full Charles was denied access to the group on the claim there was no further room for party members. It does seem odd that the expedition should have filled so quickly since Charles apparently jumped at the opportunity on first notice, but such was the case. There is certainly no reason for thinking that Mudge would have slighted him by denying him access to the party as Charles had more experience than the average student, and was appreciably more experienced than some of the actual members of the party.

In either case, Charles was to write his letter to Cope, and for the next four years his immediate allegiance was to remain with Cope. During those years, however, his expeditions and that of the opposition would meet and socialize over the clean water at Buffalo Park Station as previously mentioned. Since Mudge worked for Marsh, and Sternberg for Cope, any great animosity between the parties would certainly have come to the fore at that juncture.

Sternberg's allegiance and fondness for Cope are very obvious throughout his writings, and there is an equal void of opinion when Sternberg talks about Marsh in *The Life of a Fossil Hunter* and his other writings. There are no real direct references to Marsh beyond the casual business references one would expect, and this leads one to believe that Sternberg's opinion of Marsh most likely matched that of the other men who chanced to work for Marsh, and it most likely would have been an unfavorable one.

Marsh's deficiencies were legion when it came to his personal handling of financial affairs, and most particularly in respect to the prompt wage-paying of his various collectors. Most of the major figures who worked for him at one time or another were to speak out in rather substantive language when speaking about Marsh's failures both as an employer and scientist. Williston in particular was quite upset with Marsh, and well stated the case for it when he wrote to Cope after leaving Marsh's employ:

> I wait with patience the light that will surely be shed over Professor Marsh and his work...The assertion of Professor Marsh that he devotes his entire time to the preparation of his reports is so supremely absurd, or so supremely untrue, that it can only produce an audible smile from his most devoted admirers. I have known him intimately for ten years. During most of the time while in his employ I never knew him to do two consecutive honest day's work in science, nor am I exaggerating when I say that he has not averaged more than one hour's work per day.[1]

Williston must have been equally upset at an earlier date when he learned that his own brother, Frank Williston, who had been collecting for Marsh also, had switched allegiance and joined with Cope. Perhaps at the point when he was to write this diatribe against Marsh he could better understand his brother's defection.

Certainly one must admit that Williston is coming off a rather negative experience, and is therefore hardly objective, but we must also keep in mind that such allegations came from more than one corner. The talented James Bell Hatcher, who had performed great and marvelous services for Marsh, was treated very shabbily by Marsh, and Hatcher had to often wait at out-of-the-way junctions for late-mailed checks to finally reach him. Hatcher had been guilty of belittling his boss also, but was more discreet than Williston. Ironically it would be the very vocal Hatcher who would turn out to be the discreet one in this regard. Considering the vastness of Marsh's inherited wealth it is hard to conceive of him holding up the paltry funds that were often involved, but apparently this occurred with regularity.

The deCamp's in their book *The Day of the Dinosaur*, were not to be easy on Marsh either for they were to state: "In some ways, O.C. Marsh resembled the evil scientist of fiction. Many colleagues detested his egotism..."[2] It is not without some foundation that these stories of Marsh's excesses and lacks were to abound, for he was certainly a complex man that always bore watching. If there was an advantage to be taken one could be assured that O.C. Marsh would do so, and with alacrity.

It has been said as well that Dr. Thomas Condon of Oregon State University so mistrusted Marsh that he gave strict orders that Marsh would only be allowed access to his fossil collections when there was a guard around because he believed that Marsh had been pocketing fossils.[3]

Condon had good reason to be mistrustful of Marsh for Marsh had borrowed a number of important John Day fossils (some that were to be part of Marsh's famous horse series) that Condon had collected, and then Marsh refused to return them even though Condon persistently asked for their return. Marsh's refusal simply took the form of his not answering Condon's requests for the return of the specimens. Marsh was never to return the fossils himself for he was to die first. It was thanks to the generosity of Dr. Charles E. Schuchert of Yale that they were finally returned to Condon some thirty-five years after they were "borrowed."

William Berryman Scott was to avow his personal dislike of Marsh and did so with vehemence: "Indeed I came nearer to hating him than any other human being that I have known and his hostility to me had a really detrimental effect upon my career."[4] Scott also added the names of Dr. Guyot and Dana as being among those who particularly disliked Marsh. Scott also acknowledged that Professor Leidy told him in private that he was quitting paleontological work because he didn't want to be infected by the Cope-Marsh disease.

In Marsh's defense, however, one should perhaps relate the example of George Bird Grinnell, who was so enthralled with Marsh that he was to work for three solid years without pay as an assistant, beginning in 1874 at the Peabody Museum, because he knew he could learn much from Marsh, and there was no fund from which he could be paid.

Professor Erwin Barbour, former paleontologist at the University of Nebraska and another former Marsh worker, was to write as well castigating Marsh as a "scheming demagogue" among other things.[5]

If one remembers that Marsh would no longer lead expeditions after the expedition of 1874, and that he only made checkups on his collecting parties it would seem that maybe there was a great deal to what Williston had to say. Here again, one would expect that Sternberg would have said good things about Marsh too had he felt them about the man. Charles never failed to pass on a compliment in the case of Cope. Would he in reality have done less for Marsh?

Although there appears to be no extant record of Sternberg's full thoughts on the Sternberg Quarry situation, it is clear that this particular incident did not set well with him. Since the Sternberg Quarry was discovered by Charles, and first worked by him, and then later denied him, and the name of the quarry changed to the Marsh Quarry it would seem that there was plenty of room for resentment. However, one should also keep in mind that Charles was first and foremost a man of God, and a Christian in the truest sense. As such he would not belittle Marsh or any other man, and would "turn the other cheek."

Perhaps an inkling of Sternberg's true feelings on the matter can be gleaned in reading between the lines. In the 1905 *Transactions of the Kansas Academy of Science* Sternberg was to write:

> In 1882, while in the employ of the museum of comparative zoology (sic) of Harvard, the writer discovered the famous locality at Long Island, Phillips county. In 1884 he was employed to explore this deposit for Professor Marsh, of the United States Geological Survey, and with the assistance of J.B. Hatcher, who was at that time an enthusiastic student from Yale, and who made the first collection of his life there, they collected about ten tons of the bones of rhinoceros, representing many individuals. Professor Marsh leased this quarry, and at once it took the name of the Marsh Quarry.[6]

Later, in the same article Sternberg was to note that he still felt the quarry's name precedence was rightly his for he was to comment that he had discovered in the most recently completed season a complete lower jaw set of a mastodon "at the Sternberg Quarry."[7] Certainly no one could blame him for this light dig at Marsh, since by all rights it should have continued to bear his name as its discoverer. It is still noteworthy that Charles was to remain resolute in not wanting to appear full of ill-feeling on the matter, and his actions prove as much.

One gets a truer picture of Charles' displeasure over this takeover by Marsh in a letter fragment found at the American Museum of Natural History in New York. It appears to have been written around 1904, although the first page of the letter is missing. Charles wrote: "Thank you for sending me your memoir and a photograph of the specimen of the rhinoceros that was collected in 'my quarry' and not Marsh's, for ever the laborer is worthy of his hire."[8]

The incident involving Marsh's leasing of the Sternberg Quarry was apparently prompted by the fact that the owner of the land, Anthony Overton, had a change of heart in reference to the use of the land. Overton, who had been satisfied with being hired out as a helper at $1.50 per day, suddenly came to realize the worth of the fossils being taken out from his property and decided in August of 1884 to stop allowing Charles and his crew to dig there. Charles was therefore obliged to write Marsh, which he did on Aug. 9, 1884:

> ...the man who owns the place [Overton] has at last come to the conclusion that the bones are valuable & refuses to let me dig anymore. The only way is to purchase the place. I think it would be a paying investment to purchase it if it can be done at the usual price...[9]

Before he had a response from Marsh, however, James Bell Hatcher had stepped in and negotiated his own deal with Overton without consulting Charles. Sternberg was livid and wrote a scathing letter to Marsh:

> I wish to write you on a subject which if not fully settled I will resign my position on the 1st of Sep (sic) & collect either for myself or Prof. Cope. Mr. Hatcher without consulting me made arrangements with Mr. Overton to collect there & pay him $50 per mo. for his services. I to be left out & not work there. Now if after two years work I discovered the only rich formation in this deposit in the state am to be turned out of the locality by an

assistant & all the work given to him I have had enough of it.

The arrangements I wish are either Mr. Hatcher must be sent to Colorado or elsewhere or you must say that if any work is to be done on Overton's place I am to have charge of it...[10]

It is not exactly clear what actually transpired from this point. Charles did indeed resign from Marsh's party, but not until late October, and then presumably not for the reasons stated in his demand letter. He claimed in an October 26 letter to Marsh that his wife Anna was deathly sick and that he therefore needed to return home immediately.

In a final footnote to this incident, Charles was to write to William Diller Matthew of the American Museum noting that "...I am indeed surprised that the locality I discovered in 1882 and which Marsh got posession (sic) of in 1884 is exhausted. I thought it underlaid the whole quarter section of Overton's land."[11] He clearly remained upset about Marsh's handling of the quarry situation, perhaps until the very day he died.

To understand how difficult it must have been for Charles and the other fossil hunters who were forced to switch sides during this period, we need to go back to the beginning when Cope and Marsh first met in order to see how the famous feud developed.

Marsh and Cope had first met in Germany while Marsh was a student there, and Cope was on a tour of Europe. Although they were to share good times after this point, and might even have been considered friends, Marsh was to downplay these facts in a retrospective consideration of their first meeting:

> My acquaintance with Professor Cope goes back twenty-five years, when I was a student in Germany at the University of Berlin. Professor Cope called upon me and with great frankness confided to me some of the many troubles that even then beset him. My sympathy was aroused, and although I had some doubts of his sanity, I gave him good advice and was willing to be his friend.
>
> During the next five years I saw him often and retained friendly relations with him, although at times his eccentricities of conduct, to use no stronger term, were hard to bear. These I forgave until

the number was approaching the Biblical limit of seventy times seven, when a break occurred between us and since then we have not been friends.[12]

For them to have had the kind of difficulty from the beginning as Marsh implies seems quite difficult to believe, and one is led to the conclusion that Marsh's remembrance is tainted with the bad blood that had subsequently developed between them. Robert Plate in his book on the feud entitled *The Dinosaur Hunters* indicates that his research has shown that Marsh was to visit Cope at Haddonfield in 1868, and that the two of them had enjoyed a week's walk through the "marvelous marls" where the first known dinosaur unearthed in America had been found.[13]

To further dispel Marsh's letter comment one should also be aware that Marsh was to name one of the Mosasaur species he had discovered after Cope; the *Mosasaurus copeanus*. This is certainly ironic considering their later arguments over the rights to name species. One must also admit it does appear to be an odd thing to so honor an individual when you claim never to have liked him.

The split between Marsh and Cope was a gradual one, and the speculations from commentators as to the "final straw" are myriad. Without question, however, is the timeframe. It was sometime around 1870-71 that they finally parted company, and the deepening of the mistrust escalated into outright animosity soon after.

By the time Charles began his career as a fossil hunter the more obvious rivalry had already begun. Marsh had been in the Kansas chalk as early as 1870, but it was not until 1871 that Cope was to arrive on the scene. In Marsh's mind, however, Cope was an interloper who was trespassing on his realm (since Marsh had already led two parties there) and from this point forward there was real out-and-out war, with not even the pretext of professional courtesy.

Before going further it would be beneficial to take at least a limited look at Marsh, and the power that he already commanded in the West when Cope made his initial foray into the Kansas territory.

Othniel C. Marsh may have been many things, but he was certainly never to be accused of timidity. If he was a coward it did not show in his masterful handling of some rough situations. He was also wise enough to always bring along plenty of cavalry protection, something Cope could ill afford financially, or manage politically. Marsh was to gain at least three different Indian names thanks to his personal idiosyncracies; one being a Pawnee appellation, "Heap Whoa Man" assigned to him because of the nervous quirk that caused him to react in fits and starts, and "Bone Medicine Man" due to his affection for fossil bones. He was also to earn the Sioux sobriquet "man-that-picks-up-bones" (wicasa pahi huhu) that is such a good example of the Indian's sagacity in assigning meaningful names to individuals.[14]

Marsh, although somewhat of a dandy, did consort with some fairly rugged individuals in his western adventures. One such character was Luther North, who was to gain fame for his exploits in the U.S. Army. Luther, along with his brother Frank, a Major, managed to acquire a reputation as great Indian fighters. The North's unusual responsibilities helped to build those reputations since they used Indians to fight other Indians. They commanded an elite squad of Pawnees who distinguished themselves on many occasions.

North, a scout at the time, was to tell of his initial encounter with Marsh in a letter to an uncle:

> I started back that night [Sept. 1870] and when I got there the next day I found that Prof. Marsh of Yale Colledge (sic) wanted me to go out with a party of men into the Bad Lands of Colorado to get fossils for the Yale Colledge Museum there was five of us in the party and we had a small tent one team and three saddle horses we started out the 28th Oct. and got back on the 10th of Dec. had a nice trip and got lots of Fossils the night before we got back to the R.R. we laid out in ten inches of snow with no wood to build a fire and the thermometer eighteen degrees below zero. How would you like that that (sic) kind of a life. I froze one of my feet a little but was all right otherwise. I staid (sic) at Frank's a few days and then come home and bought into this livery Buisiness (sic) in May of last spring Prof. Marsh wrote me that he wanted me to go with

his Assistant on Genl. Custars (sic) Black Hills Expedition and I started for Fort Lincoln on the 3d day of June and met Mr. Grinnell (Prof. Marshs Asst)...[15]

This expedition was the first of the expeditions (1870), and it was to play a prominent role in the lives of some other great fossil men. An article in *Harper's* entitled "The Yale College Expedition of 1870," was the impetus for William Berryman Scott, Henry Fairfield Osborn and F. Spier wanting to take their first trip to the western fossil beds.[16]

In the spring of 1874 North went on the expedition as described, as assistant to Grinnell, and under the overall command of Custer. North claimed "of course I knew nothing about geology, but was glad of the chance to go." He does not personally indicate whether in his mind the expedition was a success or not.

Buffalo Bill Cody, another of those rugged individuals with whom Marsh was to have contact, was also to describe the beginning of the 1870 Marsh expedition. He claimed that he was to initially be the scout for the party, but due to an Indian problem had to bow out. He says:

> The day before the Professor [Marsh] arrived at the fort I had been out hunting on the north side of the Platte River, near Pawnee Springs, with several companions, when we were suddenly attacked by Indians who wounded one of our number, John Weister. We stood the Indians off for a little while, and Weister got even with them by killing one of their party. The Indians, however, outnumbered us and at last we were forced to make a run for our lives...The General wanted to have the Indians pursued, and said that he could not spare us to accompany Professor Marsh.[17]

Cody was not overly impressed with Professor Marsh's geological knowledge for he was to say with obvious tongue in cheek:

> As he was starting his bone-hunting expedition...He gave me a geological history of the country; told me in what section fossils were to be found; and otherwise entertained me with scientific yarns, some of which seemed too complicated and too mysterious to be believed by an ordinary man like myself; but it was all clear to him.[18]

George Bird Grinnell hints at this light-hearted scorn of Codys in his article *An Old Time Bone Hunt* when he describes a campfire sequence where Marsh was giving a "campfire lecture" on the subject of Geology to his 1870 party:

> Buffalo Bill, who had ridden out with us for the first day's march, was an interested auditor and was disposed to think that the professor was trying to see how much he could make his hearers believe of the stories he told them.[19]

Marsh's overall contribution to the building of science in America, and his contribution towards the growth of western geology cannot be overstated. While his personal relationships were certainly lacking, and his temper may have often been out of bounds, he still must be commended for what he was able to accomplish. The only question, I suppose, lies in determining how much credit you actually give the man, and how much to the organization (even though it was he who formed it into a powerful scientific juggernaut). For even here one must admit that the organization is generally an extension of the man.

The fact that Marsh never married, but remained lonely and largely friendless is of some interest. While he can not be labeled as having been anti-social, since he had a deserved reputation as being a dandy who had cut a wide swath through the ranks of the New England women of his social circle, it does seem to say that he likely preferred the single life. Perhaps his simple devotion (no matter how misguided at times) to the study of fossils in America overrode all other considerations. While similar in their devotion to fossils Marsh in his solitude could not have been any more different than Edward Drinker Cope who was a warm, gregarious and loving family man, albeit a little eccentric at times.

The great naturalist Humboldt well understood (as did Charles Sternberg) that a life totally devoted to fossils, and lacking devotion to family and friends, was not a life worth living. In such a misdirected lifestyle the job would certainly suffer. It is this same malady that seems to have contributed to Howard Hughes' downfall. The importance of maintaining a wholesome family life was the subject of a letter that Alexander Humboldt had written to his good friend Louis Agassiz, and it certainly has merit when we look at the Cope-Marsh feud: "It is not enough to be praised and recognized as a great and profound naturalist; to this one must add domestic happiness as well."[20]

If there is a recognizable weakness in Marsh's character it would have to be related to this very fact. Marsh had a decided problem in not being able to form or sustain any personal relationships. This undoubtedly was the cause, or was the main contribution to the cause, of the strains that developed between Marsh and everyone that knew him for he was always to remain a man aloof. Loyalties just don't prosper under such circumstances.

When Cope was to make his flanking move into the western territory Marsh was furious, and his own reaction was mirrored by that of his organization. Again, one cannot say that this was necessarily by explicit command, for it may have only been by the implicit legitimization of drastic action against the opposing parties. One of the more obvious "dirty tricks" type of behavior was the old salt scheme:

> ...In a variation of a miner's trick, for instance they [Marsh party members] salted claims.
>
> This happened to Cope one day. For hours he had watched Marsh's men scratch and dig about a certain spot. When they departed empty-handed, Cope inspected the dig. He found a piece of the base of a badly weathered skull, and some loose scattered teeth. Assuming that the Marsh party had passed the fossils up as worthless, Cope gladly took them. Later, on the basis of the skull and teeth, he described a new species.
>
> Not until years later did Cope learn he had been hoaxed. Confronted with a similar skull with the teeth still in place, he saw that the teeth were quite different from those he had found with the Bridger skull. The pranksters had deliberately mixed parts of different creatures to confuse Cope.[21]

This action was certainly less detrimental than the breaking up of actual fossils, but it could have been equally damaging for the scientific community. It had the same possibility for misdirecting the science of paleontology as did the

Piltdown forgery later. Since the Piltdown debacle was to hang on, confusing even the most eminent scientists, for some 40 years it is not hard to see how damaging this "light-hearted" fun could actually have been. When Cope discovered he had been used he commendably admitted his mistake, thus avoiding any further misrepresentations.

Although each man had their own kind of forum (Marsh the Geological Survey and Cope his magazine) the real feuding began as almost a genteel sort of innuendo. As time passed, however, and as tempers were aroused the facades began to crumble. This bickering in print was such that each felt they had to continually top the other. This is very evident in Cope's articles in the *American Naturalist*. A review of a few different kinds of articles is enough to give one the flavor of what was going on.

In writing about the family Menodontidae of the perissodactyles presented by Marsh Cope at first lauds the woodcuts and then goes further:

> The publication of pictorial scientific papers is praiseworthy, but something more than pictures is necessary to make a paper scientific. And if one does not examine the easily-accessible types of the work of others, he is very apt to make publications which savor strongly of plagiarism. In any case, however, while a man may do what he pleases with his own money and take the consequences, it is a public scandal that a scientific bureau of the United States government should permit its money to be used in the way indicated in this paper and in others by the same author.[22]

In an earlier issue Cope had castigated Marsh in a somewhat different mode when discussing Marsh's *The Dinocerata:*

> In connection with the description of the brain of the dinocerata, Professor Marsh gives us much new valuable information as to the brain characters of a number of extinct ungulates. He, however, fails to give Professor Lartet credit for the proposal of the general theory of brain development in the mammalia with the progress of geological time.

And then further:

> Professor Marsh corrects by implication a good many errors made by himself several years ago when criticizing the work of another author on this group. Thus he adopts the species *Loxolophodon cornutus* Cope, and no longer considers it identical with a species subsequently described by himself.[23]

In each of these above mentioned circumstances one can only describe Cope's actions as petty. They have the ring of the petulant child, and could only have served to heighten the bitterness of their ongoing fracas.

Edwin H. Colbert in his book, *A Fossil Hunter's Notebook*, relates one of the "lighter-side" stories addressing the ridiculous nature of the feud, as told him by his mentor Dr. Erwin Barbour:

> ...Osborn and Scott came to Yale to study some of the fossils that Marsh had collected. They were Cope men, and that was anathema to Marsh, yet Marsh could not deny them entry to a public institution. However, he instructed Barbour to cover up and hide all of the specimens that he did not want Osborn and Scott to see. Then during the whole time of their visit the famous Professor Marsh, director of the Yale Peabody Museum, skulked among the nooks and crannies of the building, his feet encased in carpet slippers so that he would not betray himself, and from around corners and behind storage cases spied upon the two...[24]

That Cope could laugh at himself can be seen in the humor that was displayed when he was to name a species as *Anisonchus cophater*. Osborn, who related the incident in his biography of Cope said that he was curious as to what the Greek name might mean and therefore had consulted his Greek dictionary only to have Cope explain that the derivation of the word was a union of the English words "Cope" and "hater." He had named the species in honor of the legions of Cope-haters who were all around him.[25]

The feuding between Marsh and Cope was to become more pitched as each day passed. The unsatiable appetites of these two men did nothing but continue to gnaw at each of them. Cope had his "Marshiana" file in his drawer, frequently threatened to use it against Marsh, and in the end finally did, although without gaining any satisfaction. Marsh meanwhile was to build political clout through favor with John Wesley Powell and Clarence King (for whom California's famous King's Canyon is named). Much to Cope's chagrin Marsh was elected President of the National Academy of Sciences. Cope's dwindling wealth, due to bad investments and heavy expenditures, made him less able to compete against Marsh, yet in the end he was able to bring down Marsh by contributing to the circumstances that led to Marsh's firing in 1892. The actual firing was prompted by the more authoritative Louis Agassiz who had also been aroused.

In the summer of 1892, Cope was to return to Texas, and with William Fletcher Cummins he was to collect vertebrate fossils from the Llano Pliocene beds. The interesting fact here is that Cope "volunteered his services in return for field expenses"[26] for he no longer could be free with his money as he had been in earlier days.

Even after all these incidents, however, neither was to find any satisfaction. In all cases the setbacks each caused the other were temporary, and the hearts of the two men were beyond any softening. To be sure both men went to the grave regretting that they had so little affected the course of the other, and had much to answer for in terms of their behavior.

It is a real shame that these two great men were to fritter away so much valuable time in their feud. One can only imagine what each might have been able to accomplish if they had expended those same energies in the name of their science. But perhaps it was the feud that fed the men, and maybe their accomplishments were actually motivated by their extreme distaste for the other. That is perhaps an unanswerable question.

In all this entanglement and intrigue, however, Charles Sternberg was to remain loyal and as aloof as one could be given the set of circumstances. That is indeed admirable!

Although it is perhaps a bit of a digression from the intended purpose of this chapter it seems somehow appropriate to say somewhere that Sternberg was critical of Marsh and Cope in terms of some of their scientific proclamations, and was never simply a yes-man. That Sternberg had opinions that ran counter to these men and vocalized them is without question, and this seems as good a place as any to mention some of these differings.

Sternberg was to vocalize his criticism of Marsh and Hatcher both in an article for the Kansas Academy of Science. The occasion of the criticism was the finding of a great specimen of the duck-billed dinosaur *Trachodon*. Charles' criticism was mainly that snap-decisions were often made by the paleontologists when they relied too much on speed and not enough on accuracy:

> This discovery proves that all the existing mounts made up from disassociated bones by Marsh and Hatcher, and the other authorities who copied after them, were wrong; and instead of its being a kangaroo or frog, it is a water animal, and was web-footed, with a skin as thin as ours, covered with minute scales no larger than a snake's. So, as I have said so often before, the most carefully prepared restorations made by our highest authorities fall to pieces when the complete skeleton is found of the animal in its normal position.[27]

Charles' observations, certainly credible enough at the time and given the set of circumstances, seem not to be in keeping with present thought as it relates to dinosaurs since the present generation of paleontologists seem to favor the "hot-blooded" lizard theory of dinosauria that characterizes the creatures as forerunners of modern birds.

Cope too did not escape Sternberg's criticism even though we are aware of the great love and respect he held for the man. In another article for the Kansas Academy of Science he was to take Cope to task for his incorrect conclusions regarding a tortoise skeleton, *Protostega gigas*. Cope had apparently made mistakes in thinking that the creature was twice as long as it was wide, and had therefore positioned some of the bones incorrectly. Sternberg, where scientific truth was involved, could be very uncompromising:

> One mistake leads to another. The professor [Cope] thought the skeleton lay on its back...It is useless to try and understand how he could have made such mistakes when he had so much of the skeleton present. For many years his description remained uncorrected, though in 1876 my party sent him a nearly complete skeleton.[28]

That Sternberg was convinced that the main problem was the speed with which they handled publication was obvious and is highlighted in his book about the Red Deer area of Canada:

> In my own experience in the field I have proved too often that I was mistaken, to doubt that other scientific men might also be. I could write a book about the mistakes of scientific men but will not burden my pages with them, except as I discover facts absolutely different from those commonly accepted, as in the case of my *Chasmosaurus* under discussion. In the past, men have been too anxious to publish results before complete skeletons have been found, and almost invariably, when one is found, it does not bear out its own individuality, the expectations of its author.[29]

There is no question about the fact that Cope and Marsh both played integral parts in the life of Charles Sternberg, and each contributed greatly to his later success, and yet we must also recognize that Charles did not fall into the same trap that they did in fighting each other. Sternberg was not willing to compromise either his Christian or scientific principles regardless of what was at stake.

NOTES

1. Elizabeth N. Shor, *Fossils and Flies*, Univ of Oklahoma Press, Norman, 1971, Pgs. 118-119

2. L. Sprague deCamp & Catherine C. deCamp, *The Day of the Dinosaur*, Doubleday Co., NY, 1968, Pg. 216

3. Dan Cushman, "Monsters of the Judith," *Montana the Magazine of Western History*, Vol 12, #4, 1961, Pg. 21

4. William Berryman Scott, *Some Memories of a Palaeontologist*, Princeton Univ Press, Princeton, NJ, 1939, Pg. 58

5. Walter H. Wheeler, "The Uintatheres and the Cope-Marsh War," *Science*, Vol #131, 1960, Pg. 1176

6. Charles H. Sternberg, "The Loup Fork Miocene of Western Kansas," *Kansas Academy of Science Transactions*, Vol XX, Part I, 1905, Pg. 72

7. Ibid, Pg. 74

8. AMNH Ltr fragment, n.d., (To Henry Fairfield Osborn) [Written after 1900 and possibly as late as 1923], From the Archives of the Dept. of Vert. Paleo.

9. Yale Univ Ltr, Aug (?) 1884 (To O.C. Marsh), Othniel Charles Marsh Papers, Manuscripts & Archives

10. Yale Univ Ltr, Aug 12, 1884 (To O.C. Marsh), Othniel Charles Marsh Papers, Manuscripts & Archives

11. AMNH Ltr, Apr 17, 1905, From the Archives of the Dept. of Vert. Paleo.

12. Robert Plate, *The Dinosaur Hunters; Othniel C. Marsh and Edward D. Cope*, David McKay Co., NY, 1964 [From an 1890 Ltr], Pgs. 60-61

13. Ibid, Pg. 90

14. James H. Cook, *Fifty Years on the Old Frontier*, Pg. 228

15. Donald F. Danker (ed.), *Man of the Plains: Recollections of Luther North 1856-1882,*, Univ of Neb Press, Lincoln, 1961, Pg. 307

16. *Harper's New Monthly Magazine*, "The Yale College Expedition of 1870," Pgs. 663-671

17. Don Russell, *The Lives and Legends of Buffalo Bill*, Univ of Oklahoma Press, Norman, 1960, Pg. 167

18. Ibid, Pg. 167

19. George Bird Grinnell, *Blackfoot and Cheyenne*, "An Old-Time Bone Hunt," Chas Scribner's Sons, NY, 1961, Pg. 6

20. Edward Lurie, *Louis Agassiz: A Life in Science*, Univ of Chicago Press, Chicago, 1960, Pg. 78

21. Robert Plate, Pgs. 133-134

22. Edward Drinker Cope, *The American Naturalist*, Vol XXI, #10, Oct 1887

23. Edward Drinker Cope, *The American Naturalist*, Vol XIX, #7, Jul 1885

24. Edwin H. Colbert, *A Fossil Hunter's Notebook*, E.P. Dutton, NY, 1980, Pg. 49

25. Henry Fairfield Osborn, *Cope: Master Naturalist*, Princeton Univ Press, Princeton, 1931, Pg. 583

26. Walter Keene Ferguson, *Geology and Politics in Frontier Texas 1845-1809*, Univ of Texas Press, Austin & London, 1969, Pg. 152

27. Charles H. Sternberg, "In the Niobrara and Laramie Cretaceous," *Kansas Academy of Science Transactions*, #23-24, 1911, Pg. 73

28. Charles H. Sternberg, "Protostega Gigas and Other Cretaceous Reptiles and Fishes from the Kansas Chalk, *Kansas Academy of Science Transactions*, 1905, #19, Pg. 123

29. Charles H. Sternberg, *Hunting Dinosaurs in the Bad Lands of the Red Deer River Alberta, Canada*, (2nd ed), Privately Published, Jensen Printing Co., San Diego, 1932, Pg. 76

CHAPTER TWELVE
STERNBERG THE PROFESSIONAL

A degree does not a successful man make, nor does the lack of a degree make for a man that is necessarily doomed to mediocrity or worse. Nowhere does that statement find more meaning than in the life of our subject, Charles H. Sternberg, for he was to prove that it is not only higher education that makes the man.

While Charles Sternberg may not have been "degreed" in any institution of higher learning (although he did receive an honorary degree from Midland College in Atchison, Kansas), he certainly was so in the proverbial "school of hard knocks." While it may seem trite to say so, Sternberg may actually have had an advantage over some of the other collectors and paleontologists of the day for he was to hardly ever leave the collecting field, and it is on this field that the most can be learned about many aspects of the science.

While it may be true that only long and diligent research can finally turn up the less obvious truths of past animal life, it is also equally true that it is of major benefit to the scientist to be right there as a specimen is being pulled from its earthen grave. The fine points and distinctions that help to ascertain exact strata placement, previous habitat, relationships to other paleo-objects, and the like are more difficult to ascertain if one is not there from the beginning. Paleontologists of today suffer by comparison, as their field time is appreciably lower, and their studies much more specialized. Field men of Sternberg's day were only as specialized as were their particular fossil haunts, and they collected anything handy, even though they had a decided penchant for the large and splashy dinosaurs that would mount well and provide good press, thereby benefiting future funding.

While it was more common in the early days of the science than it is now for paleontologists to spend substantial fieldwork time, even the most prodigious paleontologists and collectors then could hardly have spent even a small portion of their overall time in the field. Such fieldmen as the great Barnum Brown and the prolific James Bell Hatcher, both giants in the area of fieldwork, still could not match up with Sternberg because they had other duties to attend to as well.

The rigors of Sternberg's professionalism were substantial and the avenues the family often had to take to extricate their fossil treasures were plenteous. They frequently had to be innovative to get the job done. Examples of these extra hardships sprinkle Sternberg's writings: digging steps into the hillside in order to provide footing so that they could cart in necessary supplies of plaster sacks and water containers into inaccessible areas, building roads into backareas so that their fossil sleds could be pulled out by their horse teams, digging wagons down into the hillside in order to drop their heaviest fossils comfortably onto their wagon beds, fashioning lever and pulley systems, tripods and other structures to assist in hoisting heavy loads with limited manpower, and all sorts of other labor saving methods.

The Sternberg men were always willing to do what it took to bring a new fossil to light. Charles was in a sense an earthen-journeyman, and he felt truly called to the profession of fossil-hunter. He said it was something "to which I was born."[1]

While we have taken note that Charles was obviously much more professional than the average "layman," he was equally less professional than the studied paleontologist who had been coached in the anatomical make-up of animal life, and had been fortunate enough to be allowed access to the collections of major museums and universities. The most evident difference between Charles' life and the life of the degreed collector lay somewhere in the fact that the research-oriented man had less to worry about in reference to day-to-day existence, and hence could spend more time in thinking about his finds. Sternberg had little time to dwell upon past victories, but had to be after his next fossil (and thereby next meal) without any delay, or he and his family would have starved. At one point he described a particularly bleak financial period as "staving off the wolf from the door," and more than once he had to be bailed out of situations by others because he didn't have ready funds to extricate himself.

In all this, however, Sternberg still was clearly well above the average collector because he had an uncanny knack for discovering fossils, even in areas where others had looked and failed to find fossil material. While this special, innate ability is indeed rare, it is at the same time hardly singular. There is a certain breed of individual who seems to fit this mold. The Leakey's, both father and son, perhaps are one of the better examples of this rugged ability, as they themselves learned early that there was a distinct advantage to using native workmen to scour the African gorges because these men had a propensity for fossil-finding that could not be learned. The success of the Leakey's in great part has been because of their own abilities to turn these native workers into true contributing "scientists." The number of excellent fossils found by these men has been astounding, and the face of geology today has been greatly altered by these men and their native wisdom.

Henry Fairfield Osborn of the American Museum had recognized this very same native ability in Harold J. Cook, son of the famous Agate Springs Ranch owner. Cook, who had honed his collecting skills on his father's own property (site of the Agate Springs Quarry), clearly had a penchant for fossil collecting and proved so. Osborn was able to cultivate this quality in Cook, and felt deeply that this was a trait worth developing in Cook:

I can cite a number of cases where men in Yale, John Hopkins, and other colleges have had rather unorthodox training, but they are among our greatest scientists. I realize that an orthodox schedule has to be set up as a pattern toward which to work, but I don't think we should permit ourselves to be restricted by such arbitrary patterns. When we run across people of exceptional interest, we should have intelligence enough to give them every opportunity to do whatever they can.[2]

Since Cook excelled in his education and in fieldwork it would seem obvious that Osborn's faith in him was not unfounded. It would also appear that Osborn would most likely have held the same kind of opinion about Sternberg judging from his views on Charles' talents. It is perhaps then fitting that Charles was to dedicate his book, *A Story of The Past*, to Osborn. His choice of tribute is worthy of note:

> To
> my friend
> and the friend of paleontology
>
> PROFESSOR H.F. OSBORN, L.L.D.
>
> President of the American Museum
> of Natural History, New York City
>
> With the hope that it will lead the
> reader to visit this great museum
> and others in Europe and America
> where are beautifully preserved the
> animals and plants of other days
> and also with the hope that it may
> lead others to give their lives to the
> work of a fossil hunter in the great
> fields of the west where the harvest
> truly is great and the laborers few

In general, Sternberg seems not to have been greatly hampered by not having had a degree in one of the sciences. This might be seen as another of those parallels that can be drawn between he and Richard Leakey. And there are certainly other equally appropriate modern examples such as Jack Horner and others. Sternberg's constancy of work, his lasting friendships with some of the greats of paleontology, and the prestige with which his geological writings were received (particularly by the scientists of his home state of Kansas) all seem to point to no real or major affect in not having an "education" in paleontology.

However, that does not mean that things might not have gone considerably better for him had he taken the time to attend college as part of his professional training. Nor does it imply that there were not rocky times for him when the "smooth-sailing" of his early days were all but forgotten. With all of his talent Sternberg could not escape entirely the in-built fencing that academia manages to build, and as such he was always something of an outsider. On the other hand, by not taking out time to attend classes Sternberg probably saved at least two or three years that he managed to use profitably in fieldwork, didn't have to write laborious papers on the fossils he found, and could concentrate on his popular writings which gave him great pleasure.

Sternberg makes, however, one casual mention in his later autobiography, *Hunting Dinosaurs in the Badlands of the Red Deer River, Alberta*, of one incident that caused him much distress, and makes it all too clear that not being a member of the "elite" was the reason for his being mistreated. The trouble began because of his desire to collect dinosaurs in the Belly River series, on the American side of the border. You might recall that this is the same series that Sternberg was to work on the Canadian side of the border, and it is evident that his desire was to extend his knowledge of the prolific strata.

The specific section that he wanted to explore had been previously combed by C.W. Gilmore for its superb dinosaur material, and Sternberg had heard of its excellent reputation. The reluctant puzzle-piece that would refuse to fall into place was that this particular fossil site was located on a Blackfoot Indian Reservation. It was therefore "government" controlled in the sense that the Department of the Interior presumably had final say as to whether or not anyone was allowed or denied access to this property. Sternberg was refused entry to the site, and stated: "as I had no authority to visit and collect in this rich field, I was obliged to give it up...owing to red tape."[3]

This brief statement is all that appears in his autobiography, but in fact we find that he was much more incensed over this matter than he lets on in his book. By studying other material one gets a clearer picture as to how devastated Charles was in this refusal. In an unpublished letter to C.W. Gilmore, written from Ottawa, Charles had requested that Gilmore provide a letter of introduction to the Agent, so that his entrance to the Belly River series might be facilitated.[4] The tone of Sternberg's letter makes it evident he expected no trouble in completing this task and was merely trying to facilitate entrance to the area. He certainly did not expect being rejected. Gilmore complied with a suggestion instead that Charles contact a Mr. S.T. Mather, and judging from Gilmore's off-handed reply he too saw no real difficulty in Charles getting his wish. When Mather, however, rejected outright Sternberg's plea to enter the field Charles was overtly distraught. Mather's rejection prompted the following letter response to Gilmore. This is reproduced in its entirety because to slash anything would be to reduce one's understanding of the situation:

Steveville Alberta Canada
July 9, 1916

Mr. C.W. Gilmour (sic)[5]
National Museum Washington DC

Mr Dear Sir:

The enclosed correspondence will explain itself According to your suggestion I applied to the Hon S T Mather and with the result that he refused to allow me to collect fossils in the Blackfoot Reservation Because as he says I was not regularly engaged by a Public Museum (I am now engaged by the British Museum of Natural History) Now the way Mr. Mathers (sic) puts it (Though for the life of me I cannot understand it) American Antiquities according to the enclosed act means Fossil Vertebrates as well as ancient works of man If that interpretation is correct and I go on Government land to collect (a thing I have done all my life to the great advantage of Science) fossils I am subject to arrest unless I have gone through a regular form and receive a permit after proving I am employed by some University or other place where they have a Public Museum, I had intended to go over the ground and may if I can, this fall but I don't want to run into an arrest by Government Officers.

I have never been so outraged in my life, I was coming home once more [from Canada] to my own beloved country to spend the rest of my life collecting principly (sic) in my home country and under my own flag, and the Assistant Secretary says in so many words "You cannot ply your life time vocation in your own country" A pretty welcome home I call it.

Now I am asking you as a friend who knows something of what I have done in this line to see the Secretary of the Smithsonian Institution or any other man in authority in Washington to have this matter straightened out So I may go anywhere I please on the Public Domain Adding as I always have to human knowledge, I cannot believe that when I have endured so much, Suffered so much to augment human knowledge, and have enriched so many Museums I will be told I must keep out of the field by a man who evidently doesn't know the difference between a fossil vertebrate and a cliff Dwellers cave.

I cannot believe you would ever have written me of the Milk River field or advised me to apply for a permit if you knew the kind of an answer I was going to get Please preserve the papers I send you and when you are through with them return them to me here I am anxiously waiting to see this matter in better shape[6]

        Faithfully yours
        Charles H. Sternberg

Apparently Gilmore was not able to persuade whatever powers to intercede, or made no attempt to, as Sternberg was never able to make the survey he was desiring. It seems exceedingly clear that the bureaucrat Mather had no intention of allowing Sternberg access to the field under any circumstances, and also likely that the lack of academic credentials was either the actual reason or an excuse for the refusal. There is no reason to believe that matters were anything other than what they seemed.

An excerpt from the *Carnegie News Service Bulletin*, Vol. 2, #9 entitled "Disclosures of Ancient Life in the Grand Canyon" mentions Mather in relation with Gilmore:

> Through the cooperation of Stephen F. (sic) Mather, then Director of the National Park Service, who approved the plan, Charles W. Gilmore of the United States National Museum was detailed to visit the locality, prepare the exhibit, and make a collection of footprints for the museum.[7]

Mather, actually Stephen Tyng Mather, was a former assistant to the Secretary of the Interior, and he had been appointed by President Woodrow Wilson as the first director of the National Park Service in May 1917. Mather, who was a prominent early figure in the establishment of the national park system in the United States, had been instrumental in the formation of Yellowstone National Park. Sternberg would have undoubtedly bristled at the text for the memorial plaque erected at Yellowstone in honor of Mather's service for it said in part "There will never come an end to the good that he has done."

This incident occurred in the summer of 1924, fully eight years after the rejection of Sternberg's plea to enter the Belly River series. Certainly Sternberg would have been mortified to find that Mather was actually promoted into a higher ranking government position where he would have had even greater control over fossil hunting on government lands in the country.

In Mather's defense one must give him his due by pointing out that it was primarily due to his diligence and shrewd handling of park business that we today can enjoy so many miles of undisturbed national land for recreational purposes. His legacy also undoubtedly has served to protect a vast number of important fossils over the years from destruction by unqualified persons. It does not, however, make the amateur fossil hunter happy since he cannot collect fossils on most of this land due to the dictums of the government. It is indeed a two-edged sword.

While we have spoken to Sternberg's professionalism, his hard-won self-education and his worthiness of distinction (for his many contributions to the young science), it would seem only fair to mention that there were at least a few detractors. In the interest of balanced reporting we should at least mention this opposing views.

One detractor was John Bell Hatcher. Hatcher was a man with a reputation. His reputation was reputedly well earned, and aside from his spotless professional accomplishments one would have to say that it was not a pleasant reputation. He was considered a man very difficult to get along with, and a man better well left alone. In fact, it is apparent that Hatcher preferred being alone. On his numerous successful forays into various fossil fields he generally managed to get off some place by himself. For instance, on his first Patagonian expedition he left O.A. Peterson (his brother-in-law) to collect by himself while he visited other sites alone. He did much the same thing on his second trip to the same area with A.E. Colburn (who was to serve primarily as a taxidermist). Hatcher's reaction to Barnum Brown's arrival on the second expedition was to leave for the interior, spurning Brown's company. It is evident that he simply preferred collecting by himself, and equally evident that he was not a sociable individual, but a loner in most ways.

Although there are stories of an unsavory nature that have surfaced about Hatcher (particularly relating to his penchant for gambling) they appear to be just that; stories. Considering his loner instincts it would seem only right that stories of a negative nature would have eventually surfaced because others would have been put off by his behavior.

All this is by way of introduction to Hatcher's appearing on the scene of a fossil site discovered by Sternberg. In 1882 Charles was to make his greatest discovery, that of the Sternberg Quarry in Long Island, Kansas. It is perhaps unfortunate that he was to work for O.C. Marsh after this time, for it was to end up that the quarry was to have its name changed to the Marsh Quarry. This was most likely due to the direct intervention of Marsh who was known to garner whatever distinction he could regardless of the ethics involved. Marsh was to lease the quarry from the land's owner for a period of five years after its discovery. This quarry was literally filled with the bones of early ancestral rhinoceros, and they were indeed legion. Charles was head of a party collecting there for Marsh in the summer of 1884 when Marsh was to send the initiate

James Bell Hatcher there to work under Sternberg. Hatcher was fresh to the field, having studied under the prestigious James Dwight Dana, and had no previous field experience. While Hatcher was to immediately prove his worth as a fossil hunter and collector, he was at the same time to immediately again show that he was not a pleasant companion.

Although no one can really say as to whether or not Hatcher's response was purely vindictive, was a ploy to gain either recognition or job enhancement, or was simply an attempt to make a name for himself with Marsh, it is clear that he was no friend to Sternberg. He was to write to Marsh belittling Sternberg's capabilities:

> I [Hatcher] found a set of rhinoceros teeth together the other day and Mr. Sternberg sent them to you by mail yesterday. I hope he packed them all right. Yesterday I found all the toe and foot bones of one foot together...he broke some of them a little getting them out...[8]

Url Lanham in his entertaining and lucid book, *The Bone Hunters*, documents the circumstances which led up to Hatcher's comments:

> Hatcher arrived on the scene in July [1884], burning with desire to prove himself as a bone hunter. Within two days Hatcher began to display the trait that was to be manifest for the rest of his short life --he was a difficult man to work with. He wrote Marsh that he would not be able to send a report on his work unless he were given permission to work independent of Sternberg's group, and ventured the opinion that Sternberg was doing a poor job of excavating fossils. Soon he was asking Marsh not to judge his work by that of Sternberg...[9]

What gall the man had! Considering that he had only a few days experience in collecting, and was working for one of the most experienced and prestigious fossil collectors alive, you would have thought that he would have taken time to learn the craft from Sternberg, rather than criticizing from the outset. The fact that he was working in one of the world's foremost fossil sites at the time, and the fact that his co-worker had actually discovered it, apparently meant nothing to him. If it were not for the fact that Hatcher was to so quickly become a magnificent talent in the field himself we would not even bother with his comments. As it is we still cannot sympathize with a man so callous and insensitive, and must wonder how it was that some of the rougher elements of the West or Patagonia had not seen fit to provide him with a bullet for his behavior. It probably speaks to Hatcher's toughness.

But, we must remember too that it is Charles Sternberg we are discussing, and he too was not an ordinary man. That he was to rise above Hatcher's criticism is evident, and it bespeaks the kind nature of Sternberg that he could look with favor on such a rascal as Hatcher. Sternberg was to extol Hatcher's merits rather than to deride him for his treachery when he discusses Hatcher's work at the quarry:

> That year, 1884, in which I explored the quarry at Long Island, was a memorable one, not only because we secured a large carload of rhinoceros bones, but also because we had with us Mr. J.B. Hatcher, who afterwards helped to build up three great museums of vertebrate paleontology,--the museums of Yale and Princeton and the Carnegie Museum...A bright, earnest student, he gave promise of a future even then by his perfect understanding of the work in hand and the thoughtful care which he devoted to it. I have always been glad that I had the honor of being his first teacher in the practical work of collecting, although he soon graduated from my department, and requested me to let him take one side of the ravine while I worked the other.[10]

One certainly suspects that Sternberg's real associations with Hatcher could hardly have been this pleasant, and that Sternberg magnanimously places Hatcher in a better light than he truly deserves. This response from Sternberg shows him for the kind and considerate individual he really was. It also seems to show that Sternberg's christianity went beyond mere preaching and into the realm of practice.

Of further interest in this same passage is the language of description. Sternberg's word choice evokes for us a picture of a man who was very proud of his own personal accomplishments, and the expertise that he had been able to attain. It is equally clear that he was well aware of his own "station" and the limitations of his own scientific contribution. He gave Hatcher his due, and was able to appreciate Hatcher for his major accomplishments.

As a further footnote to understanding the Hatcher incident it is helpful to consult Charles' letter of July 8, 1884 to O.C. Marsh. Obviously Charles was attempting to give full due to Hatcher for he was to state:

> We reached here last eve to find that Mr. Hatcher had reached here Sat. & started at once for Oberlin. He however returned last night & -- has had his first lesson in collecting Loup Fork beds...I am much pleased with Mr. Hatcher & think he will be a most valuable addition to the party...[11]

In a magnanimous gesture Charles was to say that on all of the items being collected they would be labeled with his name and that of Hatcher and the other helper, Mr. Russ. Considering the import that Charles was to place on receiving due recognition for his contributions to science, this was a great gesture on his part.

Hatcher was to again show his true colors when he was to accuse Charles of stealing a foot (presumably a rhinoceros foot). This caused Charles to write to Marsh on Nov. 22, 1884 accusing Hatcher of treachery and claiming that this was a "piece of impudence beyond belief."

It is ironic that in March 1904 James Bell Hatcher was to spend three days with Charles Sternberg on behalf of Princeton University. Hatcher was there to review Charles' fossil trove for purchase, and he did purchase a number of good items while there.

Although Hatcher can be considered the prime detractor Samuel W. Williston was guilty on that same score. Since Sternberg considered Williston a friend perhaps that makes his less caustic remark more notable. If Charles had known of Williston's remark he would most likely have been deeply hurt, as it is clear that he respected Williston highly, and most assuredly felt the respect was mutual. Williston's comment is recounted in Elizabeth Shor's fine book about Williston entitled *Fossils and Flies:*

> ...Sternberg & one assistant is down on the Smoky...I ascertained

his plans but kept my own counsel. He doesn't know what a pterodactyl looks like & hardly what a saurian is. He has had directions from Cope to collect all vertebrates--and we will take pains to leave him plenty of fishes....[12]

Charles' professionalism was also tempered with compassion as is indicated in his hiring of H.A. Patrick Disney to assist them in 1914. Disney had been originally referred to W.E. Cutler by A. Smith Woodward of the British Museum, but Disney found that he could not work with Cutler. Disney was to note to Charles that he could "learn nothing" from Cutler who was working an area around Calgary, and Charles felt sorry for Disney's situation and therefore took him on staff for the season to the benefit of both.

Although there always seem to be some detractors around, wherever one goes, it always seems that the good comments generally outweigh the bad. In Sternberg's case the negative comments are rare, and the positive comments are substantial. As a final note to this chapter it seems only fitting to end with an encapsulating comment, in this case by the great fossil hunter, Edwin H. Colbert. Colbert touts Sternberg's honesty in his book, *Men and Dinosaurs*:

...he [Sternberg] worked hard & conscientiously, and, with the best interests of his clients at heart. The museums that purchased his fossils without doubt received full value and more for the specimens that were sent to them from the field by Charles Sternberg and his assistants.[13]

What better epitaph could be written about a man than to say that "he gave full value?" There are none around that could claim with any conviction that Charles Sternberg wasn't a hard and capable worker who gave full value. Mounds of fossils in a myriad of museums speak directly to this diligence.

NOTES

1. Charles H. Sternberg, *Hunting Dinosaurs in the Bad Lands of the Red Deer River Alberta, Canada,* Privately pub, World Company Press, Lawrence, 1917, Pg. 125

2. Harold J. Cook, *Tales of the 04 Ranch*, Univ of Neb Press, Lincoln, 1968, Pg. 201

3. CHS, *Red*, Pg. 127

4. USNM Ltr, May 5, 1916 (To C.W. Gilmore from Ottawa, Ont.)

5. Charles H. Sternberg used to live on a Gilmour Street and therefore this mistake is easily explained.

6. USNM Ltr, July 9, 1916 (To C.W. Gilmore from Steveville, Alb) - [Sternberg's lack of punctuation and misspellings clearly indicate his extreme agitation]

7. "Disclosures of Ancient Life in the Grand Canyon," *Carnegie Institution of Washington News Service Bulletin*, Vol II, #9, 1930, Pg. 65

8. Charles Schuchert & Clara M. LeVene, *O.C. Marsh: Pioneer in Paleontology*, Yale Univ Press, New Haven, 1940, Pg. 208

9. Url Lanham, *The Bone Hunters*, Columbia Univ Press, 1973, Pg. 199

10. Charles H. Sternberg, *The Life of a Fossil Hunter*, Henry Holt & Co, NY, 1909, Pgs. 132-133

11. Yale Univ Ltr, July 8, 1884 (To O.C. Marsh), Othniel Charles Marsh Papers, Manuscripts & Archives

12. Elizabeth N. Shor, *Fossils & Flies*, Univ of Oklahoma Press, Norman, 1971, Pg. 78

13. Edwin H. Colbert, *Men and Dinosaurs*, E.P. Dutton & Co., NY, 1968, Pg. 184

CHAPTER THIRTEEN
EX-PATRIOT AMERICAN

Sternberg's staunch love for America is without question. His patriotism is to be found in all of his various writings and letters, and he is unabashed in his praise for his country. Some might find it hard to reconcile this vociferous preaching of love of country with the fact that he was to voluntarily give up life in the states and was to spend a significant portion of his life as a Canadian resident. He was, in fact, to spend his final earthly hours in that country rather than in the United States.

This is not really incompatible, however, when one considers the obvious factors that culminated in his decisions to move to and from Canada. It is also noteworthy, and perfectly in keeping with his character, that these main reasons dealt with the things most important to him: his children and his career as a fossil hunter. There is no question that Charles would have refused consideration of this great opportunity if either family or science were to suffer, but since they weren't to suffer it was to him a godsend.

We get a sense of the vehemence of Charles' decision to move northward, leaving his beloved home in Lawrence, in a passage from *Hunting Dinosaurs in the Bad Lands of the Red Deer River, Alberta Canada*. This passage also provides for us an accurate picture of the emotional depth of this decision when he considered leaving the scene of his happiest moments in life:

> As my readers will bear witness, I have seen my choicest treasures collected and prepared during the past fifty years, leave my hands for ever, to add to the glories of museums I shall in all probability never see. When the opportunity came, however, so suddenly and unexpectedly--the opportunity of a lifetime -- to crown my last days with a monument that only time's ravages or the vandal hand of man can efface in that growing Dominion of the North that promises to be one of the great countries in the boundless Western Hemisphere, it seemed to me like a call from heaven. Though the ties of nearly a lifetime, that bound me to many a dear friend at Lawrence, Kansas, must be severed, though I must leave the protecting folds of my father's flag and mine, and live under a flag that has waved a thousand years--under a monarch, in fact--I, a republican of republicans! [1]

This is an arresting picture of turmoil as Charles Sternberg was to decide leaving America for Canada, but it is also in truth somewhat melodramatic when one considers the close physical proximity and the close natural ties between Canada and the United States. Also of note is the inherent fact that Sternberg could and did travel back and forth on trips with some frequency, in each direction, both before and after the move. It was, therefore, not as traumatic an affair as he made it sound. However, one can easily see what would have appealed to Sternberg about Canada, looking beyond what he said in this passage, since we know he had a great naturalist's sense of the pristine and primitive, and Kansas was now showing growth that was beyond the imaginings of those with even the most prophetic outlook.

Canada was primitive at this date in the same way that Kansas had been primitive just a few years before (and now wasn't), and this simplicity would have been very alluring to the likes of the Sternberg family. One gets a good sense of the allurement of this arcadian existence by reading the autobiographical, *Covered Wagon Geologist*, by the early midwestern geologist Charles Newton Gould:

> The sky was a blue dome. The horizon was a circle stretching away in all directions for unnumbered miles. There was no timber;

nothing to meet the eye but grass-covered plains. Roads were little more than trails wandering across the open prairie. Homes were chiefly dugouts, sod houses, and claim shanties. This was western Kansas in the eighteen eighties.[2]

Gould also touches on a point that must be considered: that he found the settlers in primitive countryside to be the "happiest, jolliest, most hopeful people" he had ever known. He attributed this elation to the fact that "where there is hope there is happiness, and these people had much for which to hope."[3] One suspects that Charles Sternberg felt that this ingredient of the early Kansas he so loved was being lost and it would have been a major factor to him, perhaps even the final straw.

To further delineate Charles' feelings one need only look at a letter that he was to write to A. Smith Woodward of the British Museum of Natural History:

> I cannot help but congratulate you on the purchase of these fine specimens, every year it is becoming harder to find good specimens in the Chalk. The country is being settled, and farmer boys roam over the beds, and out of curiosity destroy the fossils...[4]

And then in another letter to Woodward that same year he stated:

> I wish to tell you that in my estimation it is a question of a short time when Kansas Chalk is so thoroughly exhausted that it will not pay to send parties out. I sent my son to a locality we had explored 2 years ago and got a good many fishes. He worked there four days and did not even find a fish. The country has been settled within a year by hundreds of families and every boy of them is born vandal (men too), he will pick up and destroy all surface specimens for the fun of the thing...[5]

In addition to this we cannot overlook the obvious (since he was to highlight it for us), this being the opportunity he would have there to perform his work under the umbrella of a prestigious, yet young institution that he could help to mold into a world-class institution. Since necessity had forced him to be a loner in his profession (not in the Hatcher sense) he would certainly have found a radical change in receiving a regular paycheck.

Obviously too, he had an acquaintance with the geology of the Canadian provinces and recognized their relationships (at least as well as any one else at the time) to their southern counterparts in the United States. He must too have seen the wealth of fossil horizons from which to work, and knew that it would be of major benefit to him and to science for him to make this change in station.

As a member in good standing of the Kansas Academy of Science he would have had occasion to visit Canada for meetings before this date, and the border was often traversed in both directions for all sorts of reasons: from economic to humanitarian to scientific. Undoubtedly the size and variability of Canada appealed to him as well.

Charles' son Charles Mortram Sternberg (always referred to as "Charlie" by his father) and Levi Sternberg were to stay and become Canadian residents with full citizenship and with positions of great eminence in the northern scientific realm. Since they remained in Canada it was a natural drawing card for Charles at a later point when he was to lose his wife to illness. After her death in San Diego he was to return to Canada, and was to remain there for the few years that remained of his life.

Although it need not take up much of our time we might mention that Canada provided its own set of problems for the roving American, and was not a total sanctuary from all of the civilization woes to be found south of the border. For instance, the buffalo and the climate were already affected by the ravages of men:

> The atmosphere then was much cleaner than it is now, and you could stand on a little rise and look way off, far away into the distance, and in the spring the prairie would look sort of white. You know what that would be from? The bones of buffalo. The prairie would give off a whiteness from these old bones and they would be everywhere, just everywhere.[6]

Even though Canadians and Americans have always shared common problems and goals, and have a commonality of interest (and to some extent language), things were not always great for the American traveling north. Even though most Americans themselves were still pretty new (being immigrants or the sons of) and could as easily be called Italians, Irish, Germans or Swedes, the Canadian settlers felt that the intruders were "yankees" and "revolutionaries" and there was some conflict that was natural. Whether Charles himself ever encountered such animosity we'll most likely never know since it's not mentioned in his writings. Considering his gentle ways, however, it would seem likely that this was not a major hurdle for him as he generally managed to blend in wherever he went.

Open expanses, few neighbors, primitive countryside and great fossil potential were the real draws for Charles, and the timing was excellent for just such a move. So, with regrets he made the wise move and headed for Canada, and the world of paleontology was to gain.

Simple and basic was the byword in the Canadian move since the equipment with which they were to work was just that. The museums of Canada were not as wealthy as their American counterparts, and therefore they had farther to go both in terms of manpower and equipment. Charles was to describe the rather sparse assemblage of mounting material as being "boxes and barrels, and there was not a tool in sight."[7] Because of this lack Sternberg was forced to do what he always did well, improvise. He explains how they built an anvil and a sandtable, and gradually managed to make a workshop from scratch.

Not only the equipment was primitive, but so was the museum complex. Charlie Sternberg said in an interview that "we were working in the basement of the museum at the workshop. We opened the workshop in 1913. When we came here the yard was just a mass of rocks. The building had been finished but the yeard hadn't been cleaned up and it was anything but a pleasant looking place."[8]

Charlie was to further comment on the survey's lacks in the same interview:

> I had a team and wagon and father had the camp outfit. The Canadian government had no horses or no outfit so they paid us a small rental for the use of our camp for a couple or three years. Then I said I don't want to rent these horses anymore. Soon after that we got a car and we were working out of cars...[9]

Clearly the Sternbergs were a perfect match for the Canadian Survey in this regard for their versatility, and ability to manage under the most adverse circumstances, was indeed a necessity.

The local Lawrence paper was to highlight the Sternberg's move to Canada, envisioning them as becoming a "subject of King George." Both George and his father returned to Lawrence at the same time in 1912, to "pack their specimens here preparatory to leaving Lawrence and plunging into the Canadian wilderness." Both son and father were, however, going in different directions, as George was immediately joining Barnum Brown's American Museum team in Alberta to hunt dinosaurs and Charles, although reporting to the same general area, was to work for the National Museum of Canada. There is irony here surely since while Charles had been hired to protect the Canadian dinosaur heritage, son George had been hired by the opposition Americans to remove dinosaur material from the country. As we shall see this proved to be an area of some conflict before George's tenure was over.

The Canadian move was in reality a family move, for it was not just Charles and George who moved north, but the whole of the family. The initial move northward was cause for their settling in the city of Ottawa, in Ontario province. Once there the family got down to serious business, serious fossil business, with the Canadian Geological Survey, and things began to happen. Although the Kansas finds had been very fine, and Charles' contributions to the young science in Kansas were especially important, he and his sons were to make even greater contributions as a result of their move to Canada. Charles' dinosaur finds had been limited to this point, and his main finds had consisted of fishes, fossil leaf impressions and invertebrates of all types. This would quickly change since almost everything he would collect in Canada would be dinosaur material.

There is no mistaking that Charles felt comfortable in Canada and generally enjoyed life there. This is not to say he didn't miss life in the states, only that his Canadian sojourn was not without its pleasures. He was to comment that "I have learned...a man is as much a man amidst the snows of the Lady of the North, under the Union Jack, as under my own beloved Stars and Stripes." It is obvious he still felt a closer kinship to the country he had left, and yet Canada was both promising and pleasant to him. He went further to say that Canadians and Americans shared much as "our hopes, our ideals, our aims are much the same."[10]

It appears this sharing of mutual interests was important to him, for he comments upon it more than once. A man with real solid roots, such as those his family had in Kansas, would tend to be very cognizant of the similarities and dissimilarities of the two countries, and would need to find them compatible or he simply couldn't have lived there. Canada was acceptable because it was so like America; the kinship was both necessary and obvious.

Life in Canada for the Sternbergs was much different than life in the states, however, principally because they had regular jobs with a specified salary and therefore did not have to worry about what the next field season would bring, from a financial point of view. At the same time, however, it should be noted that son Charles was very critical of his father in respect to his lacking a good fiscal business sense. Charlie claimed his father was so preoccupied with wanting to collect the best of specimens that everything else, including making a living, became secondary. Charlie didn't blame Brock, who hired the team, but his father since Brock had apparently asked Charles how much he wanted per month to collect. Charles answered that $100 per month would do just fine, and a deal was struck at that amount. According to Charlie his father should have asked for a more realistic rate of $250 per month. This perhaps accounts in part for why Charles and his wife Anna found themselves fiscally troubled in later life while living in San Diego.

Son Charlie also commented that he was paid at a beginning rate of $35 per month by his father, and that he should have been making closer to $150 or $200 a month himself. It is not clear whether this was paid from what Brock paid for services or whether this came directly from Charles' share.

Their boss at the Canadian Geological Survey was Dr. Lawrence M. Lambe, a foremost paleontologist with expertise in Cretaceous reptiles. Charles was given the title of "Head Collector and Preparator of Vertebrate Fossils," and had both Charles M. and Levi working for him at the start, and George was to join in at a later date. While Charles was pleased initially with his title we will see later that it became something of an irritant.

The Canadian Survey was being revamped at this time by its director, Reginald Walter Brock. Brock has been credited with great forethought in his having placed great emphasis on the hiring of men who would make a real and lasting impact on the survey. He decided that the best way to accomplish this would be to select only the ablest and best qualified of men. Although Brock himself was to emphasize the educational backgrounds of his selections, it is interesting that the unschooled Sternbergs were chosen because of their expertise in their field. It is indeed a tribute to their talents as fieldmen that they gained their positions, and Brock in the process was again proven right.[11] Obviously Brock felt that Brown (and other foreigners) were taking away too much of Canada's fossil wealth, and he needed the Sternbergs in the worst way. Since it was the Americans who were taking away the bulk of the fossil wealth it is ironic that Americans were turned to in order to stem the tide. It is clear, however, that it made real sense to look south for the answer, since there was clearly no one in Canada at the time who was capable of fulfilling that role. Loris S. Russell was to remark, for instance, that Lawrence Lambe could not even be relied upon in this area since he was "less informed on the art."[12]

One of the first major efforts that the newly formed group was to undertake was to explore the Edmonton beds which were to prove so prolific in terms of the dinosaur remains they would eventually disgorge. It was in these beds that Charles and his family were to really make a name for themselves in an international way, and it was these finds that were to prove that their move north was indeed the right one.

The Red Deer area had first been worked in 1884 by the Canadian geologist Joseph Burr Tyrrell. Tyrrell had quickly recognized the expanded amount of dinosaur material, but personally had neither the time, funds nor expertise to pursue the matter then. Such was to prove fortunate for both Barnum Brown, and Charles Sternberg.

This general vicinity, called the Red Deer area because of the river of that name, was already being worked by the famous Barnum Brown in an expedition for the American Museum. Brown, also a Kansan, was some twenty years younger than Charles, and had gotten his training initially under the tutelage of Samuel Williston. He then studied fieldwork under J.L. Wortman before gaining his affiliation with the great museum. The Canadian Survey had hired the Sternbergs in hopes they could take major finds from this area thus saving these "home-grown" trophies for the museums of Canada. They were not to be disappointed.

Since Brown had already been in the area for sometime before the Sternberg's arrival it was already an established fact it was good dinosaur ground, and so it was simply a matter of ferreting out the fossils from their once watery graves. Charles was to describe these magnificent beds in this way:

> The Edmonton beds are brackish water origin. On top is a great bed of oyster and clam shells. Below the principal bone-beds are about 200 feet of greyish clay (that crumbles under the feet), interlaid with dark shales and seams of coal. Many of the clay beds have hard iron concretions scattered through them. As these are practically indestructible, they remain scattered over the surface, the other material having been carried away by the water. There is a bed of massive sandstone within a hundred feet of the top, and it weathers out into table lands. Below, the soft clays form conical mounds, often capped with grey sandstone that is fluted by weathering. The rain water becomes so thick with clay that it never settles but gradually evaporates into mud.[13]

Sternberg's description of this fossil territory lacks the lilt we normally find when he is picturing fossil lands for the reader, even badlands, and this was perhaps because he was less impressed with the terrain's looks. But what the terrain lacked in looks it certainly made up for in what it was to deliver.

The area had come to light as good fossil territory thanks to the intervention of an Alberta farmer by the name of J.L. Wagner. Wagner had visited the American Museum in 1909, and had mentioned to the staff there that it was an excellent badland area, packed full of immense fossils. Barnum Brown, never one to overlook any good possibility, was to visit there a few months later, and saw the worth of Wagner's claims. At the first opportunity Brown, on behalf of the American Museum, staged a full scale expedition into Canada, and the area immediately proved its great worth by disgorging varied and substantial dinosaur material. In 1911 alone Brown's party was to take out a total of 104 cases of valuable bone. Brown had quickly recognized the value of approaching the collecting task via the waterways, and hence had constructed his barge which was to prove so fruitful.

Following in the footsteps of Barnum Brown the Sternberg expedition was also to construct a river barge that would allow them to survey the area with more speed, and allow them access to areas that would otherwise have been inaccessible. The walls along the river canyon (Dead Lodge Canyon), where some of the best bones were to be found, were almost perpendicular and in some cases rose to as much as five hundred feet above the river. Dead Lodge Canyon, now incorporated in the area designated as Dinosaur Provincial Park, was a vast boneyard of dinosaur material. The canyon received its name from an Indian tribe that had been annihilated by a smallpox epidemic.

Because the Sternbergs too were to attack the Red Deer by means of a barge, and were hunting too for dinosaurs (which heretofore had been Brown's private realm) Brown was apparently much chagrined at their approach.

Brown was to explain that his barge, constructed to his specifications, was a flat-boat, 12 feet wide by 30 feet in length and he likened it to a "Western ferry-boat." He indicated that it was constructed upside down, right by the river's edge, and that special caulking was applied to make it watertight. The Sternberg's used an oakum and coal tar to caulk the bottom of their vessel, and they apparently suffered some anguish in the process for they related having major discomfort to their eyes. Brown's barge was oared with two great oars on the gunwale, each of them stretching out 22 feet in length, and it seems likely that Sternberg's rendition would have been somewhat similar.[14]

Brown was to provision his barge for the entire season, so his crew had to be on constant vigilance to protect their larder against loss by any means, as well as making sure they had an ample supply of plaster and wood for preserving their finds. They also needed to have on hand the right tools for extracting any conceivable kind of fossil find. The weight maintained on the barge must therefore have been substantial.

Sternberg was to outfit his barge in much the same manner. He indicates that he had a row boat made for him, and purchased a small five-horse power boat in Drumheller, but he also notes that he and his sons built a flat boat that was to measure 12 by 28 feet. Upon this flat boat they were to pitch two tents, one to serve as the kitchen and the other for sleeping. His son Charlie was to be the motor boat operator and he provided the power by means of pulling the flat boat along after him.

The treacherousness of the Red Deer waters was certainly of some cause for alarm to the various parties. On more than one occasion the Sternbergs were called upon to assist in locating drowned persons, and in seeing to their disposition. Because of the remoteness of the area men had to rely on one another, and because the Sternbergs had an innate concern for others, they always quit work to help in these instances. In his field notes of 1915 Charles was to indicate that the "strong under current along the island" was the true culprit in these drownings, and he described how they would transport the recovered bodies to the authorities by lashing the bodies to a board and floating them down river.[15]

Because Brown had the financial backing of the more wealthy American Museum it is likely that he could outfit things with less concern for the money involved, but Sternberg had advantage in that he was working for the mother-country whereas Brown in essence was considered a foreigner. Each of the two parties therefore had both strengths and weaknesses, and where they would lack in one area they would gain in another. Son Charlie was to comment on this very point: "If it hadn't been for this competition, the Director would have said we won't send out this year, we are scarce of money...So if it hadn't been for Brown working there in the first place I wouldn't have been employed -- at least not at the time."[16]

On this first expedition the Sternbergs (minus George) were to be accompanied by a Jack McGee, who almost ended tragically in a freak accident. Apparently the ferry-man some miles below Drumheller had stretched a long string of barbed wire across the river for some unknown reason. Charlie saw it as he approached and yelled to his father and McGee to watch out as it hung low upon the water and their oars were well above that height and would certainly have hit. McGee had his back to the wire, and stood up at the wrong time for the wire whipped past the boat taking McGee's hat rather viciously with it. Sternberg comments that there was only a mere six inches that saved McGee's head from becoming a permanent fixture at the bottom of the river.

On the trip down river they were to stop at a small town near the mouth of Berry Creek, and had hospitable lodging at Steve Hall's Hotel. Here they were to learn that George had recently left the Brown encampment of the American Museum and would be joining up with their party. He was now also a member of the Canadian Geological Survey.

Steve Hall's Hotel, or Steve's, as it was generally known, was similar to an oasis on the desert. Steveville, the only town in the immediate vicinity was named for Hall also, although this town no longer exists. He was obviously a man of some prestige. The badlands begin about one mile east of where Steveville once stood, and are some 25 miles in length, with their width ranging between 6 and 10 miles. Another of George Sternberg's scrapbook articles (name and date of paper not shown and therefore unknown) entitled *New Variety of Stegosaurus Discovered in 'Bad Lands' of the Red Deer Valley* by a Joe F. Price was to recount a conversation with Steve:

At Steveville we met 'Steve,' after whom the 'Bad Lands' receive their name. 'Steve' keeps a stopping place. He does not profess to know much about Dinosauri but he knows the modern history of the country and recalls the first trip of Barnum Brown and his party into the country some fifteen years ago, and can relate stories of the principal discoveries which have been made since then.

Soon after their encampment Charlie was to discover the remains of a Cretaceous dinosaur in excellent condition, and so they moved their camp to be near this site. The camp was close to the small town of Steveville, on the southern side of the Red Deer. Charles was to comment on how pleasant it was to enjoy some of the luxuries of clean water and abundant wood for fires that were so notably absent in the Kansas Chalk. They did, however, suffer some deprivation in the form of the pesky mosquito, for like most river environs they filled the air in the early evenings. As protection they had to use both smudges and netting. Brown's expedition had the same complaint, Brown claiming the mosquitos had "flocked to anything that moved."[17]

Brown was to describe the camp delicacies culled from the area as being highlighted by the excellent fish. He claimed that pike, pickerel, sturgeon and a type of fresh-water herring named "gold-eyes" were the most prevalent.[18] After having to scrounge for food in inaccessible areas in previous camps the Sternbergs appreciated the little added comforts that this provided.

The Sternberg camp, at various times, had the delicacy of baked fish (ling) and wild duck and often added berries to their menu. Charlie, for instance, noted picking buffalo berries on days too rainy to work deposits, and they made jelly in the camp from them. He also frequently mentioned their having Saskatoon berries.

Dinosaur Country, a book by the Canadian Renie Gross, recaps well the work done by both parties in the Red Deer. Gross acknowledges that both parties got along quite well in the first two seasons with members of both camps visiting the other. This may have been in part thanks to George having been a member of the rival party. As early as 1913, however, Brown was becoming edgy over territory and it seems his relationship with George was ruined by suspicion. Brown suspected George was leaking information to his father's party, and even went so far as to write this to his superiors. Gross notes however that Brown "neglected to mention that George had been equally candid with him about the activities of the Ottawa party."[19]

In explanation of Charles Sternberg's own feelings on territorial rights one only has to read his field notes for 1914, for he takes William E. Cutler (from Calgary) to task, "he has not done a very honourable thing coming into my own camp."[20] Cutler was a transplanted Englishman who was homesteading in the badlands. He learned his craft through Barnum Brown, but because he lacked the social graces he was generally not accepted by other collectors and professionals. He died of malaria at Tendaguru in Africa while on a major dinosaur dig. Some months prior to his death Cutler had been assisted by L.S.B. Leakey.

In 1914 things deteriorated substantially to the point that one of Brown's assistants, Peter Kaisen, was to write in his journal that the Sternbergs could not be trusted because they did not keep their word. This comment was the result of a difference in interpretation of "bone ethics," and over who had rights to collect where. Since Brown was a foreigner, but had prior claim, it was a touchy situation since both parties felt they were in the right. Gross claims the real disagreement, however might have been more personal:

> Perhaps the nub of the disagreement was a bit of chagrin on Brown's part which erupted over George Sternberg's discovery of a very fine skull and partial skeleton of a new type of horned dinosaur. It was the most complete skull of a ceratopsian found up to that time and it was retrieved from an area where Brown had prospected earlier. Footprints assumed by the Sternbergs to be his were noted a short distance from where the skull was eroding from the earth.[21]

Charlie was to comment on this same incident in later years. According to him it was he who was to find the Brown-missed fossil, and this partially by accident. Charlie was working on a steep embankment when he slipped down the embankment some 25 feet or so. After scrambling back up the embankment he found an exposed dinosaur skeleton of a duck-bill only a few feet away. He did comment that Brown's footprints were only eight feet away, but said also that "anybody could have missed [it]."

The same sort of incident was to occur between Charlie and his own father. In 1917 Charles had combed a particular site with his party and had found nothing. Charlie came along after and found a very fine fossil specimen. He noted how upset his father was at not having found the specimen on his survey. He wasn't, however, angry at his son, only at the situation.

In testament to Brown's prowess, however, was Charles Sternberg's fieldnote comment of July 8, 1915 in which he stated, "I spent the whole day on Sand Creek only to find wherever there was a prospect it was dug out by Brown's party."

It is certainly clear that Brown, a proud and competent collector, would have been very upset over even the hint that he had missed such a find. Even if one assumes the footprints were those of Brown (rather than another party member) reason would tell any bone hunter that terrain frequently changes and what can be seen today might have been invisible yesterday. It is a shame that the two famous collectors couldn't have gotten on better, although they did seem to mend fences in later days.

As a result of the wonderful carnivore dinosaur find found so early by Charlie, which took two members of the party a full six weeks to excavate, the season was ensured a true success. Charles was so elated with the find that he wanted to make certain that it would be mounted in the most advantageous manner. He thus at seasons end obtained permission from the director of the survey and took a trip to Pittsburgh where he was to discuss the specimen and its mounting possibilities with Dr. W.J. Holland and Earl Douglas (who would work the famous Dinosaur Quarry in Utah on behalf of the Carnegie Museum). He then traveled to Washington where he visited Dr. C.W. Gilmore and Dr. J.W. Gidley of the National Museum, and from there went on to Princeton where he consulted with William Berryman Scott. He also was to visit New York again where he met with Henry Fairfield Osborn, Walter Granger, and Barnum Brown after which he made a final visit to Yale where he met with Dr. Lull.

Each of the men consulted were in agreement that the specimen should definitely be slab-mounted, and so that was the way in which the project was concluded. Dr. Lambe of the Canadian Survey was to call the specimen *Gorgosaurus libratus* because of its fierce look. Precedence, however, has led to its renaming as *Albertosaurus libratus*. This dinosaur might be described as a diminutive *Tyrannosaurus rex* (as they were very similar in appearance), although the *Gorgosaurus* was only a mere 35 feet or so long, and was more gracile in its build. It, however, has the same fierceness characteristic of the Tyrannosaur family.

On the same trip, and in another area close by, Charles was to discover two reasonably complete skeletons of a duck-billed dinosaur which Dr. Lambe was to name *Stephanosaurus marginatus* ("the crowned lizard"). This same dinosaur has been renamed *Lambeosaurus* since that time, obviously in recognition of Dr, Lambe. One of these specimens was particularly praiseworthy because it was to have a beautiful skin impression preserved.

At another site Charles was to discover further duck-bill remains; again a pair of skeletons. Discovered near Loveland Ferry, about 10 miles from their famous locality of Dead Lodge Canyon, these skeletons provided useful information and bones, although the completeness of the skeletons was not as good as they would have hoped. If there were ever any question as to whether or not Charles was truly a scientist rather than a simple collector it should be answered in those writings where he attempts to capture the long ago scene that would explain the remains of today:

> The other contains three caudal, or tail vertebrae, and the whole column in front, with arches, front and hind limbs, except that one hind foot and one fore foot were missing. A very fine head was found pressed back against the back bone, showing that the animal had died in the water, when the gases raised it to the surface and the pressure of so large a body against the head, forced it back.

> When the gases were liberated, the body settled in a mud bank, where it became covered over, and lay buried, through all these ages, undisturbed until the recession of the bluffs carried away the tail. Underground channels destroyed the two feet.[22]

He was to further describe the nature of the bones themselves as being "enclosed with a heavy coating of bog iron" and notes that the bones were poorly petrified since the bone itself was still spongy-bone, and the spaces were not filled with rocky material. Sternberg recognized this and took precautionary measures to see that the friable bones were not destroyed in the preparatory stages by filling the spongy bone with repeated applications of a shellac-like substance called ambroid. Anyone ever having prepared fossil bone of this type will recognize the necessity, and difficulty, of working with this kind of fragile bone.

The interest here is two-fold. It is obvious that Sternberg thought extensively about what he was seeing and mulled over the habitat in much the same way a taphonomist would today. In a sense he might be considered one of the precursors of that line of science. He also again shows his innovation and forethought. Always finding newer and better ways to preserve the bone at hand. This is not to say that he was always right, because he certainly erred with some frequency, but he did make a concerted effort to identify the setting as best he was able.

Sternberg was to first travel northward to Canada early in the year 1912 and by his claim had amassed over five car loads of the dinosaur material available from the Edmonton beds by 1916. This is really an exceptional feat when one remembers that in the year 1917 Sternberg would celebrate the fiftieth anniversary of his first fossil adventure. As a frame of reference one could compare this to Brown's claim that the American Museum expeditions collected in four years 300 large cases or three and one half car loads of fossils. That has to be considered a sizeable undertaking no matter how it is viewed. Brown claimed that two-thirds of his finds were exhibition quality, and that same percentage would probably have held true for Sternberg. The value of these specimens is even greater when one considers Brown's claim that fully two-thirds of the specimens were also new to science. It is clear what a windfall this trove was to the scientific community.

When they began work in the Belly River series of the Cretaceous they were to make even more startling discoveries. Charlie was the darling of this expedition and made the most discoveries. Charles gave his son credit for this exceptional talent, and mentioned one incident in particular. Charlie had apparently taken a short trip to assist a teamster moving a load of the bones Charlie had been laboring over when he found another duck-bill skeleton, and then later walking to a site where he was taking out a carnivore he was to espy a trachodont eroding from the hill. It was definitely Charlie's area in the largest sense.

The 1913 expedition alone was to unearth fossil members of the family ceratopsidae, trachodontidae, theropoda, stegosauridae and plesiosauridae, and a variety of mammalia were also in the collection. If you add to this the further accumulations made by the American Museum encampment the wealth of the area seems truly staggering.

Sternberg pays tribute to Brown in his autobiography, and in that tribute we find his lack of malice concerning the success of one of his competitors:

> On July 19th [1913] Mr. Barnum Brown went down the river with his scow, motorboat and rowboat [sound familiar?] bearing his party of five men and all his outfit... Consequently, with his five collectors, all first class men, filled with energy and enthusiasm, with such a leader and hunter, it is little wonder that he secured that year a great collection, now being mounted in the American Museum.[23]

In his enthusiasm over the new discoveries (the duck-bills) Sternberg tended to give some of their other discoveries a back-seat. Such was the case with their horned dinosaur finds and they deserve better.

While at Steveville Charles was to make a startling discovery as he was traversing a ravine stretching through a flood plain. In so doing he was to come across the horn cores of a horned dinosaur. The find was providential as it related to this particular specimen, for portions of the skull had already eroded out and had been lost due to weathering. There was enough, however, for Sternberg to recognize it as new to science. In a frenzy of elation they used a scraper pulled by a team to remove some of the overburden. When they got down to the finer work with small picks and awls they were to find what Charles said appeared to be mud cracks. These "mud cracks" on closer examination turned out to be skin impressions; a series of polygonal scales arranged in a "beautiful mosaic pattern." The real beauty of this find lay in the fact that this was the first discovery ever made of skin impressions for any member of the horned dinosaur family. Sternberg was obviously delighted at the find, and indicated that it proved again to him that "the wisdom of man is foolishness to God."[24] His reason for saying such lay in the totally mistaken thick-skinned hypothesis that was prevalent at the time.

Once the dinosaur was unearthed and properly prepared it was given the name of *Chasmosaurus* by Dr. Lawrence Lambe. It received its name for the chasms that appeared to be gouged out of the crest and skull. Since part of this skull was damaged beyond repair when it had weathered out from the rock Sternberg initially thought there would be no value in trying to mount it. Upon reflection, however, he realized that his 1913 find of the same animal was relatively the same size. He then was able to use parts from each and made casts of them so that he was able to create from the two partial skulls two complete specimens fully capable of being upon public display.

This persistence, and this ability to innovate with the barest of tools was indicative of Sternberg's style in the field. His motto, had he actually had one, would most likely have been something like "untiring effort will accomplish results in the fossil fields." Since this is a quote from his autobiography we have no trouble in believing that this was indeed his credo.

The wealth of the season's finds had to be transported back to the workshop, and the drayage duties fell to Levi. It was his job to haul the fossils from the Denhart upper camp down to the connecting branch of the Central Pacific Railway near Swift Current and Bassano in Alberta. This was a distance of between 35 and 40 miles over exceptionally rough terrain.

In 1914 particularly, and in other seasons as well, the Sternbergs were to work the area around what was called Happy Jack Ferry. Happy Jack Ferry was located about 12 miles below Steveville on the Red Deer. The previously mentioned Dead Lodge Canyon was below Happy Jack Ferry.

The ferry was named for a famous character who dominated the locale because of his singular ways, Hansel Gordon "Happy Jack" Jackson. Jackson, a hard-drinking loner, seemed very much a precursor of Harry Truman, the 83 year old hard-bitten owner of the Spirit Lake Lodge, who perished during the Mt. St. Helens eruption because he refused to vacate his property. Jackson, a lean and handlebar moustached cowboy, was a seasoned veteran of cattle trail drives in Texas and Mexico who had made his way north to Alberta in 1903. He became boss of the large cattle spread that came to be called the Mexico Ranch, the ranch being owned by the brother of Admiral Lord Beresford of Britain. When Beresford was killed in 1906 his land reverted to the government. Happy Jack in 1908 was able to claim 160 of those acres along with the original buildings for the ranch. He was to remain there on the property until his death in 1942.[25]

Happy Jack seems not to have been too happy at all judging from the observations of his infrequent visitors, and from the terse journal entries that recount how he spent his time. He was always a loner, and clearly was consumed both literally and mentally with the art of drinking.

According to Charles' fieldnote entry of Aug. 18, 1914 George had to stop over at Happy Jack's in order to sharpen his field chisels and picks. The hardness of the rocks and the concretions in the vicinity damaged their working utensils very quickly and so they frequently required repair. Due to this fact it is reasonable to assume that George made the trek to Happy Jack's with some frequency.

George also made an important discovery of a complete dinosaur skeleton on Happy Jack's property. In *Hunting Dinosaurs in the Bad Lands of the Red Deer Alberta, Canada* Charles indicated that George was to find a large

Corythosaurus in Mr. Jackson's pasture.[26] Both Charles and his son Charlie were in Montana when this particular find was made.

In actual time Charles was to spend about 10 years on Canadian soil (five early and five after the death of his beloved wife), versus the 83 years he spent on American soil. In these few short years in the north, however, he was to make some outstanding contributions to the science in terms of his actual finds, his surveying of possible fossil sites and the discarding of others, in bringing his expertise to teach others in Canada, and in publicizing the area in local newspapers and magazines which brought more prominence to the fossil sites. This made a major difference to Canadian paleontology and as such they are each remembered for their contributions to the country in many earth history books.

George, on the other hand, never really liked Canada. He did spend about 10 years time there, in spurts, but never really enjoyed being there. He was happiest in Kansas. George vacated his position in Canada in 1918 when there appeared to be no immediate prospects for fieldwork there that year. In this sense he was certainly very like his father. Since George both lived and breathed fieldwork, he just could not remain confined to the office.

Charlie was to spend a lifetime in Canada. After going there in 1912 he was never to return to the United States except for occasional visits. He seemed a Canadian through and through. His contributions to Canadian science are well documented. His service with the Canadian Geological Survey lasted until 1948 when the National Museum of Canada was to take over responsibility for vertebrate paleontology, and Charlie moved with the transfer. He was to retire officially from government work in 1950, but was to remain active as a consultant until much later. He was instrumental in founding the first Canadian park of its kind, called Dinosaur Provincial Park, at Drumheller.

Levi, the youngest of the sons, was also a Canadian for all intensive purposes. He too was to remain in Canada for the rest of his life. He was to marry in the country and die there too.

The Canadian move for the Sternbergs was certainly a profitable move for all. Canada gained an expertise in excavating and mounting dinosaurs that it lacked before their service and secured the lifelong services of two of the sons. Charlie, George and Levi were to acquire further experience in being able to collect the massive dinosaurs that were less plentiful in areas formerly worked in the United States, and they gained from learning further how to mount these beasts. They also profited by being able to visit all the Canadian locales and therefore better understand the relationships between the various beds in Canada and their southern counterparts. When Charles and George returned to the United States they too were more well-rounded collectors than they had been before their Canadian sojourn. Clearly the Canadian venture was a worthy effort for all involved.

NOTES

1. Charles H. Sternberg, *Hunting Dinosaurs in the Bad Lands of the Red Deer River Alberta, Canada,* Privately Pub, World Company Press, Lawrence, KS, 1917, Pg. 2

2. Charles N. Gould, *Covered Wagon Geologist*, Univ of Oklahoma Press, Norman, 1959, Pg. 26

3. Ibid, Pg. 27

4. BMNH Ltr, Sept 10, 1906 (To A. Smith Woodward from Lawrence, KS), Courtesy Natural History Museum Archiver, London

5. BMNH Ltr, Dec 3, 1906 (To A. Smith Woodward from Lawrence, KS), Courtesy Natural History Museum Archiver, London

6. Barry Broadfoot, *The Pioneer Years 1895-1914,* Doubleday & Co., Toronto, 1976, Pg. 109

7. CHS, *Red*, Pg. 27

8. Interview with Charles Mortram Sternberg by Free-lance Broadcaster Laurie LaMaguer, Courtesy of Tyrrell Museum, Dept. of Paleo.

9. Ibid

10. CHS, *Red*, Pg. 3

11. F.J. Alcock, *A Century in the History of the Geological Survey of Canada*, Canada Dept of Mines & Resources, Ntl Mus of Canada, #47-1, 1947, Pg. 61

12. Loris Russell, *Dinosaur Hunting in Western Canada*, Contribution #70, Life Sciences, Royal Ontario Museum, Univ of Toronto, Pg. 11

13. CHS, *Red*, Pgs. 36-37

14. Barnum Brown, "Hunting Big Game of Other Days," *The National Geographic Magazine*, Vol XXXV, #5, May 1919, Pgs. 407-429

15. Fieldnotes Courtesy of the Saskatchewan Museum

16. Interview with Charles Mortram Sternberg

17. Barnum Brown, Pg. 427

18. Ibid, Pg. 426

19. Renie Gross, *Dinosaur Country; Unearthing the Badlands' Prehistoric Past*, Western Producer Prairie Books, Saskatoon, Saskatchewan, 1985, Pg. 95

20. 1914 CHS Fieldnotes, Jun 8, 1914, Courtesy of the Saskatchewan Museum

21. Renie Gross, Pg. 97

22. CHS, *Red*, Pgs. 62-63

23. Ibid, Pg. 73

24. Ibid, Pg. 79

25. Michael Klassen, "Hell Ain't a Mile Off: The Journals of Happy Jack," *Alberta History*, Spring 1990

26. CHS, *Red*, Pg. 87

CHAPTER FOURTEEN
AMERICAN MUSEUM MYSTIQUE

Anyone with even the remotest interest in fossils cannot help but have a great appreciation for the contributions to the field made by the American Museum of Natural History in New York. The stature of this museum in the field of paleontology is unsurpassed. While there are certainly other institutions of great repute in the United States and the world, this great museum can boast having had the greatest of collectors, the greatest of staff paleontologists, the greatest overall discoveries, the greatest of fossil collections in the world and the greatest institutional history of the science. It may not be as old as the Academy of Sciences in Philadelphia, nor as various as the Smithsonian complex, but it simply must be considered the greatest paleontological assemblage to be found anywhere in the world. Even the fantastic British Museum of Natural History pales in comparison. Although the American Museum only celebrated its centennial in 1969 it has been able to amass a fortune in relics, artifacts, fossils and other natural history items in this relatively short lifespan, and it is without peer.

The great museum was the brainchild of Professor Albert S. Bickmore. Bickmore, a scientist of modest repute, had worked (apparently with some dissatisfaction) as an assistant to the multi-talented Louis Agassiz in Cambridge before striking out on his own. Bickmore did not like the restrictive methods that Agassiz used in handling his employees, and therefore was willing to take the gamble that was to lead to great things both for he and the American science. It was Bickmore's intention to form a national-type museum in the greatest city in the country, New York.

Bickmore, who was of modest means, had a true penchant for fund-raising, and he was able to use it to good advantage in amassing the funds that were required to start the museum venture. Without this talent his efforts would have been totally in vain. Bickmore was to prove successful in raising capital (even though opposed by the notoriously powerful New York political machine of Boss Tweed), and the American Museum of Natural History was underway.

Before one can appreciate the museum fully one needs to get at least a rough idea of its holdings, for it is the collections that ultimately make the museum. This is harder to ascertain than it might first seem since the greater part of a museums holdings are always hidden away in nooks and crannies away from public view. The American Museum is said to house in excess of 50,000,000 bones. Included in this figure are somewhere over 330,000 fossil vertebrates. These figures are astounding, and come to a total that would exceed the number of bones (alive or dead) to be found in a good sized American city.

While the greater part of the museum's collection is not open to the general public, and is only accessible to the scientist, it does cause one some wonder to consider that the regular display areas cover in excess of 700,000 square feet. Since there is so much stored away in the back storage areas of the museum it cannot fail to impress us with the sheer scope of the work begun by Bickmore.

Because of its great prestige there is a generally acknowledged mystique that has built up around the American Museum that is totally unlike that of any other in the United States. Even museums of fame in other sciences or the arts cannot lay claim to the same "awe responses" that the rock facade of the famous museum can inspire. There is a hushed sort of reverence that seems to emanate from the pages of the writings of the many paleontological figures who have chosen to write about its hallowed halls, and this reverence almost defies description.

This mystique is perhaps best put into words by Roy Chapman Andrews. Andrews was initially a worker at the museum, later the idolized leader of the United States' Central Asiatic Expedition to the Mongolian Gobi Desert in 1921, and was to eventually rise to a directorship at the museum. Andrews describes in rather spirited language his first impressions of the great institution:

> ...I arrived too early in the morning and walked twice around the vast pile of buildings. Then I sat down on a rock just outside the entrance to the park at Eighty-first street. I looked at the museum and wondered what kind of place it was; what sort of men I should meet there...Although Osborn, Chapman, Lucas and others were my youthful gods, I had little conception of their personalities.[1]

In another of his books, *Under a Lucky Star*, he elaborates on just how important it was for him to be allowed to become a part of the museum. He describes his intrepid visit for the personnel interview with Dr. Herman C. Bumpus, then director of the museum:

> At last, he [Dr. Bumpus] said, regretfully, that there wasn't a position of any kind open in the museum. My heart dropped into my shoes. Finally I blurted out, 'I'm not asking for a position. I just want to work here. You have to have someone to clean the floors. Couldn't I do that?'
>
> 'But,' he said, 'a man with a college education doesn't want to clean floors!'
>
> 'No,' I said, 'not just any floors. But the museum floors are different. I'll clean them and love it, if you'll let me.'
>
> His face lighted with a smile. 'If that's the way you feel about it, I'll give you a chance. You'll get forty dollars a month.'[2]

I am struck when reading this passage about the similarity of circumstances between Roy Chapman Andrews and Charles Sternberg in this particular area. Both were allowed to join organizations (where others would have failed) because of their youthful enthusiasm. In both cases the old veterans were moved by the sincerity of these young men's pleas for work.

The drawing power of the American Museum has not been restricted to Andrews alone, particularly when it has come to wanting to be associated with this most prestigious of institutions. George Gaylord Simpson, long considered a paleontological giant, relates in his autobiography, *Concession to the Improbable*, that he was approached by a young George Whitaker who was attempting to gain a position. When Simpson explained that nothing was currently available except for a secretarial position, Whitaker adroitly countered with the fact that his wife would take the job, and that in the interim (until something opened up) he would work for free until funds were available to pay him. Who could deny such a brash and confident young man? Certainly not Simpson, and his confidence in Whitaker wasn't proven wrong. Whitaker was on the paleontological staff of the museum for many years and served the institution well.[3] Whitaker, for instance, was to provide invaluable service to Edwin H. Colbert at the famous Ghost Ranch site in excavating *Coelophysis* skeletons.

The mystique of the museum is touched upon also by Edwin H. Colbert, who was associated with the great museum for over 40 years:

> Already in my young career as a paleontologist the name of the American Museum carried with it a magic aura. To paleontologists interested in the study of backboned animals the American Museum was then the foremost place in the world housing such fossils, and I think it is fair to say that this is still the case. In those days the American Museum, Osborn, Matthew, Granger, Gregory, (slightly later) Simpson, Brown were to me magic names representing paleontological colossi.[4]

Sternberg himself was employed as an independent collector at varying times for the American Museum, and for Cope who had helped to supply the museum. Charles' pride in the fossils he had collected, and that were housed in that institution, is very obvious throughout his writings. He had a special pride for the contributions made to the museum by his three capable sons. George's "mummy dinosaur," however, held particular sway in his father's heart because it is the finest example of fossilized dinosaur skin to have been found anywhere in the world. Charles' many references to this particular find are frequent, and one suspects he had more pride in that than the greatest of his own discoveries.

Douglas J. Preston has called the fossil the 'oldest and greatest mummy of them all" and characterizes the fossil scene one encounters at the prominent display case as follows:

> Lying on its back, with a gaping rib cage and grinning skull, is a fossilized trachodont, sometimes called a duck-billed dinosaur. The specimen looks like a partly decomposed carcass -- one can almost smell it -- but 65 million years of entombment have turned it completely to stone...Skin, tendons, and shreds of flesh -- all fossilized cling to the trachodont's fossil bones. The animals head is twisted behind its back in a grotesque arc...The trachodont mummy presents a sharp contrast to the carefully articulated, gleaming black skeletons of other dinosaurs in the hall, which gives an impression of monumentality, stiffness, and formality.[5]

While son George was to actually find this famous mummy it was in reality Charles who recognized the stature of the find, and who was to work towards seeing it got the recognition it certainly deserved. Professor Osborn of the American Museum was quick to make claims upon the fossil and they readily came to terms. Not only did Charles oversee the collection of this most famous of fossils, but he was also to travel to New York City at the expense of the museum in order to watch over and assist in the unpacking of the specimen so that the skin impression would not be ruined in the unpacking and preparation. This was another case where the museum was first, and much of this would have to be attributed to its stature.

In March 1912, the beginning of a pivotal year for Charles and his family, he decided to take a tour of the "teeming east" with his oldest son George. The trip was intended to acquaint them with the collections of the greater United States museums, as well as allowing them to visit interesting geological sites along the way. One of their major stops was at the American Museum:

> It would be useless for me to attempt to describe the wonders of the American Museum at 77th street and Central Park West, in New York City. There is no museum on our continent to compare with it as far as I know, and I have visited nearly all. I have rarely been able to spare the time to visit any part of it, except that of

Vertebrate Paleontology, neither have I time now, to describe their most noted specimens, and since Barnum Brown has added six car loads of the wealth of Dinosaur material, from the Edmonton and Belly River series of the Red Deer, Alberta, no man can measure the wonders of her 'Animals of the Past.' How grand for science, to have such a man as Professor Osborn its President, a man who has given his life and wealth to augment its riches...I was proud indeed when I entered her walls to know that the nucleus of those vast collections was the "Cope Collection," and to remember that I had been a contributor to that collection for seven years of the best, if not the most fruitful years of my life...But what pleased me most were the more perfect specimens of a horned and duck-billed dinosaur from Wyoming, and the great fish *Portheus*.[6]

Son George was also employed by the American Museum for a time, as an independent collector working under Barnum Brown. Charles acknowledged that he considered Brown "the greatest collector of extinct reptiles"[7], so he must have been very proud of his sons connection with Brown and the American Museum. Brown, who can be considered (with little room for dispute) the most prolific collector of fossils ever, was long on field work and short on office work and publication. He was Sternberg's kind of man. Brown's brevity in letter-writing was only exceeded by his brevity in fossil description. Although Brown was much more the dandy (he often wore fur coats on expeditions, and always dressed fashionably) he definitely had qualities that were similar to Sternberg, and it is easy to see why George Sternberg would have enjoyed working for him.

If we could not already ascertain Charles' great feeling for the American Museum we would have only to look at a comment made in the transactions for the Kansas Academy of Science for the year 1905. It is certainly with a mixture of both reverence and awe that he exclaims: "Like the older Agassiz, we should uncover our heads when we enter a paleontological museum, for we stand in the presence of the wonderful works of the creator--works which only recently has man looked upon with intelligent eyes."[8]

How could one doubt upon reading this that Sternberg almost felt the museum was as close to a church in terms of its affect on him as an individual.

The monolith that would inspire such a depth of emotion is indeed a rare one.

Considering the fact that Charles Sternberg was never rich, and was often what one might label as poor, it is of especial note that both he and his third son, Charles Mortram, were both associate members of the museum. Although the cost was not that high it does say something about his priorities. He simply loved this great museum!

In later years the National Museum was to become more prominent in the lives of the Sternbergs, particularly since George became a favorite of C.W. Gilmore, and was asked to participate in a number of different and substantial expeditions around the United States. The feeling that the Sternberg's had for the American Museum was, however, never to wane and it was to always remain "the Museum."

With all the positive notes one can cite about natural history museums, and most certainly the American Museum, it must be said that from Charles' point of view there were negatives as well. The same museums that sought him out when he had valuable and ostentatious fossils ignored him as well when he did not have items for which they had a need. The business side of these relationships could be cold to the point of cruelty as it was on more than one occasion.

Particularly disheartening for Charles was his being turned down in May 1906 by the Carnegie Institute on his request for a life annuity. He had hoped to be able to be less concerned with daily expenses and more able to devote himself to fossil work, but such an opportunity was never afforded him.

Charles was to describe further some discouragement in his thoughts on museums, for he could not divorce himself from dwelling upon how few people were actually affected by the wonders to be found in the natural history museum. He felt that most people were unfortunately fed by a temporary curiosity that was more than satisfied by a brief visit. Charles was particularly concerned that they had "little conception of the enormous energy the collector and preparator have expended in heart-breaking months of exploration and nerve-trying labor in the shop."[9]

While the American Museum was always a favorite of Charles' he still in the end did not feel it right to give one museum direct advantage over another. In a letter to A. Smith Woodward of the British Museum of Natural History he was to state:

> I can only say you are mistaken about my asking you to compete with any museum. I send out to all paleontological museums similar letters to the one I sent you and it is the museum that accepts my terms first, that procure my material. I can not keep special specimens for any museums unless I receive positive information that in case the specimens are satisfactory they will accept my terms...
>
> For many years owing to poverty I was obliged to allow museums to set the prices on my material. This kept me in such financial straights (sic) that I could hardly get means to carry on my work which is a life-work. I consider that even a fossil hunter is worthy of his hire. I have after year(s) of struggle outlived the condition referred to, and my methods of doing business is due to years of experience from my point of view, and I never charge more than the cost to me in time and labor put upon the material, fully realizing that science in this cannot pay what skilled laborers receive in other lines of human effort.[10]

The reason that Charles may have felt so comfortable in dealing with individuals from the American Museum is touched upon by Edwin H. Colbert who was to comment upon the type of scientist drawn to that institution: "It was no place for people who wore gray flannel suits, all parted their hair the same way, and behaved in general exactly like one another."[11] Although one perhaps could not number Henry Fairfield Osborn in this group because of his personal affectations one would have to allow him in because of his ability yet to befriend people such as Cook and Sternberg. Charles found him to be a very influential and helpful friend who stood behind him at all times. If for that reason alone Osborn was much like the Institution for which he stood.

NOTES

1. Roy Chapman Andrews, *Ends of the Earth*, Garden City Pub Co, Garden City, NY, 1929, Pg, 290

2. Roy Chapman Andrews, *Under a Lucky Star*, Blue Ribbon Book, Garden City, NY, 1945, Pg. 22

3. George Gaylord Simpson, *Concession to the Improbable*, Yale Univ Press, New Haven & London, 1978, Pg. 133

4. Edwin H. Colbert, *Digging Into The Past*, Dembner Books, NY, 1989, Pg. 129

5. Douglas J. Preston, "Sternberg and the Dinosaur Mummy," *Natural History*, Vol 91, Jan 1982, Pg. 88

6. Charles H. Sternberg, *Hunting Dinosaurs in the Bad Lands of the Red Deer River Alberta, Canada*, Privately Pub, The World Company Press, Lawrence, KS, 1917, Pgs. 24-25

7. Ibid, Pg. 128

8. Charles H. Sternberg, "The Loup Fork Miocene of Western Kansas," *Kansas Academy of Science Transactions*, Vol XX, Part I, 1905, Pg. 71

9. CHS, *Red*, Pg. 65

10. BMNH Ltr, Nov 6, 1903 (To A. Smith Woodward), Courtesy Natural History Museum Archiver, London

11. Edwin H. Colbert, Pg. 421

Levi Sternberg, Father of Charles Hazelius Sternberg
Courtesy of Paul F. Cooper, Jr. Archives
Hartwick College, Oneonta, NY

George B. Miller, Father-in-Law to Levi Sternberg
Professor and Administrator at Hartwick College
Courtesy of Paul F. Cooper, Jr. Archives,
Hartwick College, Oneonta, NY

Levi Sternberg, Son of Charles Hazelius Sternberg
Courtesy Royal Ontario Museum Archives

George Fryer Sternberg
Courtesy of Glenbow Archives, Calgary, Alberta
Neg. # NA-3250-13

Frank Sternberg (Brother to C.H. Sternberg) and wife Nellie on occasion of Golden Anniversary, October 16, 1933, San Francisco
Courtesy Ellsworth County Historical Society, Box 144, Ellsworth, KS 67430

William Sternberg, Brother to Charles H. Sternberg
Courtesy Ellsworth County Historical Society, Box 144, Ellsworth, KS 67430

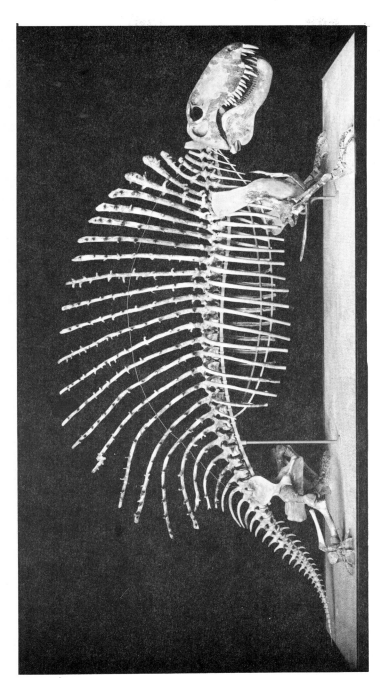

The so-called "Naosaurus." A composite skeleton from the Permian of Texas. Neg # 35338
Courtesy Department Library Services American Museum of Natural History

CHAS. H. STERNBERG,

MANUFACTURER OF

Sternberg's Silicious Soaps,

LAWRENCE, KANSAS,................188....

My Fossils have been Sent to:
The U.S. National Museum where are thousands of my plants.
To the British Museum of Natural History.
The Palaeontological Museum of Munich where I have 85 species of vertebrates.
The American Museum of Natural History, New York, where are the results also of 8 years service in the field as assistant of the late Prof. E.D. Cope.
The National Museum of France at Paris.
The National Museum of Germany -Berlin Senckenberg, Natural History Museum Frankfurt-on-the-Main.
The Museums of the University of Tunigen, Roemer at Hildersheim, Yale, Harvard, Chicago, Toronto, Princeton, California, Minnesota, Iowa, Ohio, Michigan, Cornell, Poukeepsie, Massachusetts Institute of Technology, Kansas, and many other Museums and Universities throughout the world.

The Laboratory
of
Charles H.
Sternberg,
617 Vermont
Street,
Lawrence, Kansas

Member of the American Association for the Advancement of Science.
Member of the Palaeontological Society of America.
Member of the Society of American Vertebrate Palaeontologists.
Life member of the Kansas Academy of Science, etc.
Explorer and collector of fossil plants and vertebrates.
Author of the Life of a Fossil Hunter.
Price $1.60 or $1.75 postage paid.
75 good specimens of fossil leaves from Ellsworth Co., Kansas, $10.00; 25 specimens $5.00; fine collection Fossil Vertebrates especially for students' use, $25.00.
Fine show specimens also for mounts.

Diospyros rotundifolia, Lesqx.

Office of
Charles H. Sternberg,
Laboratory 806 Vermont Street,
Lawrence, Kansas.

....1904.

Modern reproductions similar to letterhead designs
used by Charles Hazelius Sternberg

CHAPTER FIFTEEN
TORPEDO AND CYCLONE

If one stays at any occupation long enough they are certain to see a great number of things change, as well as being assured that they will be beset by a multitude of problems. They will also most likely catalogue a great many volumes of stories relating to their work that they will dole out to their grandchildren during their remaining retirement years. Charles Sternberg certainly gathered his volumes, but probably didn't have too much time for story-telling to his grandchildren because he kept working right up to a very advanced age.

We the readers, however, are the beneficiaries of these "true life stories" and they are stuff that build legend. Since Charles was on the fossil fields for over 60 years there was much that happened to him, and these things do bear recounting. These stories help additionally since they clearly humanize the profession and make it much easier for the average man to relate to the oddities of this most misunderstood science.

Charles managed, for instance, to be victimized twice in a major way by forces outside his control, and whereby major fossils were destroyed in the process. The first was through the auspices of a cyclone or tornado, and the second by the nose of a ship's torpedo. In both cases the losses were substantial, the damage irreversible, and the real loser was mankind.

The first incident of destruction occurred in the most unlikely of places; Sternberg's own laboratory. His laboratory was located in a large two-storied brick building at 617 Vermont Street in Lawrence. Sternberg rented a portion of the lower building from a Mr. Constant, a carpenter, who apparently either lived or worked there also. Charles had been a business tenant of the building for some time before the tornado struck.

In the Spring of 1911 a cyclone struck the building with enough force to push in the brick walls of the laboratory, totally destroying the building and most of its contents. Considering the massive damage to the building one would have expected that the majority of the buildings contents would have been totally obliterated, but there was at least a glimmer in all the destruction. Sternberg only mentions one specific item being destroyed, but it was a major one; the huge skull of the *Triceratops* that had been recovered the season before. This was a massive piece, as he indicates that the skull itself was over seven feet in length. It had been collected from the Laramie Beds in Wyoming, and only after a great deal of effort and ordeal. This fact made it a particularly hard loss for him, since it meant a major loss of income and a decided loss to science.

Charles and his sons had been preparing the skull during the winter months and it was intended to go to the Victoria Memorial Museum in Ottawa, Canada. Sternberg himself was away, and his son Charlie had spent the fateful day in preparation work on the skull, and had just gone home for the evening when the tornado hit. The full impact of the disaster is mentioned by Sternberg in an autobiographical account: "The weight of the brick falling on the skull not only crushed it so badly that it could not be restored and had to be thrown away, but it drove the heavy tailor's table it was on through the floor."[1]

But all was not a total loss even in this devastation for he also had a valuable 14 foot skeleton of *Portheus molossus* in the east end of the building that escaped damage. Apparently it was situated such that when the roof caved in the rafters and other debris fell in such a way so as to actually protect it from destruction. It was in perfect condition after the accident and was to be later mounted in the

Victoria Memorial Museum.

In an unpublished letter to Dr. C.W. Gilmore of the National Museum dated Sept. 27, 1910 Sternberg had described the *Triceratops* skull when he was offering it for sale. He indicated that it had been collected in the summer of 1910 on Crooked Creek, Converse County, Wyoming in the Laramie Beds. It was 11 inches longer than the "type according to Dr. Bull," and it was almost complete except for some of the right horn core and the mandibles. He explained that the scalloped ridge was in very good shape, and described the whole as being "beautifully preserved." He offered it to the museum for $1,000. Such a costly fossil, and one so difficult to collect and prepare, would have been a major loss to him as an independent collector.

The most striking fact about this whole incident, however, lies in the fact that Sternberg continued to maintain a clear perspective on things even at such a great loss, for he said "...the thought that my son had so narrowly escaped with his life made me reconciled to the loss."[2]

The second incident where a major work of his was destroyed through a calamity was later, during the course of World War I, and again the circumstances were well beyond his control. Charles and son Levi were to unearth two separate duck-billed dinosaurs from the Red Deer formation that were of exceeding quality. Both of the specimens were described as being 32 feet in length, so there had been a vast amount of time spent in excavating them. Since the Sternbergs were then working for the British Museum they duly shipped the two excellent dinosaurs to the east coast at season's end, so that they might be transhipped to Britain. The specimens because of their size filled 22 boxes of considerable dimensions, and their total weight exceeded 20,000 pounds. Unfortunately (and it was a substantial loss to science) enroute to England on the steamer *Mount Temple* they were to meet up with a German submarine which torpedoed the steamer and sent it's valuable cargo to "Davy Jones' Locker." The only consolation one can derive from this disaster is that the Sternbergs weren't on the steamer. One shudders to think what could have been had they been commissioned to mount their finds.

In a letter to A. Smith Woodward dated Jan. 21, 1920 Charles refers to this particular loss, and to the fact that he had been prepaid for this work. He felt bad about the museum's loss as well (since it was normally their procedure to cover the freight) and he attempted to repay them later as he was able.

There is also another less notorious fossil loss that occurred to Charles when a freak accident obliterated a fossil turtle: "A beautifully complete turtle was smashed to smithereens when the table on which Sternberg placed it collapsed."[3]

Although they may be the same incident described somewhat differently Charles does recount an incident involving a turtle that appears to be different. He mentions securing a number of exceptionally large turtles in the Upper Miocene of Phillips County. The turtles were grouped together in death in a manner that suggested to him that they had been traveling together when they had been overcome by a freak sandstorm that was to cover them wholly and thus preserve them for posterity. Of the 20 in number collected, the finest was of grand size and condition, so that he was particularly interested in photographing it and therefore took it to a carpenter's shop to obtain some help in moving a yet larger specimen. In the process a helper attempted to move the perfect specimen by himself, found the weight too great to hold, and the fine specimen was dropped and damaged so greatly it could not be saved at all.[4]

As freakish as these destructive accidents were, however, they must be considered in the same light as the more positive events; for these events were equally spectacular and out of the ordinary. They make wonderful stories for grandchildren as well.

We always like to think that we are masters of our own destiny, and that such applies also to those "fabulous fossil collectors." We like to believe that the great fossil finds are the result of hard work by the experienced fossil detector, and that it is only through the sheer drudgery of years of fallow searching that the inevitable day arrives when paydirt must be hit. But, such is not always the case. Whether one's background might call it luck, or Providence, or God's handiwork, there do occur those times when fossil finds literally fall in one's lap. In Sternberg's case he always gave God the credit for his finds, whether they came hard or easy.

Both Charles Sternberg and his sons were the recipients of such providential finds, and they were the kind of finds that would make Ned Buntline want to come back from the grave, although in these cases he wouldn't have to exaggerate.

One suspects, Sternberg being the professional he was, that he was perhaps down deep embarrassed at the ease in which some of these fossils were found. Interestingly enough it is only in his "popular" writings that these less than professional glimpses are mentioned. Since he was all too aware of who his audience was when writing it would certainly appear that it was only before the unprofessional audience that he would let his guard down. This is not to imply, however, that his articles ever got too technical, because they didn't, but some were certainly more technical than others.

In a 1929 article for *Popular Science Monthly* Sternberg capitalized on these "fluke finds." His title for the article, *Fossil Monsters I Have Hunted*, was deliberately intended to grab the reader in the same modern advertising sense, and certainly must have considering its amateur readership.

He begins by reiterating an incident that occurred in the Wyoming badlands:

> ...I heard, seemingly at my feet, the deadly warning of a rattlesnake. I leaped to one side, slipped, fell, slid over the edge of the crag, and came to a sudden stop sitting down on what apparently was a brown boulder, about ten feet from the top. The 'boulder' was the shoulder bone of the largest dinosaur ever unearthed. It was nearly eighty feet long...[5]

Although we can surmise from the brief description that he was talking about the finding of a giant sauropod, most likely *Diplodocus*, we are not given exact data relating to time, place or species to allow proper identification. A more professional readership would not have stood for such omissions, but it was in keeping with the article and what was trying to be conveyed. As a footnote to this incident I have to note that I recall nowhere Sternberg mentioning collecting any sauropod bones, at least none that he was to write about.

He then describes in the same article the finding of a huge Kansan fish, *Portheus molossus*:

> I dropped the geologist's pick I was carrying and the sharp point accidentally struck a slab of sandstone, flaking off a chip a foot long and half as wide. Beneath, dark brown against the pale yellow sandstone, was the tooth-filled snout of a huge fish. Careful examination revealed it to be a *Portheus molossus*.[6]

Sternberg does manage to put in the expected disclaimer, indicating that not all fossil collecting is luck, and that the collector must "follow certain definite rules." He also admits that the true harvest is gathered by the "man who looks over the most square feet of ground." Since Sternberg prided himself on diligence and organization the "methodical labor" was extremely important to him, and he couldn't quite feel comfortable with this "luck" business. As a staunch Christian he would have grown up feeling that the word "luck" was really improper anyway for he would have felt it more proper to acknowledge it "God's grace."

Although it is perhaps not so dramatic he next describes a Canadian expedition and the gratuitous location of a dinosaur skeleton just outside his tent. He explains being "walled in by rain" and therefore jumpy over being confined and not able to hunt the canyon walls for their treasure. He continues, saying:

> ...After the rain had passed and I stepped from my tent, I saw on the face of a nearby cliff, clear cut as with a knife, the complete and almost perfect skeleton of the great and terrible *Tyrannosaurus rex*, king of all carnivorous dinosaurs.[7]

He then mentions an incident whereby a casual "flopping" down to rest brought a quick jumping up due to a piercing kind of pain that came from something on the ground. Turning his eyes from the damage that it had done to his trousers he was surprised to find that the culprit was "the very end bone of the fossilized tail of some prehistoric monster." He was to find, after unearthing the specimen, that it was a particularly fine example of the duck-billed dinosaur,

Trachodon. Of especial note on this "fluke find" was the fact that it turned out to be one of his most famous finds, for the skeleton was to have fossilized skin still clinging to its bones, "well filled out, as if by the rounded muscles of life."

The last incident mentioned in the article, and one of the most bizarre, involved him stooping to pick up a fossil assemblage of gastropods. Laying his geology pick aside to study the find he found when he repicked up the tool to continue searching that it had conveniently hooked a "long-buried skull" of an ancient 30 foot beast (the fiercest of the mosasaurs) called a Tylosaur.

Charles was not the only Sternberg who was to have benefit of such interesting surprises for his son George was to also experience a bizarre incident. This particular incident was to occur on George's Patagonian expedition in 1922, Patagonia, one of the world's most famous fossil lands, has always been a draw for interested fossil collectors and so George jumped at the opportunity when Elmer S. Riggs offered him a spot on his expedition. Patagonia, after all, had been the home of the famous Ameghino brothers, Florentino and Carlos, and had too hosted James Bell Hatcher, F.B. Loomis and many others. It was too great an opportunity to overlook.

This expedition was under the auspices of the Field Museum in Chicago, and George was accompanied by a fellow collector named John B. Abbott. It was on this expedition that George was to make his interesting discovery, and it was also this expedition that appeared to be the "final straw" that finally led to his first wife divorcing him, largely on account of the great amount of time he spent away from the family.

In the interest of not destroying the clarity of George's story I take the liberty of quoting from him at some length:

> We had been at work for several months with splendid results. Then, one day when the tide was at its lowest ebb, I had followed it to its very limit to find just as it was beginning to return what promised to be the finest skull of a large *Nesodont* we had so far discovered. But the salt water was already covering the specimen, and knowing how useless it would be to try to dig it out, I left it and returned the next day to collect it. But much to my surprise

only the very top was uncovered by the water. I realized I had waited too long and that there was no hope of finding another chance to secure it for a whole month, when the receding of the highest tide would uncover it.

Our work in this vicinity had been finished and orders had been given to move camp up the coast some eighteen or twenty miles. I hated to leave my prize, but there was nothing left for me to do but to go. However, I hoped to return later. During the next high tide I arranged for two saddle horses, one to carry my bed and food and to bring back my specimen, and the other as a mount. I arrived on the spot to find a misty rain falling. But the tide was going out and I knew that far out there my prize of last month would soon be uncovered by the fast receding water. I tied one horse to a bush along the shore and mounting the other one, followed the tide as it receded far out over the shallow ocean floor. Finally I saw the top of my find. I hastened into the shallow water and started to work. Soon it was entirely uncovered and the water receded several yards beyond it. My horse turned his tail to a chilling damp wind and stood facing me. 'What a splendid companion is a faithful horse when one is out alone,' thought I.

Finally the skull was ready to load. I rolled it from the hollow from which it had been dug and getting it beside my horse I started to load it. But much to my surprise I could not raise it above my waist. What was I to do? It was already pitch dark and I knew the tide was beginning to turn. I tried several times to load it on the horse's back, but without success. Finally in despair I straddled my pony and rode with all speed to the shore where I remembered a small piece of broken life boat had been washed up high and dry. I had no trouble finding it, and tying my rope to it, I dragged it out to the specimen. I knew I would never relocate the skull unless it would be by following the edge of the returning tide. There was nothing to guide me to it, and in the darkness I could not see it from my horse. We followed the water's edge until finally my horse stopped. I thought this strange. But I got down to look more closely. And there, not two steps from us in the very edge of the water, was my find. I soon had it strapped on that battered piece of a life boat and safely towed to shore. The next day by getting my horse close to a small bank I loaded it on his back and took it back to camp.[8]

In another serendipitous event in 1882 was the previously mentioned finding of the famous Sternberg Quarry [Marsh Quarry] of Phillips County. Charles was to describe this find in an early article in *Popular Science News*. His indication was that "out of sheer absentmindedness" he allowed the team to take a wrong route near Prairie Dog Creek and didn't realize his mistake until it was too late to go back. An "inviting clump of trees" beckoned as it seemed an ideal place to camp. The site indeed proved to be ideal, and the fossils even better. If he had not taken the incorrect route his greatest fossil site discovery would never have occurred.

In another incident that reads almost like fiction father and son were to collect parts of the same fossil vertebrate, but years apart. In 1913 Charles was to uncover a skull of the elaborately frilled dinosaur *Styracosaurus albertensis*. The find was a particularly fine one and it became a main attraction with the National Museum of Canada.

Twenty years later son Levi was to collect in the same area and was to find additional bone uncovered. Apparently a small stream had shifted course over the years and in the process the remainder of the skeleton of the same dinosaur was to come to light.

Another case of long delay was recounted to me by Tim Tokaryk of the Saskatchewan Museum of Natural History. Tokaryk notes that he was given the chore in 1984 of opening a small crate that had been housed for years in the back of the museum's collections. Upon opening the crate he was surprised to find several Hadrosaurian postcranial bones that had been collected by George F. Sternberg some 68 years earlier. The bones were carefully wrapped in a Toronto area newspaper dating to 1916. Tim noted further that a small femur which was also part of the crated material is presently still in active use as a "touchy-feely bone for the kids."[9]

It is clear from the different episodes listed here that the Sternberg hunters often were blessed with gratuitous and serendipitous finds and circumstances, and that at rare times they benefitted without the usual hard work. If nothing else these stories must have made for good conversations with the grandkids.

NOTES

1. Charles H. Sternberg, *Hunting Dinosaurs in the Bad Lands of the Red Deer River Alberta, Canada*, Privately pub, World Company Press, Lawrence, KS, 1917, Pg. 10

2. Ibid, Pg. 10

3. L. Sprague deCamp & Catherine C. deCamp, *The Day of the Dinosaur*, Bonanza Books, NY, 1985, Pg. 231

4. CHS, *Red,* Pg. 177

5. Charles H. Sternberg, "Fossil Monsters I Have Hunted," *Popular Science Monthly*, Dec 1929, Pg. 56

6. Ibid, Pg. 56

7. Ibid, Pg. 57

8. George H. Sternberg, "Thrills In Fossil Hunting," *Aerend*, Kansas State Teachers College, Hays, KS, Vol #1, #3, 1930, Pgs. 151-153

9. Personal Communication, Apr 11, 1990

CHAPTER SIXTEEN
GEORGE MILLER STERNBERG

Charles' older brother George was always his mentor, and assisted him in myriad ways in his fossil hunting pursuits as well as offering him good counsel on other matters. Judging from the informations extant on the matter they seem to have been close (particularly in the days when the younger brothers Charles and Edward lived at the ranch near Ellsworth) while not being close in a physical sense since they were always widely separated in terms of age, occupations, social status and in home sites.

George, as noted before, was born at Hartwick Seminary, Middleburgh, New York just as Charles was. George was born on June 8, 1838, which made him substantially Charles' senior, and it is quite likely (considering that George was to teach some subjects at Hartwick) that he might even have been responsible for a good deal of Charles' early education. In either case George seems to have been a rather bright child even at the earliest ages, and seemed destined for greater things. The fact that he became his own man and earned his own way at the tender age of 16 says something about his capabilities. It also is indicative of the family traits of responsibility and self-motivation, which can be found in a good number of the children and grandchildren.

George was to actually begin earning his living at the young age of 13 by working for the sons of an early entrepreneur by the name of Elihu Phinney. Phinney, a multi-talented man, was publisher of the *Otsego Herald* (a weekly journal), a prominent landowner, an associate county judge, and founder of a thriving bookstore operation in Cooperstown. It was at this bookstore that George was to make his living. If this were a normal bookstore all interest might end here for it wasn't a traditional bookstore of its day, but an innovative and history-making establishment. Although it would perhaps be difficult to measure the value of such an experience upon a young impressionable boy such as George was at the time, one would have to guess that the experience would have been invaluable. Ralph Birdsall in his history of Cooperstown, *The Story of Cooperstown*, indicates that the business was established in 1849 by the Phinney family at the corner of Pioneer and Main streets and that it became "famous for original methods of conducting business." Birdsall, obviously impressed with the business and its history as it relates to the town, says:

> Large wagons were ingeniously constructed to serve as locomotive bookstores. They had moveable tops and counters, and their shelves were stocked with hundreds of varieties of books. Traveling agents drove these wagons to many villages where books were scarcely attainable otherwise. The Erie Canal opened even more remote fields of enterprise. The Phinneys had a canal boat fitted up as a floating bookstore, which carried a variety beyond that found in the ordinary village...[1]

Although George did not go to work for the Phinney brothers at their Cooperstown bookstore until a year after the major end of the business had been transferred to Buffalo it still managed to make a major impact on his life. His wife Martha says that George was to recollect to her that as a result of working at the bookstore he "devoted every leisure moment to reading fiction, books of exciting character giving him the greatest pleasure."[2] The excitement culled from these early books of adventure most assuredly had a deep affect on George that would lead to fulfillment in later adventures, and that would have translated as well to his younger brothers. Perhaps George lived out these adventures vicariously by cultivating adventures too through his younger brother Charles.

The Phinney brothers for whom he worked, Elihu and Henry, like their father before them were prestigious men in the Cooperstown area and probably too had their influence on George. Both these able men had gone to school with James Fenimore Cooper, the author.

George was to eventually graduate from the New York based College of Physicians and Surgeons. He was to receive his M.D. there in the spring of 1860, and from there he was to immediately enter private practice. George was to later expound upon this early period as a doctor in a speech given at a testimonial banquet in his honor in 1902:

> When I graduated in medicine in the College of Physicians and Surgeons in this city my ambition did not extend beyond the hope of securing a living practice in the country. My first venture was at a little town on Long Island [NY], where a vacancy was supposed to exist owing to the recent death of an old and highly respected physician. Apparently I was not able to fill this vacancy for my professional shingle was displayed for several months and I did not receive a single professional call. Not being appreciated in this conservative neighborhood I moved my base of operations to Elizabeth City, New Jersey, and was getting a little practice when the war tocsin sounded and my future career was determined by the favorable verdict of an army medical examining board as to my qualifications for duty as an army surgeon. [3]

How ironic it is that this small Long Island town in its apparent mistrust of the "new doctor on the block" should in reality turn up their noses at the future Surgeon-General of the United States Army! How often we in our shortsightedness as humans find ourselves guilty of letting our prejudices rule so that we make such stupid mistakes. But how fine too is the telling, after enough time has passed, where we can mention the incident without personal embarrassment.

George Sternberg became an assistant surgeon with the United States Army on May 28, 1861. His first tour of duty was with the Army of the Potomac where he was to serve under General George Sykes. George's career with the army was to prove to be both long and profitable, and the country itself was to gain from his brilliance, indeed the world was as we shall see.

George was to prove his fitness for military service early, and excelled in his military career. He was to participate in the first Battle of Bull Run, an early but major battle of the Civil War that was to be won by the "enemy," the Confederate Army. This battle, which took place on July 21, 1861, was to prove costly to Union troops, and George Sternberg as a surgeon was to gain a great deal of battleground experience that same day since there were so many wounded and dying to care for. In the course of the mop-up, and while caring for a portion of the 280 wounded Union soldiers temporarily housed in a church, however, he was to be captured by Confederate troops under the command of the victorious Generals Beauregard and Johnston.

In a report by the surgeon in charge of this medical unit, William S. King, to his supervisor he was to laud the actions of Sternberg and his fellow medical officers, saying that "the conduct of these officers is worthy of all commendation."[4]

At the time of this episode one of George and Charles' other brothers, J. Fred, was close at hand. J. Fred was serving under the command of Col. Dixon S. Miles with the 8th Infantry Volunteers, 5th Division, 1st Brigade, and he saw service too at Bull Run. Another of the Sternberg brothers, Theodore, was to serve in that war as well.

George was able to escape within 10 days of being captured and was able to quickly reach Washington, since it was only about 30 miles from the scene of the battle. He managed to make it to Washington within about 48 hours of his escape, but not without some peril and incident. In his biography *George M. Sternberg* written by his second wife Martha she reconstructed the escape based on his recollections:

> ...Dr. Alexander having recommended a decoction of the bark as an application for some of the wounds. The sentry said there was plenty of it in the woods about half a mile north of town. I then started in the direction of the woods, and as soon as I reached it, I gave up that leisurely pace which I had taken so far for a more

rapid gait. I came to a stream that I thought to be Bull Run. I took off my shoes and stockings and forded it...⁵

After crossing the stream he was to encounter further obstacles, and it is a measure of his resourcefulness that he was able to find food and navigate through hostile territory and find his way back to his own forces without benefit of compass or guide.

Following his escape he was sent to Rhode Island for hospital duty for a few months until around 1866 when he was to be sent to the west to assume the post-surgeon position at Ft. Harker, near Ellsworth, Kansas. It was here that he was to pause for a while and where he was to buy his ranch. It was to this ranch that he was to bring his younger brothers Charles and Edward, and eventually the whole family.

While at Ft. Harker George was to first engage the enemy that was to eventually earn him his greatest fame: cholera. In this first of many bouts with this communicable disease he was to suffer a great loss in that his wife of less than two years, Maria Louisa Russell, was to contract the deadly disease and die within a short 12 hours of displaying symptoms of the dread pestilence. George's later fame was due in great part to the lessons he learned about communicable diseases at Ft. Harker, and this beginning was to ensure that he had a personal stake in working towards the eradication of cholera, yellow fever and the like.

Cholera was certainly a virulent pestilence at this time. D.B. Long was to comment, for instance, that the outbreak experienced at neighboring Ft. Riley was to produce a death rate of a "hundred per week," and he counted himself lucky to have gotten out of it unscathed.

George, a captain in rank and an assistant surgeon at this juncture, did not after this point relish life at Ft. Harker. As one can imagine, losing a precious loved one so soon after marriage would hardly ingratiate one to the area that was to spawn the evil that was to destroy her. So it was that about a month after the death of his wife, and after the disease had subsided in the area, that he was to take leave and return eastward. This tragedy probably played some part in his later giving his property over to his father Levi.

Although it is not mentioned in George's biography by his second wife Martha, he did play a part in saving his younger brother Charles' life while at Ft. Harker, and Charles evidently placed great import on this action of unspoken love by his older brother.

Charles had apparently been injured by a bullet to the forehead when out riding alone on a buckboard or wagon of some sort. Miraculously he was saved by the very fact of being in a wagon as the horse knew the way home, and consequently was to wend its way home without the help of its unconscious occupant. Charles describes the wound to his head as "slung-shot," and indicates that he was robbed in the process. It was apparently due to his being asleep "at the wheel" that he could not be more specific about what had actually occurred. This was not uncommon of him as he makes mention of his sleeping at other times, but without incident.

The post-surgeon, Dr. Fryer, was immediately called and also his brother George. Charles was to put it this way:

> My oldest brother, Dr. Sternberg, for years Surgeon-General of the Army, was also sent for, and I found him lying on a mattress by my side when I regained consciousness two weeks later.[6]

That his brother was to be at his side during this critical time meant a lot to Charles, and it does say something about George's devotion to his own family.

It was most likely at Ft. Harker, as post-surgeon, that George was to first begin to appreciate the sciences of paleontology and archaeology, which then translated down to Charles as well. George seemed to have contributed most specifically to Charles by suggesting sites to investigate, by referring good manpower for crews (such as the referral of J.L. Wortman), by introducing Charles to noteworthy people of science (such as H.F. Osborn), and by just generally being available to assist when called upon or needed.

In a letter of June 17, 1908 George was to write to Charles and made reference to one such help:

> When I met Mr. Osborn (Henry Fairfield) at the Cosmos Club I had a letter from you in my pocket which I showed him. At the same time I gave him some account of your work and told him of your difficulties and of your perseverance in the prosecution of your life work. I am glad to know that he has proved to be such a good friend.[7]

In an earlier (Dec. 16, 1901) letter to the Director of the Paleontological division of the National Museum, F.A. Lucas, Charles was to write:

> Prof. Osborn kindly sent me a letter commending my work to President Francis (Exposition President). the President became acquainted with my work through my brother Surgeon General George M. Sternberg, who will also I believe be glad to reccommend (sic) my work, which has been continued for many years and will last until I am no longer able to enter the fossil fields.[8]

George had apparently also attempted to influence his younger brother Charles to become a surgeon. George had urged Charles to allow him to put him through medical school, and further had indicated that he would use his influence to see Charles would be placed in a respectable position in the U.S. Army. Charles, however, had other ideas and refused because he "might use the time necessary to get an education in the work of adding material to human knowledge."[9]

It is obvious, therefore, that Charles benefited greatly from the occasional intercession of his older brother, and that it generally came in the form of introducing him to the right people. There is certainly no question that becoming an acquaintance of Henry Fairfield Osborn was a feather in the cap of any individual interested in paleontology, and George had accomplished that.

In 1868 George Sternberg was assigned to General Sheridan's command, and in this transition he was allowed additional travel throughout the Smoky Hill Trail area. It was undoubtedly at this point that he was to hone his interest in artifacts and fossils. John M. Gibson in his biography, *Soldier In White*, indicates that Sternberg's field duties contributed to his knowledge of geology, and also allowed him to develop his avid interest in artifacts:

During this assignment he also became interested in Indian relics. Both interests were to give him much pleasure throughout his life. He took in good spirit the good-natured ribbing of fellow officers, who pretended to a great deal of concern lest he meet injury or death from the arrows of live Indians while exploring the relics left behind by dead ones.[10]

Samuel W. Williston was to attribute both George Sternberg and Dr. J.H. Janeway with being very early collectors of Kansas fossils in the vicinities of Forts Hays and Wallace.[11]

George, while stationed in the west, was to send fossils to the National Museum in Washington, as well as to Joseph Leidy and O.C. Marsh. Apparently there is even record of correspondence between George Sternberg and Othniel C. Marsh in which George was suggesting places for Marsh to visit as potential fossil sites in Western Kansas, and even suggesting the time of year which was the most advantageous. There is decided irony here, considering the feuding that was to later occur and the conciliatory role that Charles was to play in that matter.

He was also to have some form of communication with Professor Edward Drinker Cope when he forwarded fossil specimens to him in Philadelphia; fossils that had been unearthed while he was post-surgeon at Ft. Walla Walla in Washington state. Considering the impact that George was to have on the life of his brother Charles in this respect, it is noteworthy that George's wife Martha was to only lightly touch on both this incident, and the excavation of a burial mound in Florida. These are the only fossil/artifact incidents mentioned in the long biography, which dwells on the medical accomplishments for which he is most famous.

As camp surgeon Sternberg was responsible for a wide range of duties that included such diverse responsibilities as manning the infirmary and for seeing to the disposal of garbage and other waste materials. The position was one that commanded high respect, for the surgeon had to have a great deal more expertise than the average western soldier, and was quite likely the most educated man around even the officer's quarters.

The camp or post surgeon was frequently called upon to take drastic action in order to save soldiers from what we would consider today to be relatively minor wounds. Since it was not until after the Civil War that antiseptic methods were really established, the surgeon knew little or nothing about controlling infection, which meant that often amputation would have to follow an arm or leg wound. Sternberg was to be a true pioneer in this facet of medicine in America since he was a diligent researcher and wrote much on the subject.

It was not uncommon for the post surgeon also to service the medical needs of the civilian population, that settled around the forts because of the protection the military supplied. A former officer in the army was to say:

> At least seventy capital amputations were performed by the post surgeon on citizens who were buffalo-hunters or railroad employees, whilst a much greater number of frozen men were sent East for treatment. I think it safe to say that over 200 men in that vicinity lost hands or feet, or parts of them.[12]

Apparently George Sternberg had a thriving and "profitable consulting practice in the town and surrounding country" while stationed at Ft. Walla Walla. This was fairly common practice among military surgeons, although it was a bone of contention with other regular officers since the money was so good and the surgeon therefore managed better than they.[13] The army apparently recognized the fact that this was good press and contributed to a good community feeling about the army, and so it never took a position against the extracurricular jobs.

Since the surgeon was often thought of as a "scientist" because of his medical studies, he was often the recipient of fossils and artifacts from interested parties. Lieutenant Colonel Richard Irving Dodge in his memorable work *Plains Of The Great West* was to describe Professor Louis Agassiz' visit to Ft. Sanders when Dodge was stationed there. He gives one a good understanding of the respect the scientist commanded:

> He (Agassiz) had hardly been at the post twenty-four hours before (as, I am told, was usual with him) he had converted the whole

garrison into enthusiastic naturalists, and everything rare or curious was brought to him for examination and explanation.[14]

One cannot fail but to compare this remembrance with that of George Bird Grinnell and James Cook when they variously described Marsh's ability to translate "dry land" into a wonderland of great margin filled with a multitude of mysterious creatures. This same sort of infectious enthusiasm also found voice in Osborn's vignettes about Cope. All of these remembrances are of the same mold, and portray perhaps in the best possible way the strength of fascination that these great men of science were to command.

As recipient of fossils or as collector we are not sure, but we do know that somehow George Sternberg was to gather in some manner a small collection of vertebrate fossils from the area around Lake Washtucna in southeastern Washington. This area is almost due north of Walla Walla and therefore was easily reachable by Sternberg, so it is certainly quite possible that he himself made this collection. This small and fragmentary collection apparently consisted of fossil remains of the horse and camel and was later donated to the United States National Museum in Washington in 1920 by Mrs. Sternberg after her husband's death. Along with these fossil remains, the museum was to receive the artifacts that General Sternberg had collected from the mound excavation in Florida that he had led. These artifacts consisted of five stone implements, a shell spoon, and three pottery vessels that had been collected in the mound that Sternberg was to excavate while post surgeon of Fort Barrancas. Fort Barrancas, near Pensacola, Florida, was Sternberg's home for three years beginning in September 1872.

On the behest of William Diller Matthew of the American Museum, Charles was to write to Martha Pattison Sternberg, after the death of her husband, requesting on behalf of Matthew that a "small collection of fossils from Washtucna Lake, Washington," held in George Miller Sternberg's private collection and maintained at his residence on Massachusetts Avenue, be turned over either to the American Museum, the U.S. National Museum, or the

University of California as they saw fit. Mrs. Sternberg apparently decided upon the National Museum.

Following an area bout with Yellow Fever in which Sternberg as a medical expert was to play a major role, he was to chance upon a conversation in Warrington where a captain of a little sloop in the Naval Yard was expounding to a small group of men about a large "shell heap" that was near his home, somewhere on the bay. This piqued Sternberg's interest and George therefore determined to investigate this location in the hopes of determining something about prehistoric man. Examination of this mound, and visits with the area people, led to a similar mound in the same vicinity, and it appears to have been from this mound that the artifacts were actually taken. The care with which Sternberg opened and examined the mound is given special emphasis by his wife and biographer Martha Pattison Sternberg, and it further highlights how important the older brother was in honing Charles' fledgling talents when he first came to Kansas.

George Sternberg was to go on and accomplish much in his lifetime, and will undoubtedly be long remembered more for his attainments in the medical profession than will his younger brother Charles be remembered for his talents on the fossil field, but not surprisingly they seem to have had many of the same character traits that make for greatness. They were both very conscious of the other person and his welfare, even when at personal jeopardy. This is evident in many cases with Charles, but found special merit in his heroics during the Bannock Indian War, as noted before. This is well evidenced in the life of George through his diligent caring for yellow fever patients when he had not yet contracted the disease and was therefore at extreme risk. The brothers were also men that other men wanted to emulate. The numerous testimonials to George upon his death are a clear indication of this fact, and the respectful comments of Charles' colleagues make the same equally obvious. They were also both good and solid family men and devoted to their wives. While George did not have any children, he was desirous of them and even wrote a number of medical stories intended for youth.[15] Charles' family life almost needs little delineation, and its success is evidenced in his large and loving family. They certainly shared a deep and committed faith as Christians, and both practiced what they preached, as is evidenced by the many things that each did for others and for the betterment of mankind.

So it can be seen that, though separated by age and background, the brothers were similar in many respects. It was probably this distance between the brothers that allowed Charles to get out from under the mantle of his older and accomplished brother, and allowed him the freedom to make his own world. George's social station was legions beyond that of his other brothers. As a military officer, he immediately gained a certain command and respect, but this was clearly further enhanced when he was to attain the respected position of Surgeon General of the United States Army, a position which carried the rank of Brigadier General and which was to endear him to at least two presidential families. He was to serve as personal physician to President Grover Cleveland, and to the wife of President William McKinley and was invited on a number of occasions to functions at the White House by both administrations. He was even elevated to the exalted position of fishing partner to President Cleveland.

Charles' social peers were not of such station, although he knew some great men also and could call them friend. These men, however, would be the great paleontologists of the day; men such as James Bell Hatcher, James W. Gidley, Charles W. Gilmore, Samuel Williston, Barnum Brown and so on. These men, although very prominent, did not hobnob with presidents, but maintained a more earthly social status.

George's greatest accomplishments related to his medical work, and they were substantial. He is called the Father of American Bacteriology because of his work in uncovering the tuberculin bacilli in America, because of his contributions to the cure for yellow fever, and because of his work with cholera, and typhoid fever, and in advising the United States Army as to how to cope with the problem of germ transmittal through proper sanitation and disposal of wastes.

Charles does mention in his autobiography *Hunting Dinosaurs in the Badlands of the Red Deer River, Alberta* that on Mar. 25, 1912 he and his son George went to visit his brother the Surgeon-General at 2005 Massachusetts Avenue in Washington DC. He also noted that he "had not seen him for years."[16] In earlier days, however, they were more able to visit.

The closeness of the segmented Sternberg family is still very evident and finds voice in the little things that were said and done, and also in the bigger things that are more "costly" in terms of their affect on the giver. Such an example can be seen with this generous son as he was to turn over ownership of his Ellsworth ranch to his father. Martha Pattison Sternberg was to highlight this generous aspect of her husband:

> While stationed at Fort Harker, a frontier post in the early history of the state, some of the officers of the Army had secured quarter sections of the fertile land close to the post. Dr. Sternberg was especially fortunate in securing a piece of land beautifully situated on the wooded banks of a little river about two and a half miles from Fort Harker. At the time Dr. Sternberg was a Lutheran minister and president of a college in Iowa, and had not visited his son for years. Dr. Sternberg urged his father to make him a visit at Fort Harker and the invitation was readily accepted. The son was naturally pleased to show his father what he had done, and to talk with him of the plans for future development of the ranch. His father entered into the plans with zeal and interest because he really was in love with the situation, and several times he remarked what a splendid place it would be for the younger boys to develop. While Dr. Sternberg readily acquiesced in his opinion he did not think it a fitting place for his refined educated mother.
>
> Shortly after his father's visit Dr. Sternberg was ordered to take the field in the Indian Territory. In correspondence, his father had expressed a desire to possess this farm in Kansas and made an offer to purchase it, but the generous son could not think of that. In a quiet and delicate way, he made it possible for the father to own the farm, notwithstanding he was not wholly in sympathy with the project; for he was devoted to his lovely mother who, he knew, had always enjoyed refined society. He said to me [Martha] at a later date 'I could not say 'No' when I thought there was a prospect that father might lighten the burden of life that had been

his to bear for so many years of ministerial and college work on salaries never very large.'[17]

It was in July of 1866 that Levi Sternberg came out to visit his son's ranch. He made the trip comfortably in a buggy all the way from Iowa. Later that same year one of the older brothers, Frederick (J. Fred), came out and he was followed the following summer by the twins, Charles and Edward.

General Sternberg as the eldest and well-to-do of the family was apparently always family-minded and was to look out for family members in any kind of need. Martha was to claim, particularly in relation to the much younger boys, that the General always "felt interest in the success of his younger brothers, and he gave them every assistance from his own salary."[18]

Sternberg's further generosity was to be seen in his later purchasing a house in town for his mother as a birthday present, and his having given her the deed. He was very concerned even from even the earliest moment in Kansas that his mother be treated as the genteel lady she was, and this was his way of showing it. He was obviously a man of means, but also a man who was able to share that wealth with others, even when in jeopardy of personal loss. His generosity to strangers during the "germ wars" is certainly evidence enough, and it does allow us some insight into this man who was to directly have such a major impact upon his younger brother Charles.

It always bespeaks highly of an individual if their personal integrity and honesty are evident to their friends and fellow workers. Although General Sternberg received his share of criticism as an administrator, particularly over what some saw (and probably unfairly) as his maladroit handling of the medical aspects of the Spanish-American War, there were none who questioned his honesty. He was indeed a man respected for his contributions in a number of different fields. John Gibson shares an incident where Sternberg's personal integrity (in the face of substantial loss) was very evident. In this instance George was selling his patent rights on the device that he had invented for controlling the temperature of a room. The value of this heat regulator was evident to a number of different firms and there was bidding for the rights. Sternberg verbally communicated to one firm that he would accept their offer of $5000 for his patent, only to have a competitor offer twice as much shortly thereafter. Gibson states that even though the first deal was not legally-binding, that both Martha and George felt it was morally-binding and they therefore turned down the higher offer. This same kind of honesty is evidenced throughout the family, and very much so in Charles Sternberg's life.

Since George Sternberg's merits or demerits as a General of the United States Army, and most particularly in his capacity as Surgeon-General, are not germane to our subject matter we need only note that it was under his administrative hand that Dr. Walter Reed was to finally confirm the mosquito as culprit in transmitting Yellow Fever. Sternberg played a great part behind the scenes in that episode of medical history, and received high praise from Reed for his contributions. We should also note that his medical contributions were most substantial in his efforts to develop preventative measures to combat loss of life in epidemic conditions. It was he who pushed for and gained the support of necessary government officials to begin sanitation measures that would eliminate breeding grounds for the various carriers of communicable disease.

A Reed biographer relates Reed's enthusiasm when George was named as Surgeon-General, and it seems to give us a great sense of the feelings his peers had for him:

> Reed wanted to toss his hat in the air. At last, at last, it had happened! George Miller Sternberg was Surgeon General, appointed late in May, 1893, to succeed General Sutherland! It was significant of the changing atmosphere, the growing appreciation of modern scientific medicine, that President Cleveland had selected not an able executive, not a man adroit in departmental politics, not merely a good doctor, but the one man in the Corps with a distinguished reputation as a scientist. The fossil age, Reed jubilantly declared, was past.[19]

If we learn nothing else about George Sternberg we must know that he was a self-made man, who came from a modest background to attain the highest role a medical man can in our country, and yet a man who never lost sight of his beginnings and was always considerate of those less fortunate than him. He worked hard his whole life to make life better for others. It is in this area that we can recognize the "Sternberg character," for these same traits are to be found in his younger brother Charles. George had a profound affect on his younger brother, and cultivated in him the things that he saw to be of worth. Since he had the money and prestige acquired from his hard work, he wanted to assist his family wherever able. He did so, and in the process we who follow the science of paleontology owe him a great debt.

NOTES

1. Ralph Birdsall, *The Story of Cooperstown,* Charles Scribner's Sons, NY, 1925, Pg. 145

2. Martha L. Pattison Sternberg, *George M. Sternberg*, American Medical Assoc, Chicago, 1920, Pg. 2

3. Ibid, Pg. 245

4. *War of the Rebellion Official Records of the Union and Confederate Armies,* Series I, Vol# 2, Ntl Historical Soc, Pg. 345

5. Martha L.Pattison Sternberg, Pg. 7

6. Charles H. Sternberg, *The Life of a Fossil Hunter*, Henry Holt & Co., NY, 1909, Pg. 14

7. SDMNH Ltr, Jun 17, 1908, Materials Courtesy of San Diego Nat. Hist. Mus. Archives

8. USNM Ltr, Dec 16, 1901 (To F.A. Lucas from Lawrence, KS)

9. AMNH Ltr, Jan 28, 1907 (To William Diller Matthew from Lawrence, KS), From the Archives of the Dept. of Vert. Paleo.

10. John M. Gibson, *Soldier in White*, Duke Univ Press, Durham, NC, 1958, Pg. 37

11. Samuel W. Williston, "Addenda to Part I," *The Univ Geological Survey of Kansas*, Vol IX, Paleontology, Topeka, State Printer, 1898, Pg. 28

12. *American Heritage*, Oct/Nov 1984, "A Medical Picture of the United States"

13. Peter D. Olch, *Journal of the West*, Vol. XXI, #3, July 1982, Pg. 34

14. Lt. Col. Richard Irving Dodge, *Plains of the Great West*, Archer House, NY, 1959, Pgs. 13-14

15. John M. Gibson, Pg. 178

16. Charles H. Sternberg, *Hunting Dinosaurs in the Bad Lands of the Red Deer River Alberta, Canada*, Privately Pub, World Company Press, Lawrence, KS, 1917, Pg. 22

17. Martha L. Pattison Sternberg, Pg. 18-19

18. Ibid, Pg. 20

19. L.N. Wood, *Walter Reed: Doctor in Uniform*, Julian Messner, Inc., NY, 1943, Pg. 154

CHAPTER SEVENTEEN
THE FAMILY

The Sternberg family had its roots in the French/German Palatinate. What was known as the Palatinate was situated in that strip of land that was then variously controlled by the French kings and the German princes, on the Rhine River (Pfalz). This strongly contested land is somewhat analogous to the modern situation that Israel and its neighbors are in with the Gaza Strip. People who live in such hotly contested areas find it impossible to live in peace, and always are forced to either take one side or the other, or suffer the consequences from the opposite faction. It was because of this unrest, and the religious persecution that went with it, that the great migration out of that area occurred. This great exodus provided the United States with a substantial number of hard-working settlers.

The Sternberg family before its migration to the New World had the name of Von Sternberger, or Starenburger. In variant spellings, however, the family name can be found in America as Sternberg, Sternbergh, Sternberger, Sternburger, Sterenberger, and Staarenberger. Undoubtedly there were other spellings as well. It was perhaps in the spirit of starting again, and Americanizing themselves, that the name was shortened most frequently to Sternberg. In either case the name Sternberg has come to real fruition in this country.

The reasons behind the mass exodus from the old country are manifold, but the last straw seems to have been related to the fact that there was a change in area government that occurred with the death of Philip William of the House of Neuberg in 1690. Upon his death his son, John William, came to power, and like so many cases that have occurred through history he became a religious persecutor when he brought back Roman Catholicism to the area by edict. The populace, which was mainly Lutheran and Reformed rebelled, wars followed, and emigration or annihilation became the options.

The French, under Louis XIV, were to attempt to wrest the lands of the Palatinate from the Germans in 1689 by sending over 50,000 troops into that land against the less well organized German princes. The great cities of that region were all laid waste. Louis, although the winner proved to be the loser as well, since his treasuries were depleted by the vast expenses of the army under his command. In the end, with the signing of the Peace of Ryswick in 1697 he was forced to restore all the land that his troops had taken.

The Lutheran population, which could not accept the doctrines of the Roman Catholic Church imposed by John William, decided to emigrate to the New World, by way of England. The Sternbergs, who were Lutheran like most of the emigres, decided to try and make a new life in America. They felt that life in this primitive and wild new country had to be better than life under the strictures of John William. Sternbergs were most likely in the group of Palatines that landed in the Port of New York in mid-year 1710, which was the largest emigration of the three, in which about 3,000 new Americans disembarked with hopes of finding a freedom they had never known before.

Part of the exit agreement made with the British was that the Palatines would serve the crown by procuring tar and pitch from the pines of the Hudson for shipment back to the mother country. In return the Palatines would be granted a 40-acre parcel of land for each family, that would eventually become theirs. The plan was to have the majority of the Palatines settle in Schoharie, because it was erroneously thought that the trees of this area would prove substantial producers of pitch. After numerous delays, generally caused by political maneuvers on the part of the governor, the intended exodus to Schoharie from New York City finally took place in 1712, although even then not with the blessings of the crown. Schoharie, because it had been promised them, had become a rallying cry for the Palatines, and it indeed was in their minds the "promised land." When the governor failed to act they simply made the move on their own.

Schoharie was a wild and virgin land at the time. Since it was outside the band of English settlements, it was considered to be on the advanced edge of the province, and therefore had an additional appeal to the Palatines because of its isolation. They equated isolation with freedom since they had come from a land tightly controlled. The Palatines were not to find the land easy to live in, however, and the hardships were substantial. Because of more political maneuvering, and because of a questionable land patent a great many of the settlers of Schoharie finally left that area too in disillusionment. The majority of those who left were to find permanent residence in Pennsylvania.

Of the seven original settlements formed by the Palatines in Schoharie, one of the most prominent was called Weiser's Dorp, after one of the main early leaders. This village was to later become known as Middleburgh, and it was in this town that Charles Hazelius Sternberg was to be born.

Once the Sternbergs had finally settled in permanent towns in America they seemed to be successful. It was not so much that they were to accomplish impossible tasks, but they did show immediate promise as men of substance since they worked both long and hard, and were willing and industrious. They perhaps epitomized the good we all envision in the early American settlers.

Martha Pattison Sternberg relates in her biography of her husband George Miller Sternberg that the "first wheat" sown in Schoharie County, New York was sown by Lambert [Lampert] Sternberg in the fall of 1713 in Gerlachsdorf.[1] A later relative with the same namesake of Lampert also distinguished himself in the Revolutionary War and has been listed in the DAR's books of honor. This same Lampert was apparently a distinguished soldier in the New York militia during the Revolutionary War. He attained the rank of Corporal while serving with the

Vrooman's Regiment for Albany County in the mid-1780's. Pattison also relates that the history of Schoharie County denoted that one of the men serving on the first bench of Common Plea judges was a David Sternberg, also related [perhaps the David who was the son of the Lampert born Sept. 12, 1743].

Charles' father Levi was to translate from the German a novel called *The German Pioneers; A Tale of the Mohawk*, by the writer Frederick Spielhagen. The protagonist of the novel was Lambert Sternberg, and it was about he that Spielhagen was to spin his story. Although it is not totally clear as to why Spielhagen would pick a real individual to fictionalize, it most likely relates to the fact that Lambert was a reasonably famous man locally, who commanded respect in Schoharie, and hence he would have been a good subject.

Since *The German Pioneers* is a fictional work it is not really clear where reality ends and the fiction begins. Chapter One of the novel time-dates the action as April 1758, and whether true or no claims the young Sternberg lived on Canada Creek. The novel finds him, as with the other refuges from the Palatinate, as earning his living by helping to supply the agreed upon tar and rosin from local pine trees to the British. It is also claimed he had a brother who made his living as a fur-hunter. Other than occasional tidbits about the countryside and the people the novel is only of limited help, and then only if one can manage to obtain a copy since they are so scarce.[2]

It was also in Schoharie County that Charles' father Levi was born in 1814. Very clearly the early Sternbergs played an active and substantial part in the history of Schoharie County, southeastern New York state, and their westward development was to prove equally successful. They were a hardy and determinate bunch, and they proved over and over that they were of special mein.

But it is really not this prevenient background that we are interested in, but rather the family as it came to be with the advent of Levi Sternberg and his sons. As previously mentioned, Charles' parents Levi and Margaret were to have 11 children. The 1870 census of the County of Ellsworth, Kansas for the "Inhabitants On Thompson Creek," lists seven of the children: Theodore,

Fredrick, Charles, Edward, William, Albert, and Francis.

To the 1870 list we would have to also list George M. Sternberg since he was already out on his own, and there is a reference in Charles' autobiography of a "sister in Los Angeles" who was older. This seems to have been Rosina (or Rose) who was born Mar. 8, 1845. There was also another older sister in the family, Emily, who was born Feb. 21, 1848, and married at the young age of 18 or 19 at Albion. And the family was completed with the addition of Robert who was born Jan. 19, 1860, when his mother was 42, and who apparently died while very young at Albion.

In 1870 the census listed their ages as 56 for Levi, 52 for Margaret, 29 for Theodore, Fredrick at 27, Charles and Edward at 20, William 17, Albert 14 and Francis the youngest at 12. The only noticeable error in the census was in the misspelling of their last name. The family is listed under the name Stemberg rather than Sternberg. It is easy to see how this might have occurred when things were being translated from handwriting to typed text. The same problem occurred in the 1910 census where Charles' last name is shown as "Sternbery."

The 1870 census listed Levi as a farmer and doctor of divinity, Margaret as "keeps house," Fredrick, Edward and Charles as "working on the farm," while Albert and Francis were said to be "herding cattle." Theodore was shown to be practicing law. All of the children were shown as being born in New York.

The references to "Thompson Creek" in the census relates to its being the most specific place name near to them at the time. Thompson Creek got its name from the first known area settler, P.M. Thompson, who resided on the creek from 1858-60. He and the other few area residents were driven out in 1860 by Indian and bushwhacker difficulties, although they had homesteaded the land and built cabins. Thompson also lent his name to the river there as he apparently also bore the name of "Smoky Hill" for some odd and unknown reason.

The Sternberg children were all brought up with a clear sense of who they were and what they could accomplish. Since they came from a family that was well schooled and studied in the arts they were well prepared for life when they finally were to set out on their own.

The successes of George and Charles in their respective professions is perhaps as good a barometer as any for gauging the ultimate benefit of the family-training they were to receive early in life. Their father seems to have been both authoritative (in the old and good sense of the word) and understanding. This seems to be well evidenced in the passage where Charles describes his father's reaction to his decision to become a fulltime fossil hunter:

> My father was unable to see the practical side of the work. He told me that If I had been a rich man's son, it would doubtless be an enjoyable way of passing my time, but as I should have to earn a living, I ought to turn to some other business.[3]

His father, like Charles, seems to have been a curious mixture of talents that ranged from his being a stalwart minister, a book translator, a cattle man, a farmer, a teacher of Theology, a writer, a politician, and an administrator. His devotion to the church and his God was evident both during his years as a fulltime minister and after when one viewed him in light of his everyday behavior.

It is also evident that it was his teachings that were to make Charles the later vibrant Christian he was to become; a Christianity that was very obvious even to the casual observer and which showed itself in all his actions. Levi must also have played some part in Charles' understanding of things geological, for he appears to have had a good sense of then current geological thought. In 1868 (before Charles' bone ventures began) Levi was to publish an article in the *Evangelical Review* entitled "Geology and Moses." In this article Levi was to state his support of modern inductive science, and he claimed that the "dynamics of the past, though often, perhaps, more intense in their operation, were of the same nature as those of the present."[4] Levi's views are further explored in Chapter 20.

Levi was also a man who was able to appreciate the primitive nature of early Kansas, while yet continuing to cultivate his brilliant mind even in this rustic setting. This ability to thrive was probably one of the things that most attracted the genteel Margaret Levering Miller. Well-versed in various languages and the higher arts Margaret was certainly to prove an anomaly on the Kansas plains. This fact did not go unnoticed by her sons who were always concerned about their mother and the hardships she had to endure.

Charles J. Lyon's *Compendious History of Ellsworth County, Kansas* mentions that Levi came out in a buggy and that as he came west he "found the prevailing opinion at the then little town of Salina, and in fact all of the places east of there, that all the region lying west could never be utilized for agricultural purposes..."[5] Levi Sternberg, however, took umbrage at this notion and according to Lyon was to have written:

> In riding over the country, I became convinced that for grazing it was probably unsurpassed by any considerable portion of the habitable globe, and furthermore it was admirably adapted to the growth of cereals, and that at no distant day wheat would especially become one of its staple products.[6]

It seems obvious that Levi, even before settling in Kansas, already had a penchant for growing things, and the transition from preacher to farmer doesn't seem so unusual when viewed in this light. He certainly was proven right as far as Kansas wheat was concerned!

Apparently even while at Hartwick Seminary the Sternbergs were involved with farming. For in 1860, at the time of Charles' accidental and crippling fall, he was in his father's barn, and it contained an "old-fashioned thresher...haymows and piles of shocked grain."[7]

It appears that perhaps from the start George Miller Sternberg had planned to obtain land for his father in Kansas. D.B. Long was to write that he met Dr. George Sternberg on his return to Ft. Ellsworth (later Ft. Harker) in September and that Sternberg said to him, "You better take a claim next to mine." They then went down together the following morning where they filed two claims, "one for his father and one for myself."[8]

Levi Sternberg then was given outright the frontier Kansas farm that his son

George had owned while serving at Ft. Harker. The farm was likely to have been about 600 acres or so in size initially (judging from the known size of his neighbor D.B. Long's property) and they built upon the property a "stockade." This stockade was used for protection against Indian attacks as is related by Charles in his autobiography. D.B. Long's "dug-out fort" may have been similar. It has been described as a dug-out, "arched over with timber and earth" and having portholes from which the settlers could fire at Indians.[9]

Area residents would gather at these fortifications whenever Indian trouble would erupt and lives were endangered. We do know from Charles' descriptions that the family stockade was 20 feet long, 14 feet wide, and that it was built out of cottonwood logs laid in a trench, and then covered with split logs, brush and dirt.[10] The availability of cottonwood trees undoubtedly accounts for the ranch being known as the "Cottonwood Ranch."

As previously mentioned, George Sternberg wanted his father to have the property because he knew full well that his father had tasted little of life's luxuries while serving in his austere positions as minister and college teacher, and yet he did not give up the property without some reservations as to how ranch life would impact his mother. They all well knew the hardships that women of the plains had to endure and how frequently those hardships meant an early death.

Before turning the farm over to his father, George and his first wife Louisa did receive some benefit from the property. While they did not live on the property (since they had spacious and well-furnished quarters at Ft. Harker, complete with servants), they did manage to make maximum benefit of its fruits. Jennie Barnitz, wife of fellow officer Lt. Albert Barnitz, was to comment upon the Sternberg luxuries of "onions, radishes, green peas" and other edibles.[11]

The Sternberg farm was located closest to what is now the town of Kanopolis. Their property was south of Kanopolis, near where the railroad then crossed the Smoky Hill River. They had a quarter section located on sec. 12, twp. 16, range 8. Kanopolis was to be built at the site of the former fort, Ft. Harker, and is just four miles east of Ellsworth. Ft. Harker, once a major supply station for the U.S. Cavalry, had been closed down in 1873, and when the state's recommendation to Congress that they donate the 10,240 acres in property to the state was turned down, the fort and its property had been sold to an Ohio syndicate by Col. Henry Johnson. This syndicate, the Kanopolis Land Company "offered enough land for a city of 150,000 people."[12]

Kanopolis was never to gain the notoriety that its neighbor Ellsworth was to earn (and there are those who would think this a positive thing considering the speckled history of Ellsworth), and yet Kanopolis was begun with high expectations. Its founders wanted Kanopolis to be the capitol of Kansas, and hence the ambitious name. Since Kanopolis was situated in the center of Ellsworth County, which is in the center of the state of Kansas, and since Kansas is the center of the United States, they felt there was an inevitability of its being proclaimed the state capitol. Obviously these ample ambitions never came to pass, and Kanopolis remains today a mere roadside attraction, and that only for its past and not its future. The only real vestige of note of the days gone by in Kanopolis is the old red sandstone guard house which now houses the Ft. Harker Museum.

Previous to Kanopolis there was actually a rough city called Harker which unceremoniously had formed around the fort. It was a city primarily of dugouts. These underground homes, set in tiered rows, were structured so that the yard of one house was actually the roof of the house in the lower tier. The town of Harker disappeared with the dissolution of the fort in 1873.[13]

Once Levi made the decision to change his full-time occupation from minister to farmer he was apparently fairly successful, and the family prospered. Margaret Sternberg proved surprisingly resilient and soon expressed elation at the pristine beauty of the area. She quickly made friends in the town, as well as nurturing relationships with some women who had come from like circumstances. She therefore did not experience total alienation from her former way of life, and made the best of the situation.

Levi Sternberg was asked to take charge of a church that was to be formed in a newly established town nearby, and Margaret Sternberg was to play an active part by playing the organ and teaching in the Sunday school. This most certainly played a major part in allowing the genteel Margaret to feel more comfortable in her surroundings, and this new church undoubtedly did much to keep Levi intellectually honed.

A newspaper article of unknown origin found among George F. Sternberg's mementos quotes an early resident as stating:

> Much of this development was due to the leadership of the pastor, the Rev. Levi Sternberg. Dr. Sternberg was a college professor of German descent...He was originally a Lutheran but had adapted himself to the religious needs and preference of the people he found in his new home. Before organizing the church at Ellsworth he had established a congregation on Thompson Creek that became the Ft. Harker church...His last service was in 1889. Dr. Sternberg literally gave the church life, and nursed it to strength and maturity. It was his joy.

The same article indicates his years of ministry at this church as 1875-77, 1878-81, 1884, and 1888-89.

Wilson's county history relates the account of an early resident by the name of Luther Johnson who said:

> A union Sunday school was organized the same fall (1868);...Dr. Levi Sternberg, who lived on the Smoky Hill river south of Fort Harker, preached for us sometimes. The first church organized on Thompson creek was the Christian church, by C.G. Allen, November 28, 1870.[14]

Levi started two Presbyterian churches, the church in Ft. Harker and the First Presbyterian Church in Ellsworth. He did so because he found the area full of Presbyterians and bereft of Lutherans. A missionary organizer of the Presbyterian denomination, Rev. Timothy Hill, was to work with him in this effort. This first church in Ft. Harker eventually joined its congregation with a Methodist church and this led to the founding of the Buckeye Church, which is south of Kanopolis.[15]

D.B. Long was one of the early elders at the First Presbyterian Church in Ellsworth. This church had been founded in 1873 and its frame building used to stand on what is now part of the Ellsworth Court House grounds.

Levi Sternberg was also to serve as minister in Ellsworth at the local Presbyterian church, the First Presbyterian Church in Ellsworth. Levi was wise enough to recognize that there were more Presbyterians in the Ellsworth area than Lutherans and therefore connected himself with the Presbyterian church. Already an ordained Lutheran minister, he was to accomplish his change of affiliations by attending a local seminary and becoming thereby an ordained Presbyterian minister. The church he was to found was initially organized by Levi Sternberg with a membership of nine in 1873. Because of the strength of he and this nine, a strong group had developed by 1878, and in the following year they had completed a neat frame church at a cost of $1,500. In the census of 1880 he was still listed as a Presbyterian minister and he continued to serve the church up until about 1899.

In the halcyon days that were to follow (the same days where Ellsworth gained its infamous reputation that came with the Texas cattle business) Levi Sternberg found himself called upon to perform funeral services for perhaps the most publicized of the many deaths to occur in the town. Sheriff C.B. Whitney, (Chauncey "Cap" Whitney), himself a former veteran of the Civil War and the famous Indian battle at Arikaree and a man who had served at Ft. Harker, was shot and killed by a Texas gunman by the name of Billy Thompson (no relation to the prominent Ellsworth family). Whitney was the first marshall of Ellsworth, and hence his killing was a momentous event. Whitney was buried "in the Episcopal church yeard (sic). Dr. Sternberg preached the funeral sermon before a very large audience of mourners and friends. The services at the grave were also impressive. Dust was rendered to dust; safe from the storms, free from cares, in the bosom of mother earth, rests the body of our late Sheriff C.B. Whitney."[16]

Martha Pattison Sternberg relates further concerning the up and down nature of the family's life on the plains:

Rev. Dr. Sternberg became recognized as one of the leading educators of Kansas. Years of prosperity and adversity alternated and while more acres were added and the herds of cattle, horses and ponies grew larger, the lot of the average farmer and stockraiser in the West was at that time far more enviable. These industries brought fatigue, and at times very little profit.[17]

There is no question that Levi Sternberg was a very active man in the area almost from his inception into the neighborhood, for not only did he involve himself in the ranch, the church and the neighborhood, but he also took an active hand in the politics of his new profession:

In March 1872, the Reverend Levi Sternberg and David B. Long, both dairy farmers, had been chosen officers of a local protective association in Empire Township. At that time Sternberg declared farming and stock raising to be totally incompatible. A year later he had become a big cattleman himself and president of the Stock Growers Association of Kansas.[18]

You will recall that it was this same D.B. Long who was the "nearest neighbor" and a former hospital steward that had greatly contributed to Charles' life being saved when he was shot while riding home in a wagon. Long himself was very prominent in the area too, and like Levi, had turned from a position of strict agriculturalist to a stock man as well. A picture of him taken at the time shows him to have been a rather intense looking man with thinnish face and hair, with close-set eyes and a scraggly beard. He looked as uncomfortable in this formal picture as we imagine we would be should we have to be dressed up in similar old-time full regalia, complete with starched collar and strangling necktie.

D.B., or David Burton Long, was not in Kansas by accident, for like many of the early settlers in the Thompson Creek area, he was a former military man who had been stationed at nearby Ft. Harker. George M. Sternberg, Levi's oldest son, had become fast friends with Long when the two of them were thrown together at a hospital in Cleveland, Ohio: George as attending surgeon, and David Long as recovering patient. Due in part to George's recommendation, Long was made a hospital steward in March 1865, and later was to become chief steward at that same hospital. After proving his bravery under duress at an Indian battle in which a number of cavalry men were killed or injured, Long was transferred to Ft. Wallace. While enroute to Ft. Wallace Long stopped off to visit Sternberg at Ft. Harker (at that time still known as Ft. Ellsworth), and they both established land claims near the fort. George Sternberg had suggested adjoining tracts of land, and so it was that they became neighbors.

In 1867 Long and his stalwart wife Harriet Sage Long served together at the Ft. Wallace hospital at a time when Roman Nose and a band of about 300 Cheyennes attacked the fort. The Indians far outnumbered the soldiers at the fort and many soldiers were killed and mutilated. Both husband and wife distinguished themselves that day, but perhaps most memorably when they helped to sew back the scalp on a soldier who had been partially scalped by the Indians.[19]

After mustering out, Long returned to the claim site and began to build a frame house in September 1866 (even though his enlistment was not up until two years later) not far from the property which Levi Sternberg was to assume from his son. This frame house was the first frame house built in Ellsworth county, and Long would live there on the property until 1881 at which time he moved into more luxurious quarters on his nearby estate, "The Oak Hill Place."

D.B. Long, in an historical rendering entitled "Across the Plains in 1866," was to comment upon both the hardships and the friendships of the early 1860's in the area around Ellsworth:

> All this time the country was in as much danger from horse thieves, cut throats and murderers as from the Indians. Everybody went armed as a measure of safety. I had six horses stolen at one time and seventy-five head of cattle were driven off during a storm and were a complete loss to me. It required nerve to live on the frontier in the sixties, and yet the old timers were big hearted and hospitable. No one was ever turned away unrelieved from their doors, and to offer payment was almost an insult. Indian raids, buffalo hunts, hangings and country dances kept up the wholesome excitement and prevented stagnation of the community; everyone was poor and happy.[20]

It is certainly interesting that Long was to consider "Indian raids" and "hangings" as "wholesome excitement," but then again there is no question that such things did provide daily motivation to work, and provided impetus to the necessity of communal participation in the lives of others.

We find additional things to admire Long for, since he was to prove so adept at many things. We must also marvel at the way in which his life, and that of the Sternbergs, were to entwine. Levi and Long were to become good friends and were to serve in various capacities that complimented one another even though Levi was 26 years older. Both were to serve as officers in the local grower's protective association, and both were very active and vocal in local politics. Long was to serve further in the prestigious position of Fish Commissioner from 1877 to 1883 (stocking streams with salmon and protecting the already polluted waterways), as well as serving three terms in the Kansas Legislature for District 113 as a Republican between the years 1875 and 1877. Long was also to serve as the first President of the Kanopolis State Bank.

There is some irony in the two mens' involvement with the Stockman's Association since both Levi and David Long had been vocal opponents of the cattle industry prior to their "enlightenment." Levi was to be president of the organization. Long went on to establish the first thriving dairy in the area, and remained a staunch supporter of that industry. He had already some expertise in that field, having been reared on a farm and having made a living in the early 1860's buying and selling cheese.

Both families, the Sternbergs and the Longs, are listed as among the first and most prominent of the early settlers in the vicinity of Ft. Harker and Ellsworth. Interestingly enough, they continued to share this closeness even after death, for the gravestones of David B. Long and Levi Sternberg are in "adjoining tracts" at the Pioneer Cemetery in Ellsworth where the families are buried.

Levi was to keep active after his wife's death and would apparently travel each summer for a lengthy visit to Schoharie County to visit with family. Grace Sternberg Borst, grand-niece of Levi, was to comment that Levi would "come down to the old homestead there" for a summer stay of two or three weeks and that he would sit with her under a big shade tree and try to teach her to play chess, a game he really enjoyed.[21]

Some time after Margaret Sternberg's move into town to live in the home built for her by her son George, her death in December of 1888, and Levi Sternberg's death in February of 1896 the farm was sold, and was later translated into a cattle ranch by a large company. The younger sons had already moved on into their respective occupations and were not therefore really substantially affected by the change.

Once the family was to disperse from the family homesite, they seemed to quickly spread out across the breadth of the nation. As mentioned before, William was to become the station manager for Buffalo Station, the haven where Charles used to water his animals and commiserate with his fellow fossil hunters. His niece, Florence Ethel Sternberg, in a newspaper account, was to provide much valuable information on William, but she also indicated that he was born on the "Cottonwood Ranch, Ellsworth, Kansas" which conflicts with the 1870 census, as well as with the obituary from the *Tacoma News Tribune*, which both claim he was born at Hartwick Seminary. The obituary places his date of birth as Mar. 14, 1853, making him three years Charles' junior. It seems fairly clear that the niece was wrong in this case.

Florence Sternberg further indicated in this 1949 article in the *Berkeley Daily Gazette* that William Sternberg was to prove brave, and that it evidenced itself while he was performing routine duties as station manager at Buffalo Park for the Union Pacific railroad:

> ...When a stranger proffered at $20 gold piece in payment for a lantern, he became suspicious.
>
> Having heard that there had been a train holdup by four bandits a few nights before and that a payroll in gold coins had been stolen, he kept the stranger talking with his assistant while he telegraphed his suspicions to military authorities at Fort Leavenworth. His tip led to the arrest of the gang.[22]

William's magnanimous use of his reward is also worthy of note, for according to his niece he "said he didn't want money, but would like a vacation with pay and passes for himself and mother, Mrs. Margaret Sternberg, to visit California." The two of them apparently took the trip and visited San Francisco, where they stayed at the fine old Palace Hotel so famous in its day. This all occurred at the end of the Civil War, and again bespeaks how much Margaret Sternberg's sons thought of her, and how they strove to see to her comforts.

It seems reasonable to assume, considering the manner in which Charles received his middle name, that his brother William received his first and middle names in honor of William Augustus Muhlenberg. Muhlenberg, like Ernst Lewis Hazelius (from whom Charles received his middle name), was a great and influential churchman. Muhlenberg, son of the highly respected Henry Melchior Muhlenberg (who was considered the founder of Lutheranism in America), was also related to Conrad Weiser, founder of Weiser's Dorp which was to become Charles' birthplace of Middleburgh, New York.[23]

William Augustus Sternberg was apparently to leave Ellsworth County for the Klondike in 1898 where he evidently met with some success. He left that area and from there moved to Washington state from whence he was to go on to become a prominent citizen of the city of Tacoma. According to Katherine Rogers, another Sternberg biographer, William went to New York, graduated from law school and practiced there for a time. She also says he additionally owned a large cattle ranch in western Kansas before moving to Tacoma, although none of this has turned up in my research.

He was in any event to prove a jack-of-all-trades kind of individual, and his accomplishments were indeed praiseworthy. He was in his lifetime, among other things, a station manager for the railroad, a telegraph operator, cattle salesman, real estate businessman, banker, superintendent of a cold storage business, hotel owner (the Pierce Hotel at 919 Broadway), City Treasurer (two terms beginning in 1897), deputy Pierce County Clerk (1923-26), and first Grand Master of the Masonic Lodge known as Lebanon Lodge #104.

William never married even though he had much to offer in terms of personal traits and wealth, and he died without benefit of a mate on June 27, 1929 after a brief hospital stay. He had known that his time was short for he had written an autobiography recounting his reckless days in Alaska and his other ports of call, but it was apparently never published and its whereabouts are unknown. It would presumably have become the property of his San Francisco brother Frank who had been caring for him at the time of his death.

There is an intriguing footnote about William Sternberg in a letter from Edward Drinker Cope to Henry Fairfield Osborn. This letter, dated May 13, 1885, indicated that William had been asked to collect fossils for Cope, but had declined.[24] His refusal led to their enlisting a William Overton and son to handle the job. This is the same Overton who had railed against Charles collecting on his land, but who eventually had come to work for him. The son was to become a valued member of Williston's crew at a later point. The mere fact that Cope asked William would lead one to believe that he must have had some experience in this area, and perhaps that he had actually expressed some interest in working for Cope at some juncture. Charles, however, never mentions his brother William in relation to fossil hunting, only touching on his twin Edward's collecting abilities in fringe articles.

Like all of the Sternberg men, William had a striking resemblance to his older brother Charles and his other brothers. He had the same close-cropped hair, arching bushy eyebrows, slanted-back ears and bushed moustache characteristic of the family. One could not fail but to be struck with their noble demeanor and strength of personality.

The aforementioned Florence Ethel Sternberg was the daughter of the Frank (Francis) Sternbergs who came to San Francisco from Ellsworth in 1900. Frank was born Mar. 31, 1858. A monument in the Pioneer Cemetery at Ellsworth indicates that Frank and Nellie Sternberg lost a son, Clarence Denby, in September 1892. He was only six months old. According to the history of Ellsworth County compiled by Francis L. Wilson, Frank Sternberg had a grocery in the town of Ellsworth, right next to the bank on Douglas street in the 1890's.

Wilson's account and descriptions of the building were taken from the remembrances of Dr. Charles D. Wright, who said that Sternberg "had a delivery wagon painted like a circus wagon."[25] It appears that Frank may have taken over the business begun by his brother-in-law, Ira Warner Phelps.

Before moving out of state Frank Sternberg owned a local business where he raised chickens. The business was of some magnitude and reached its apex around the year 1890.

When Frank moved his family to San Francisco he started out working for the U.S. Government in the downtown building on Jessie street that housed the U.S. Medical Supplies Department. This particular department was responsible for packing material and medicines for transport to the Philippines.[26] It was he who went north to stay with his brother William during William's final illness, and presumably he who was to handle the estate as executor.

Frank Sternberg and his wife Nellie were to celebrate their Golden Anniversary on Oct. 16, 1933 in San Francisco. He was to apparently die in California.

Theodore was perhaps the most striking of the Sternberg men. He was more robust in appearance, and had a full head of hair even in later age (not a Sternberg trait). He also sported a heavy-bushed moustache which further contributed to his robust and hearty look. An older brother, Theodore was born in Dansville, New York on Sept. 15, 1840. He spent a great deal of his life in the military, but again showed his versatility by wearing many different hats. Theodore enlisted at Hartwick on Aug. 18, 1862 at the age of 21. He was following in the footsteps of both George Miller Sternberg and his other brother John Frederick Sternberg. Theodore's first enlistment was to last until June of 1865, during which time he served as 1st Lieutenant and Quartermaster for Co. E, 121st Regiment, NY Infantry Volunteers. Among the early battles he participated in were Crampton's Gap, Antietam, Fredricksburg, and Gettysburg.

After the Civil War Theodore was to join the rest of his family in Kansas where he was to practice law from November 1865 until July of 1898 in

Ellsworth. Not just satisfied with that, however, he was to build a sugar mill, edit a newspaper, and involve himself with the raising of chickens. His chickens, however, were strictly fancy breed. It is not clear whether he or brother Frank was the first to foster the Sternberg name in the chicken industry.

Theodore was to return to the military in 1898 where he was to serve as Quartermaster of the U.S. Army transport *Samoa*. The name of this vessel was later changed to the *Dix*. It was during this period that Theodore was to serve in the Philippines (October 1898 to 1901), and he had the distinction of serving as Paymaster General for troops stationed there. Additionally Theodore as an officer and an attorney was engaged to serve on a General Court Martial Board for a period around May 1901. Undoubtedly his legal training made this an ideal assignment.

George M. Sternberg was to visit Theodore briefly when Theodore was docked in Manila at a time when the General was passing through. During this period Theodore's wife Bertha and their daughter Lottie lived on a farm near Kanopolis. Son John, who was older, was apparently already out on his own. When Theodore returned from overseas the family moved into town.

Theodore was also to briefly command the garrison at Ft. Harker in the years following 1873, at the time of the dissolution of the fort according to information at the Ft. Harker Museum, although I was not able to corroborate that fact in official records. Theodore was to marry Bertha Margaret Schmidt (1856-1914), and their union was to produce what I was told to be the last of the Sternbergs in the Ellsworth area: Charlotte (Lottie) Sternberg.[27] Theodore was to die in 1927 and his tombstone was said to state that he was a veteran of the 121st New York Volunteers from 1861 to 1865, although the *Historical Registry and Dictionary of the United States Army 1789-1903* has him listed as serving from August 1862 through June 25, 1865. He apparently was discharged Sept. 15, 1904 as required by law because of his age.

Theodore's grave, and that of his wife Bertha, can no longer be found in the Buckeye Cemetery, West Empire Township, although supposedly that is where they are buried. The marker from which the above information was earlier taken can no longer be found.

Charlotte Margaret (Lottie) Sternberg is remembered by the Curator of the Ft. Harker Museum, Mrs. Mae Thornton, as a warm individual who never married. Mrs. Thornton recalls that Lottie Sternberg was tragically killed by a train as she was crossing the tracks on her way home from work one day. This was in 1957 while she was working as a telephone operator in Ellsworth. She too was buried in Buckeye Cemetery, a rural cemetery about four miles south of the town of Kanopolis. The house where Lottie lived was regrettably burned down not long after her death. A picture of Lottie can be seen in the Ft. Harker Museum along with other memorabilia of the Sternberg family. Although not a particularly attractive woman she did seem to have an air of nobility about her, and is remembered warmly still by some of the area residents.

Fredrick, Frederick, J. Fred, or John Frederick seems all to be one individual. He was apparently named after another famous churchman, John Frederick Ernst. Fred was born Mar. 12, 1843, died in 1928, and he is also buried at Buckeye Cemetery (next to Lottie, his niece). He was a veteran too, but trying to track his military career is a confusing maze at best. John Frederick initially enlisted on May 17, 1861 at Albany, New York, but only served a few short months until his discharge on Aug. 15, 1861. His sketchy military file notes that he was unable to perform duty for a period of 40 straight days, and was therefore discharged due to "chronic rheumatism." He apparently wanted to be a military man, however, for he enlisted again at Ticonderoga, New York on Dec. 25, 1861. At this point, however, he fictionalized his name to John F. Miller.

Sternberg (AKA Miller) was to serve first with Captain Donovan's New York State Volunteers, Co. F, 18th Regiment, New York Infantry and then later with the 5th New York Cavalry and a light artillery battery. He began as a clerk, but later served as a hospital steward. Unfortunately his military career was to take a turn for the worse, however, when he took furlough at Plattsburgh, New

York and was reported as a deserter on Aug. 31, 1864. His military records are so sparse that it is only possible to tell that he was arrested on May 1 of a subsequent year. I could find no other record that further commented on what followed.

But John Frederick's story appears not to end here. According to Wilson's history of the county "Frederick" Sternberg arrived at the Sternberg ranch in the fall of 1866, before Edward and Charles who did not come out until the following spring. Lyon's history seems to indicate that he had his own separate property for he says that Frederick "in the fall of the same year [1866]...came to the county and filed on a quarter section north of the Smoky adjoining the homestead."[28]

Frederick appears to have been the "black-sheep" of the family, at least as far as history books are concerned. There is reference made to him, rather ingloriously, in the *History of the State of Kansas*, Ellsworth County section:

> Scarcely had the people recovered from the shock occasioned by this terrible murder, when they were startled by another. This occurred in November, 1882, on the farm owned by Rev. Levi Sternberg, about five miles east of Ellsworth. The farm was worked by one of Doctor Sternberg's sons, named Fred, who had in his employ a hired man named Hughes. The forenoon of the day on which the murder was committed, young Sternberg and Hughes were out in the field gathering corn, at which they worked until noon, when they went to the house for dinner. What occurred between them, if anything, will, probably, never be known, but as they were returning to the field after dinner, and just as they crossed the bed of the Smoky, Sternberg drew a revolver and shot Hughes, causing him to fall from the wagon, and while lying on the ground, Sternberg jumped down from the wagon and shot him again, although he was dead at the time, as it was proven at the coroner's inquest that it was the first shot that killed him. Sternberg immediately surrendered himself, and is now in jail awaiting trial, and no excuse can be offered for the commission of the deed, except that Sternberg was insane, which, people believe, must have been the case.[29]

John Frederick was a young man even then, being only 39 years old at the time of this tragic event. He apparently ran the farm alone at that juncture since Charles was already married at that date, Frank was working in Ellsworth, and so forth. Scant records indicate that John Frederick was indeed declared to have been insane at the time of the crime and was incarcerated for an unknown period in a hospital in Osawatomie, Kansas. He died at Kanopolis at the age of 84. He curiously is buried in the family plot right next to his nephew John Levi Sternberg and his niece Charlotte M. Sternberg as previously noted. A casual walk through the isolated Buckeye Cemetery, which can be reached only after a trip over a long gravel and sand country road, failed to show the gravestones of their parents, Theodore and Bertha.

In the 1880 census Charles Hazelius Sternberg was listed as living in Empire Township as a farmer, with two workers, Martin L. Baily and William Livingston, working for him. It is not clear if this is on the land of Frederick, or his father, or a separate piece of property.

Edward Endress Sternberg, Charles' twin brother, is somewhat more unknown since he has not had his name splashed across the newspapers as with some of the other brothers. According to Charles Mortram Sternberg Edward had married into the Scates family of Ellsworth. He was apparently both a farmer and a circuit preacher. Being a circuit preacher naturally meant one had to have few ties, that one was constantly on the move, and so tangible and locatable references of this brother are almost non-existent.

The circuit or itinerant preacher in most Kansas denominations was given a scheduled route, and they would travel from one town to the next over this specified route eternally. At most of their stopover points they would preach or perform other religious rites in houses, since these small communities seldom could afford the luxury of a church building, and most certainly they could not afford a fulltime pastor. Edward was probably paid in farm products and other staples more often than he was in cash, and it is quite likely that he had to supplement his preaching "income" with other work in order to keep a household.

Edward seems to have been involved for a while in some sort of retail business in nearby Holyrood, Kansas and then a cheese factory (perhaps at D.B. Long's inspiration?), before he settled upon his choice of serving the church. He was to become a Methodist minister, and was to move to Oklahoma with his family.

Edward apparently farmed in Oklahoma as well. In 1906 Edward was living in a small town called Dane, which is about 80 miles southwest of Enid. While there he apparently continued to communicate with his brother Charles and almost helped him further with his fossil business. A neighboring farmer had discovered what was presumably a mastodon on his property and Edward reported all this to Charles. Charles then tried to talk Osborn at the American Museum into financing excavation of this fossil, but it appears such never came to fruition.

The Oklahoma census records for 1910 show Edward living in Seiling Township. At that point he was listed as a minister and not a farmer. It shows him as being married to an Eliza A. Sternberg who was 22 years his junior, with children Herbert L. (16) and Frank (13). This was a second marriage since it showed them only having been married for three years.

Edward seems to have lived for a period in Putnam, Oklahoma as well and appears to have died at a relatively young age as he is listed as deceased at the time of the preparation of George Miller Sternberg's last will and testament. It has been claimed that he died in Michigan.

Albert Sternberg's name does not appear in any of the other documents I have had occasion to see, nor is he listed in the will of George M. Sternberg. In the 1870 census he was listed as being 14 years old, which made him the third youngest son. He is not listed at all in the 1880 Kansas census. According to Katherine Rogers, Albert had also been involved in the grocery business at some point in Ellsworth.[30]

One of the two daughters, Rosina (Rose), eventually went to live in southern California where Charles was to later visit she and her family. She had apparently married Ira Warner Phelps of Ellsworth, a grocer, and it was with them that brother Frank was residing in 1880 according to the census of that year. When the Phelps' moved to California they changed occupations again, as they apparently owned orange groves.

The other daughter, Emily, was also older than Charles. She apparently married a man named Francis (Frank) Humlong while the Sternberg family was in Albion (1865-1866), remained there with her husband when her family moved to Kansas, and is buried there in a local cemetery. Her husband Frank was a native of Bracken County, Kentucky, who had moved to Albion and had served with Company B of the 11th Regiment of the Iowa Infantry Volunteers. He had just left the service in May of 1865. His occupation before entering the service was shown as that of a farmer. There are a few Humlongs listed in George Miller Sternberg's will who are apparently descended through this branch of the family. They are Robert Sternberg Humlong, George Arthur Humlong, and Margaret Louise Humlong.

Emily appears to have been known too as Emma, as that is the name found in various probate documents. She and Frank had at least two children, Margaret L. Humlong (later of Loveland, Colorado) and George Arthur [Arthur] Humlong, later of Eldorado Springs, Colorado.

The youngest family member, Robert Sternberg, who was born in 1860 appears to have died at either five or six years of age. He is buried as well at the cemetery in Albion according to research done by Myrl Walker.

Neither of George M. Sternberg's unions in marriage were to produce any children, and therefore he did not further contribute to the family tree. Charles, however, was to prove another story.

On July 7, 1880 Charles was to finally make up his mind that it was the proper time to settle down and get married. According to Katherine Rogers, the romance flowered when Charles was a patient at the Ft. Riley Hospital. Being 30 years of age was fairly old for a man to be getting married, but little did the happy couple realize that he still had two-thirds of his life yet to live and that they would come to celebrate their golden wedding anniversary together!

Knowing how conscious Charles was of Christian ideals, it is not surprising that his choice of a wife would be the daughter of a man of the cloth. Apparently Charles was to visit Ft. Riley as a bachelor, perhaps to visit his older brother. Ft.

Riley, a military post in West-Central Kansas was not too far for Charles to venture even while a resident of Ellsworth (a distance of some 85 miles) since the following historical note found its way into the Ft. Riley *Union*:

> Another of Chaplain Reynolds' daughters was married early in July. Miss Anna M., to Mr. Charles H. Sternberg of Ellsworth. The wedding took place in the post chapel and the ceremony was performed by the Reverend W.H. Hickox (sic) of Wakefield.[31]

The chapel in which Charles and Anna were married was described by Bishop Thomas H. Vail as a "...beautiful chapel, neatly and suitably finished, with a commodious chancel, and school room attached, and would be an appropriate model for any village church."[32]

Charles Mortram Sternberg was to state in an interview that he felt it was perfectly natural and inevitable that his dad and mother would come to know one another, for as he was to put it, "the daughter of the minister and the son of the doctor were bound to meet."[33]

Charles Reynolds had served during the Civil War with the 2nd Kansas Cavalry Regiment. He was to first become a chaplain in January of 1862. He was an educated man, since he had graduated from Columbia College in 1843. As early as January of 1865 Chaplain Charles Reynolds was noted as being in charge of the post chapel at Ft. Riley and he was described as being "one of the most erudite, eloquent, and venerable clergymen of the Episcopal church."[34] Since Chaplain Reynolds' leanings were clearly high-church in contrast to Charles' background in the more austere Lutheran faith one wonders whether there were ever conflict that resulted.

Reverend Hickcox of Wakefield was also Episcopalian in faith, having served as deacon under Pastor Reynolds before beginning his own ministries. Like Levi Sternberg, Rev. William H. Hickcox, was both a farmer and a minister, and he seemed very similar to Sternberg in terms of his missionary zeal. Since he served as an assistant to Pastor Reynolds, it seems quite logical to assume that he may have had a firm bond with the younger couple, and that it was for this reason that he was chosen to perform the ceremony.

An earlier marriage at the fort (in either May 1869 or 1870) had united another of the Chaplain's daughters, Elizabeth (Bessie), with a Lieutenant George P. Borden of the 5th U.S. Cavalry.[35] This service had been solemnized in the presence of Bishop Vail. This marriage had perhaps started a family tradition that Anna did not want to break, even though Charles himself was not a military man. It certainly would have pleased her father greatly in any event since he had founded this particular work at the fort.

Levi Sternberg and his son's father-in-law were both to serve with distinction as members of the Board of Regents of the State Agricultural College, Levi from the years 1871-1873 and Charles Reynolds from 1869-1874. This connection between the two families perhaps might have additionally brought Charles and Anna together at times prior to Charles' trips to Ft. Riley and area. The faculty of Kansas State had recommended, and the regents had conferred the honorary degree of Doctor of Divinity upon Charles Reynolds in 1868, at which time he was serving as both chaplain at Ft. Riley and a member of the board of regents.[36]

Charles Reynolds played a prominent part in early church history in Kansas predominantly because of his missionary zeal. He was born in Gloucestershire, England on Dec. 19, 1819 and did not even arrive in Lawrence until the age of 39. He also was well thought of in legislative circles, as he served with distinction as chaplain of the 4th Territorial Legislature in January of 1859 at Lecompton. This may well give an indication of his political leanings.

Charles' father-in-law and his wife were apparently well thought of by the contingent at Ft. Riley and they were frequently mentioned in area newspapers. One such mention was made in the *Union* which indicated that Chaplain and Mrs. Reynolds celebrated their 39th wedding anniversary in July of 1881 with a party, but that shortly afterward in September of the same year, Mrs. Mary Braine Reynolds was to die. They had served together at the post for about 17 years, presumably having met initially in her birth-state of New York. Further note was made in the history of the fort written by W.F. Pride that Dr. Reynolds himself was to pass away just a few years later in December of 1885, succumbing to "paralysis" at his home in Junction City (which was the town closest to the fort).

It is not surprising to find Charles being married at the fort since his brother George had served at the fort himself as post surgeon from August 1867 to October 1870.

Charles was to recount to a San Diego newspaper on the occasion of their 50th wedding anniversary a humorous incident which befell them as they readied leaving the fort:

> 'When we left Fort Riley the troops turned out to give us a sendoff,' Sternberg said. 'Yes, sir, and they fired some big guns, too. On the train we boarded was a brigadier general, who was enroute to Kansas City. With the band playing and guns saluting the bride and bride-groom, this brigadier general stepped to the rear of the train and took a bow. But the big officer came to when an old friend of mine took off a shoe and threw it me. The shoe fell at the feet of the brigadier general. Then he woke up and ran away from the crowd.[37]

Dr. Charles Reynolds, Charles' father-in-law, was also to be their pastor while they were in Ellsworth. Dr. Reynolds was to found a number of churches, but the most prestigious was probably the Sternberg's home church, Trinity Church in Ellsworth. This fine church is the oldest church building of the Episcopalian faith to be found in Kansas and it has been described as being "a little gem of gothic architecture." It is conveniently located just a few short blocks from where Charles maintained his old fossil laboratory. The church is located at Tenth and Vermont streets in Lawrence. Reynolds founded the church in August 1859, but it almost had a short existence owing to the Quantrill raiders who greatly damaged the building and murdered some of the church members. Vermont street runs parallel and next to Massachusetts street which was the street most devastated by the raiders.

Record keeping at the Trinity Church in Lawrence is quite good, and their records reflect that the Sternberg family was membered there, and that baptisms and other church rites were performed for the family regularly. Records indicate that the family became members as early as Oct. 1, 1888, and perhaps even earlier. The Sternberg family residence listed at the time on the church rolls was 1033 Kentucky, and the family was shown to consist of Charles H., Anna M., George F. Charles M. and Maude. Levi was not yet listed as he was not born until a later date.

A search of old city directories for the city of Lawrence shows the Sternbergs listed for the first time in 1883, at the Kentucky street address. Until the year 1888 they were carried at that address, but then they aren't listed again until the book for 1902-03 when they are carried at a rural route address (#2 by post office definition). Charlie was to note that the family moved into the town of Lawrence in 1905. According to Katherine Rogers Charles and his family received financial help from his older brother George.[38]

In January 1893 (yet curiously notarized in August 1896?), Charles purchased property in the west half of the southeast quarter amounting to 20 acres for a sum of $2,000. He purchased the tract from A.A. and Susan A. Greene. It seems possible the Sternbergs may have lived on the land before the actual purchase.

While in the Lawrence area the family's address was to eventually revert back to a town residence, first to 806 Vermont and then to 1315 Connecticut. Son Levi was to marry early and moved down the street from the rest of the family, taking up residence in a small thinnish house at 1037 Connecticut. Both houses on Connecticut still exist, although the area has deteriorated substantially from earlier days, and the houses are certainly showing their age.

The women in the family are less well known, and information about them in many cases is sketchy at best. It is interesting to note that while we today are used to the women in families outliving the men (and by a number of years), the opposite was certainly true then. There is no question that the rigors of life on the frontier was much more difficult for the women. While the men at least had some diversions such as the breaking of virgin ground, exploring, hunting and so forth the women were pretty much forced to wait at home, performing the day to day chores, and often suffering the doldrums of life without companionship. The life of a wife in the west was hard, and it took its toll, showing itself frequently in an early death.

The Sternberg family was to prove no exception in this regard. Father Levi outlived wife Margaret by eight years, son Theodore survived Bertha by 13 years, Charles outlived Anna by five years, and so on. Their friend David B. Long kept the tradition also, outliving his wife Harriet Marie Sage Long by a total of 30 years.

With the limited knowledge of the Sternberg women it is hard to find too much to say, but one can clearly say that they seem to have been exceptional and strong women, and women who were very supportive of their husbands. Since they were not as often written about it becomes more difficult to track their lives as with their husbands.

We have already spoken about Margaret Sternberg, wife of Levi, and mother to Charles. She was a genteel woman, born in New Jersey, well schooled in all the arts and yet strong in many ways that one might not normally associate with such a genteel woman. Her sons adored her, and often expressed concern for her hard life on the plains. She was to have another 10 children (certainly a hardship in itself) and performed many of the routine chores on the Ellsworth ranch.

This must indeed have been a hard life for a woman of her social calibre. D.B. Long was to accentuate this fact when he was to note that in 1867 at Ft. Wallace, for instance, that there was only one woman at the fort. The number of women living around Ft. Harker and vicinity was evidently sparse as well, leaving very little social graces for the women to share in.

Bertha Margaret Sternberg, wife of Theodore Sternberg, was born in St. Louis, Missouri in September of 1855. She was to die Feb. 3, 1914 and is supposedly buried along with her husband in Buckeye Cemetery. Their spinster daughter Charlotte (Lottie), born in April 1883, was the last of the Sternbergs in the Ellsworth area. They also had an older son, John Levi Sternberg, who was born in September of 1878.

Charlie Sternberg was not impressed with his aunt Bertha in the least. He indicated that one of his uncles [Theodore] "married a German woman -- a sort of low class woman and he and she never got along. This woman was a vile person, I visited them once and she cursed and everything. I guess he married her because he had to. This one was the lawyer."[39]

Francis or Frank Sternberg was to wed Nellie Sternberg. They were to have an infant son, Clarence Denby, who died Sept. 24, 1892 at the age of six months. They were also to have a daughter, Florence Ethel Sternberg who was born in March of 1888, and a younger daughter Bernice, who was born in Kansas in February 1896. I've not really been able to discover anything further about Nellie at this point.

It appears that Florence Ethel Sternberg was to also call herself Ethel F. Sternberg and in 1956 she was living in Oakland, California at 1570 Jackson street. She had attended the University of California at Berkeley. The other daughter Bernice was to marry a man named Purrington and in 1956 she was living at 1664 Ninth Avenue in San Francisco.

Anna Musgrove Reynolds Sternberg, Charles' wife, was born Mar. 5, 1853 in Brooklyn, New York and was to die Dec. 24, 1938 in San Diego, California at their residence on Arizona street. She was one of the daughters of Reverend Charles Reynolds and Mary Braine Reynolds. She was sister to Helen Root and to Bessie Reynolds Borden and also apparently sister to a Dr. Theodore B. Reynolds. She was aunt to Miss Merle Ione Scates. The Scates family was prominent in early Ellsworth County history and was connected to the Scates family because Edward Endress Sternberg, Charles' brother, had married into that family.

A nephew of Charles' was Herbert E. Reynolds. He was a sometimes helper who assisted Charles in excavating fossils, although his main profession was that of an electrician. It was he who assisted Charles on the San Diego whale recovery at Ocean Beach, and who is mentioned in the beginning of Charles' autobiography

Hunting Dinosaurs in the Bad Lands of the Red Deer River Alberta, Canada. Reynolds lived variously in San Diego and La Mesa and was to marry an Anna Frieda -- Reynolds from Topeka, Kansas. Anna was to die at the young age of 45 at their home in La Mesa. She was survived by her husband, son Herbert T. Reynolds and daughter Blanche T. Reynolds all of La Mesa. Herbert E. Reynolds was to die himself just a few short years later, succumbing on Apr. 12, 1941.

NOTES

1. Martha L. Pattison Sternberg, *George M. Sternberg*, American Medical Assoc, Chicago, 1920, Pg. 1

2. Frederick Spielhagen, *The German Pioneers; A Tale of the Mohawk*, (Translated by Rev Levi Sternberg), Donohue, Henneberry & Co., Chicago, 1891

3. Charles H. Sternberg, *The Life of a Fossil Hunter*, Henry Holt & Co., NY, 1909, Pg. 17

4. Levi Sternberg, "Geology and Moses," *Evangelical Quarterly Review*, Aughinbaugh & Wible, Gettysburg, Vol #19, 1868, Pg. 138

5. Charles J. Lyon, *Compendious History of Ellsworth Kansas,* Printed @ the Reporter Ofc., Ellsworth, KS, 1879, Pg. 36

6. Ibid, Pg. 36

7. CHS, *Life*, Pg. 4

8. George Jelenik, *Ellsworth, KS, 1867-1947*, Salina, KS, 1947, Pg. 2

9. Ibid, Pg. 1262

10. CHS, *Life,* Pg. 11

11. Robert M. Utley, *Life in Custer's Cavalry; Diaries & Letters of Albert & Jennie Barnitz, 1867-1868,* Univ of Neb Press, Lincoln, 1977, Pg. 59

12. Ft. Harker Museum pamphlet

13. Ibid

14. Francis L. Wilson, *A History of Ellsworth County, Kansas,* Ellsworth County Historical Society, Pg. 7 [excerpted from Adolph Roenigk's *Pioneer History of*

Kansas]

15. Ellsworth History Fact Sheet, (Printed by *The Ellsworth Reporter*), Courtesy Ellsworth County Historical Society (Hodgden House)

16. Francis L. Wilson, Pg. 26

17. Martha L. Pattison Sternberg, Pg. 20

18. Robert R. Dykstra, *The Cattle Towns*, Atheneum, NY, 1974, Pg. 309

19. Joanna L. Stratton, *Pioneer Women,* Simon & Schuster, NY, 1981, Pgs. 120-121

20. David Burton Long, "Across the Plains in 1866," Chpt IX in *Pioneer History of Kansas,* (ed by Adolph Roenigk), 1933, Pg. 67

21. Hartwick College Archives (File Interview)

22. *Berkeley Daily Gazette*, Aug. 2, 1949

23. Ann Ayres, *The Life and Work of William Augustus Muhlenberg*, Harper & Bros., NY, 1880, Pg. 7

24. Henry Fairfield Osborn, *Cope: Master Naturalist*, Princeton Univ Press, Princeton, NJ, 1931, Pg. 379

25. Francis L. Wilson, Pg. 44

26. *Berkeley Daily Gazette*, Aug 2, 1949

27. Personal Communication, Mae Thornton, Curator Ft. Harker Museum

28. Francis L. Wilson, Pg. 36

29. Charles J. Lyon, Pg, 1275

30. Katherine Rogers, "The Incredible Sternbergs; They Left Their Mark on Kansas," *Kanhistique*, Vol. 16, #4, Aug 1990, Pgs. 1-5

31. W.F. Pride, *The History of Fort Riley,* (Privately Published), 1926, Pg. 181

32. *Boots & Bibles*, Riley County Genealogical Society pamphlet

33. Interview with Charles Mortram Sternberg by Free-lance Broadcaster Laurie LaMaguer, Courtesy of Tyrrell Museum, Dept. of Paleo.

34. *Boots & Bibles*, Courtesy Ft. Riley Museum

35. W. F. Pride, Pg. 159

36. Ibid, Pg. 444

37. Newspaper Article of Unknown Origin, n.d. (Part of GFS Scrapbook Memorabilia)

38. Katherine Rogers, *The Sternberg Fossil Hunters: A Dinosaur Dynasty*, Mountain Press Pub Co., Missoula, 1991, Pg. 79

39. Interview with Charles Mortram Sternberg

CHAPTER EIGHTEEN
CALIFORNIA SOJOURN

Charles' reasons for moving to California have not been very well delineated, and it forces us to speculation as to the real motive or motives for the change of states. There was never any question about Charles' love for Kansas, or for his love for Canada (in a different but equally obvious way), but there never appeared to be any flagrant extolling of the virtues of California to be found in any of his writings. At least one that would suggest a real love for the land, the kind that might have been a draw for him since he placed so much emphasis on what the land had to mean to him. This does seem odd, considering the depth and breadth of geology to be found in the state of California, but it is not the kind of geology Sternberg was used to and it differed in context, so one would have to guess that it was something other than the land itself which brought him to California.

Charles had first visited California on his trip to Oregon and the John Day region while working for Edward Drinker Cope. He had passed through the state via Sacramento in the northern central valley portion of the San Joaquin and had traveled north through Redding before heading out of state. His only substantive comment about the California leg of the trip was in respect to the regal and majestic display that Mt. Shasta provided for he and his fellow travelers. There were no comments about the wonderfully scenic trip over the Sierra-Nevada mountains, or comments about the stateliness of the California state capitol, or comments about the spectacular central valley itself that is a geologic wonder to behold.

Perhaps it can be assumed, that it was the aura of "gold" that he most disliked or the fact that this territory was so foreign to his taste, since it was essentially composed of volcanics and non-marine sediments of a major difference from the Kansas chalk. Whatever the reason, it seems safe to say that there is likely something of interest to be culled from his omissions about this state as far as his autobiography was concerned. California is ultimately too spectacular to be ignored, largely because of its extreme variability. Even the people that hate it (for whatever reason) generally do so with vehemence. One can hardly believe that Charles Sternberg was indifferent to its wonders since he was ultimately so attuned to things geologic.

In his autobiography, *Life of a Fossil Hunter*, Sternberg leaves the period from the end of the Oregon/Washington period in 1879 until the beginning of the Texas Permian period in 1882 open, with little indication as to what had happened to him during that timeframe. There is, however, an inkling of his whereabouts to be found in a letter of Cope's. This letter was written to his wife from San Francisco on Aug. 25, 1879 and says, in part:

> I find Dr. Condon has just explored the region near Silver Lake where I wished to go. But he has proposed so I am told to send all the specimens to me, and so I am not disturbed. But I must see the ground and visit perhaps a new locality. Mr. Sternberg has left Oregon and come to California where I will probably find him a job.[1]

Charles had not yet married at this point, for his taking of Anna Musgrove Reynolds' hand in marriage did not occur until July of 1880, well after this time. We know also that his real move to California, the San Diego move, was not made until 1923. This interlude, or "dark period," is intriguing because we know relatively little of what occurred between 1878 and 1895, with the exception of the Oregon/Washington sojourn.

Charles' residences in California were multiple, for we know that he lived variously in both Los Angeles (Glendale) and San Diego, and had stayed near the Coalinga oil fields at McKittrick when he worked at the famous brea fields of McKittrick. It was not until fairly late in Sternberg's life that he was to work at the McKittrick site, however, for this was not until around 1925 when he was well into his 70's. The wonderful picture taken of him on the California oil fields that was used as a frontispiece to his article entitled "Fossil Monsters I have Hunted" in *Popular Science Monthly* was taken sometime prior to 1929 and shows him an old man at that point. This picture certainly strengthens our claim as to how durable an individual he was, and a drive through the McKittrick area even today suggests a primitive site that would be sometimes difficult for a twenty year old to negotiate. But then again we are getting ahead of ourselves, so we should get back to the time-frame as best we understand it lest we lose track of the nature of this journey.

The real California period is the time that he spent in California while living and working in San Diego. He moved to San Diego sometime around 1920 and bought a home in the North Park area of San Diego proper, in an area above Balboa Park. This quiet area has both blossomed and fallen on sad days since the days when Sternberg and his wife Anna lived there, but it still remains appealing, if with somewhat less flourish. The section is a part of old San Diego today, and is essentially a racially-mixed area with housing that has tended to approach the shabby in outward appearance. Sternberg's house at 4046 Arizona Street remains an unassuming small white dwelling with a typical open front porch of the type popular in that era, and has a driveway leading to a small detached garage. The house appears not to have been changed in any pertinent manner since the days when it was first built, and it most likely was either new at the time he bought there, or was built to their specifications.

One can still see today why this locality would have appealed to Charles and his wife. It is close to the spacious park that houses the city's ornate museums and their valuable treasures (and most particularly the natural history museum for which he was to work) and it was equally close to a Lutheran church, and one which catered to those people with a Germanic heritage. Surprisingly, however, it was to be the English Lutheran Church in which they were to take their membership.

This church, the 1st Lutheran Church, was located on Second Avenue between "A" and Ash Streets in the downtown harbor area. This particular church no longer exists, as is so often the case with churches in changing districts. Shopping was also within easy walking distance of the Sternberg home, and the area was obviously well-manicured in its heyday. As with people, it's often sad to see things get old, for even most wines don't improve with age, but only hint at what used to be.

In August of 1920 Dr. Harry Wegeforth of the San Diego Museum of Natural History had approached Charles about both Charles and his youngest son Levi, possibly accepting respective positions at the museum. Charles was to suggest that Levi be made Curator of Natural History and he himself Curator of Paleontology. It is not clear as to why either Levi or the museum chose not to pursue that course in respect to Levi, and why Charles' position was to eventually prove more honorary than real.[2]

Charles' initial negotiations with the museum were closely tied to his old friend and advocate, W.O. Bourne of Scott City, Kansas. In the autumn of 1920 Bourne had approached Miss Ellen Scripps about purchasing what was referred to as the Sternberg Collection for the museum, but bureaucratic red tape had held up the purchase. Ellen Scripps, who was more than willing to make the substantial commitment certainly had the means. Daughter of E.W. Scripps, the quirkish newspaper magnate, she had inherited her father's abiding interest in natural history. Scripps himself had shown an avid interest in all the natural sciences, and had been the founder of the prestigious Scripps Institute of Oceanographic Research in La Jolla.

Part of the red tape apparently involved what to do with the husband and wife team of Frank and Kate Stevens, since there was some discussion of their possibly being "kicked out" as part of this transaction. Both Bourne and Charles seem not to have been in favor of that action. Since both the Stevens' were to actively serve the museum for many years after this date it is obvious the situation was somehow averted.

Bourne had a great amount of respect for Sternberg's talents and wanted badly for the museum to ratify the agreement, but at the same time he was concerned about the details of the arrangement as well. He definitely wanted it clearly set up where Ellen Scripps would be the sole donor. In a pleading letter of Sept. 11, 1920 Bourne was to state, "...for if we go out and raise any part of it by subscription than the thing is cheapened."[3] Bourne was a very principled man.

Although Bourne had a financial involvement in this as well, it was not the kind of agreement where he expected to make any money. In June 1920 (perhaps because of Charles' poor financial condition) Bourne had advanced a sum of $1,070.04 to Charles pending consummation of this sale. A Miss Irene Punam had contributed another $250.00. Bourne had shown his willingness to lose all of those funds if necessary in his commitment to seeing a truly fine museum established at Balboa Park.

As part of the collection to be sold to the museum were a number of substantial fossils. The more significant items included mounted skeletons of *Tylosaurus, Clidastes, Equus scotti* and a *Portheus molossus*. The largest single fossil was a 30-foot long duck-billed dinosaur from Canada.

Charles eventually sold his vast fossil collection to the San Diego Natural History Museum in Balboa Park in 1920. Bob Sullivan, paleontologist at the San Diego museum of Natural History, indicated it was his personal feeling that Charles was to "bamboozle" the institution, and that the museum substantially overpaid for what they got. He also mentioned that some of the referenced material (which was perhaps the best material of all), had been stored away and somehow has been lost.[4]

Sternberg sold his collection to the museum for $3,680.00, the payment being made in "gold coin of the United States of America." [5] Some have claimed this showed that Charles Sternberg was distrustful of "newfangled things" such as paper money, and that this was evidence of the same. A look at other areas of Charles' life, however, would quickly dispel any such idea, for he was very adaptable to new ways of doing things as we have already seen. A more likely explanation is that he felt there was much more safety and value to be found in being paid in gold. Since he was in a position in this case to bargain, why shouldn't he have?

Since the museum had made the overtures to him at the same time as the sale to garner his services as curator of the paleontological collection, this presumably made it impossible for Charles to resist making the move. Charles had always felt the lack of a significant scientific title. He desired one that would properly account for his lack of college training and yet allow too for his "learned expertise." Undoubtedly "Museum Curator" would have sounded very appealing to him.

This was a very loose arrangement, according to Carol Barsi, librarian of the San Diego Natural History Museum, with the title being an "honorary" one, but one that was equally advantageous to museum and Sternberg. He was to be on staff there for about twenty years. The beauty of this arrangement was that it gave Sternberg an office and laboratory from which to work, and he was to lend them his efforts and expertise.

His office was located originally in the old Foreign Arts Building at Balboa Park, but was later moved as the area was needed for exposition.[6] Charles was able to spend about six months in the latter part of 1920 in mounting "a number of old fossil bones of extinct mammals, fishes and lizards," before the exhibit was to officially open in December 1920. An article from son George F. Sternberg's scrapbook, presumably from a San Diego paper, states: "Out at the San Diego museum he has a corner all his own where he is busy every day, studying and developing his findings and getting scientific stories and pictures ready for publishers."[7]

During his tenure as "curator" Charles was to commission the talents of the artisan J. Elton Green to fashion some small clay replicas of various dinosaurs as Sternberg envisioned them in life. Green, who was head preparator at the museum, was exceptionally talented because of his added expertise in plastic surgery, and he performed the task well. Included in these fine pieces were a *Triceratops*, a *Tyrannosaurus rex* and a *Chasmosaurus*. The *Chasmosaurus* is still on prominent display for today's museum visitor in San Diego.[8]

During this same time frame he was to find numerous ways to keep busy. An article of Jan. 19, 1931 in the *San Diego Union* shows Charles and his young nephew Herbert E. Reynolds near the remains of a great whale fossil uncovered in the San Diego area. The headlines proclaim: "San Diego Scientists Find Bones of Huge Whale..." and tell of their excavating a 500,000 year-old fossil whale, *Balaenoptera borealis*, from the sands of Pacific Beach on behalf of the O'Rourke Zoological Institute. According to the article, the bones were initially unearthed thanks to the yearnings of a hungry dog, although this certainly sounds remarkably like an apocryphal story.

This Pliocene whale was plentiful in what it left to posterity for they were able to unearth a complete skull, twelve ribs, and five vertebrae of this large beast of the distant past. The vertebrae themselves were massive and of a size that naturally inspired the local newspapermen to elaborate on the size of the whale. Charles was assisted on this dig by Dr. W.H. Raymenton and Zoward B. Smith of the O'Rourke Zoological Institute.

The O'Rourke Institute of Natural Sciences was also located in Balboa Park. The Institute was headed by a Mr. and Mrs. Patrick F. O'Rourke, and it contributed to the furtherance of science through the use of its teaching facilities. By means of various natural history specimens and related apparatus they catered mainly to the education of local children. It is not really clear as to how Charles actually interfaced with them in this instance (or why), and whether or not this was their lone joint venture as local records of this organization have not been located. When the O'Rourke Institute finally disbanded, however, it is known that its specimens and equipment were incorporated into the San Diego Natural History Museum.[9]

Son George was to visit his father and mother in San Diego on at least one occasion around this time (January 1929) and then again in early 1935. On the second visit he found things both good and bad with them and wrote as such to C.W. Gilmore at the National Museum: "We spent five days with father and mother and physically and mentally, I never saw them look better but financially, they are down to the bottom of their income."[10]

This comment by George relating to the financial difficulty of his parents may help to explain further why Charles was to so quickly leave California upon the death of his wife Anna only three short years later. This weakened financial position too must have been a very disheartening fact to Charles, since as a proud man he was always concerned over being the provider for his family, while balancing this with his need to perform a service for mankind. One also suspects that at this late date he was still reacting against his father's advice that his choice of a livelihood was not a wise one.

As early as March 1911 Charles seems to have been in financial straits, and notably so. In a letter to Dr. Fritz Drevermann of the Senckenberg Museum he was to denote that he had sold nothing in the United States that year at all and that he had "suffered great financial loss." He went on to indicate that as such he had found it necessary to borrow quite heavily from the bank.[11]

Things reached such a low in 1923 that Clinton Abbott was to purchase a "spectacular pair of whale's jaws" that Sternberg had collected in San Diego, and Abbott was to pay Charles from his own personal funds. He did this because Charles was "almost on the verge of starvation and was so urgently in need of money that I drew the check on my own account."[12]

Since Charles was sometimes curator at the museum of natural history, and found odd jobs in between working for different museums as an independent collector, it is hard to conceive of him having financial problems at this juncture because he had more stability here than ever, but there might well have been unexpected burdens and expenses that cropped up when he least expected them. Perhaps the freelancing that Sternberg did was due in part to his yearnings to get away from it all, but also because his earnings from the museum position were paltry due to budgetary restrictions at the museum and hence he needed added income.

One incident, certainly unexpected, stands out as a clear case where he would have been drained financially. An unnamed and undated article (perhaps 1926?) from the scrapbooks of George F. Sternberg relates an incident that almost caused Charles to lose his life:

> Clarence H. Danforth...killed...and Charles S. (sic) Sternberg, 76, 4046 Arizona street, was perhaps fatally injured when he was struck by a motorcycle at Richmond Street and University Avenue...Sternberg, a pedestrian was struck by a motorcycle...incurred a fractured skull, internal injuries and shock.

Sternberg was so badly injured that very few felt he had any chance of recovery. Yet from his bed in Mercy Hospital he was to arise, little worse for wear considering all he had been through. We can note that he was to still live for another 18 years! That Charles was hale and hardy has never really been open to question.

Most assuredly this incident would have caused a certain amount of financial stress, regardless of any remuneration from insurance if he had any, and this condition would have been further hampered by Anna's failing health.

The Sternbergs, both husband and wife, were married fairly late in life (he thirty and she twenty-seven) and yet they still were to celebrate their golden wedding anniversary. One of the San Diego papers was to herald the event: "Golden Wedding Biggest Event to Noted Scientist." They celebrated this blessed event on July 15, 1929. Before they were through they were to add eight more years, celebrating their last anniversary in 1938, the year of Anna's death.

Although Charles and Anna had an excellent and long life together, it was not uncommon for them to spend substantial time away from one another. A fossil hunter's wife had to expect such an existence. Clinton Abbott was to bring to light one incident relating to Charles' married life in a letter from 1922. Abbott was to mention that Charles had approached him about fashioning a workshop on the museum grounds in a building called the Standard Oil Building. Charles was to further query Abbott about his being allowed to live in one of the offices as well. During the interim Charles had inquired if Abbott had objection to his "putting a cot for himself, without his wife, in one of the vacant offices of the Standard Oil Building."[13] As much as Charles loved his wife, he often would place fossil business over personal comfort and even perhaps over their relationship.

As appealing as the San Diego area was it would only remain of interest to Sternberg while he had a home atmosphere, with a loving wife to return to after his collecting trips. When Anna died suddenly on Dec. 24, 1938 he was devastated. Anna Musgrove Reynolds Sternberg died at their residence on Arizona Street, and was buried at Mt. Hope Cemetery, with their pastor, Delmar L. Dryerson, presiding.

Anna had been suffering from Chronic Myocarditis since 1928. Myocarditis is an inflammation of the heart muscle generally thought to be caused by a virus. This is a fairly rare condition that evidences itself through extreme shortness of breath, palpitating heart, and a generally weakened condition that would curtail almost all normal movement. Her health had been aggravated by complications of Arteriosclerosis and Chronic Interstitial Respiritis. These diseases had undoubtedly helped to dwindle the family savings as well, since the medicines needed to treat these conditions would have been quite expensive.

After Anna's death there was virtually nothing to hold Charles in the area and so it was without question that he would move back to Canada to be close to his sons and their families. One suspects that he would have preferred moving back to Kansas rather than Canada, but the family connections in Canada were stronger, and most of his old Kansas friends and work-work-mates had long since passed on.

An article in 1939 from the *San Diego Union* seems to take exception to Sternberg's move to Canada for it proclaims that he moved "last January [1939]" to Kansas again to live with his son.[14] This seems consistent with an article in the *Kansas State Teacher's College Leader* which indicated that Charles visited the school and spoke to the students, and that he was "still enthusiastic about fossil hunting, although he is not able to do much actual fossil hunting now." [15] One notices that the word of emphasis is "much," and not the "none" we might expect of an eighty eight year old man.

Because of the major contribution that was made both to Sternberg's life and the general field of paleontology by the McKittrick fossil sites, it is essential that this segment of Sternberg's life in California receive special treatment.

One of the most famous fossil sites in California history is at McKittrick. This world-renowned site is noted for the fine preservation of the many articulated specimens found within its confines. Its fame is second to none in terms of the quality of its specimens.

Because of the value of the work that was done at McKittrick, and because of the reputation that this fossil site has in paleontological circles, it is essential that we describe in greater depth some of what went on there, and particularly as it related to Charles himself.

A trip through the area around McKittrick today gives one a good sense of what life was like when Charles was there, for there has been little in the way of urban development in the district. The roads are undoubtedly better, the oil derricks more modern, and the towns larger and more resplendent with modern contrivances, but for the most part the region remains both quiet and "old" in the sense of the feeling you get when driving through the area.

McKittrick itself is now approached from Interstate Highway 5 which did not exist at that time, so it certainly was more difficult to get to easily than is now the case, but once one leaves Highway 5 and drops off onto the county maintained Highway 33, things become pretty primitive in short order. Taking 33 for just a few miles brings one into the heart of Coalinga, a region long known for its famous oil fields and its fossils. Coalinga is (and was) the largest town in the area and most of the oil workers have congregated there.

The fossils of the Coalinga district are almost all invertebrates as the area is a past ocean bottom basin of great length and breadth, and only the occasional fish will be plotted on the paleontologist's vertebrate listing. Fossil gastropods and pectens can be found in proliferation in the whole of the area, and fossil hunters have traversed the general region for a good many years.

The general area around Coalinga mainly consists of Cretaceous or Tertiary rocks that have undergone a considerable amount of disturbance, and a look at the locality today would lead one to believe that its treasures would have been obvious to the earliest geologists and that it must have been very early a major area of reconnaissance, but this was not the case. It was not until around the turn of the century that this locality gained real recognition and at which point it became the subject of a study by the geological survey. The very remoteness of the location, and the barrenness and isolation of the surrounding countryside made it a "late discovery."

Perhaps this is not so hard to understand if we take note of the fact that as late as 1938 there still remained over twenty-five percent of California's soil that had yet been unmapped by the geologic survey. Much of that unmapped area was therefore relatively unknown.[16]

The only sour note on collecting fossils in the area (and unfortunately Sternberg would not have been aware of this in his day) is the presence of the pestilence known as Valley Fever that haunts the would-be hunter of fossils. Sternberg wore no protective devices and hence would have been highly susceptible to this virulence, but was probably saved by the type of deposits he was working in, since they presumably don't have the same penchant for the fungus.

Valley Fever, or Coccidioidomycosis, infects a great number of San Joaquin Valley residents who have either tilled the soil or otherwise been somehow involved in breaking up the rocks in the area where the culprit hides and performs its insidious work. When released into the atmosphere and inhaled it causes mild to severe cases of respiratory difficulty, with particular emphasis on the noticeable damage it does to the lungs specifically. Fossil hunters in the area must arm themselves with protective masks of the type that cannot be penetrated by the airborne fungi when they are digging, and even more so when they are preparing their finds in the laboratory, or they will catch this debilitating disease most certainly.

This disease is indeed a major deterrent to the casual fossil hunter, and preparation and forethought is essential when in this famous locality. Sternberg does not mention the fungus in his writings so it is not entirely clear whether or not he and his fellow bone hunters were prepared to ward off this pestilence by using masks, but I expect not. Modern medicine is much more adept at isolating these kind of nuisances than was the case 50 years ago. One wonders whether Sternberg or any of the others developed Valley Fever as a result of their work there, but Charles does not mention being stricken by this ailment.

The neighboring town of Coalinga in that day has been described variously, but finds good voice in an autobiography by an early geologist by the name of Olaf P. Jenkins. Jenkins, who traveled in the area around 1915, was to describe Coalinga as a "frontier town" and then was to say:

> I learned from authorities in the barber shop to stay away from the front street or I might get struck by a flying bottle from one of the saloons that lined the street. The labor in Coalinga was beginning to feel their oats, I judged, for linemen would not go to work unless they had on silk shirts. Always the best place in town to get educated is in the barber shop.[17]

One can hardly help but compare this to Sternberg's definition of a Kansas frontier town that we looked at earlier. If Coalinga, which was the only real town in the vicinity of any size, was described as a "frontier town," one can imagine what Jenkins would have said about McKittrick. These were not places for the meek, and only the hale and hardy found them to their liking.

McKittrick, which is to the south of Coalinga some 60 miles or slightly northwest some 35 miles of Bakersfield (another major oil and fossil locale in California), lends its name to the McKittrick formation which is predominant in the Temblor Range, and completely forms the Elk and Buena Vista Hills. This formation in spots is said to achieve a thickness of around 2,500 feet, and will alternately consist of fine to coarse conglomerates, sandy shales, and a distinctive bluish sandy clay.[18]

The general region, south of Coalinga and near the town of McKittrick, was first brought to the public's attention when the pioneer paleontologist Joseph Leidy was to describe in the 1860-70's an extinct species of horse from the Buena Vista Lake locale. This discovery preceded the now more famous, but similar, La Brea Tar Pit finds. John C. Merriam was also to describe fossils from a site near McKittrick at Asphalto, whose name cannot help but give one a true picture as to its geological character.

It was not until 1921, however, that the famous brea beds of McKittrick were brought to light during excavation for an area highway in a cut for the highway being constructed between Taft and McKittrick. As with so many things, however, this proved both a blessing and a curse, for Sternberg was to later find that the highways positioning was such that it was to destroy as well, for it tore right through the principal deposit. While the automobile may be the cause of urban blight and atmospheric haze, one must marvel at what it has done for fossil horizons. But here, as ever, the automobile can be both praised and maligned.

Charles' work at McKittrick was not to begin until 1925, well after initial efforts had been made there by the University of California. On Hanna's recommendation he began after the hot August weather since the heat would have been too much for a man his age. The crews from the university had already removed a mastodon skull and a substantial collection of bones from the area before Sternberg was invited to try his hand at the fossil wealth. Charles indicates that it was owing to Dr. G. Dallas Hanna, curator of the California Academy of Sciences in San Francisco that he was invited to work as an independent at the site.

Although he was to begin here as an independent collector, he did need to have some affiliation in order to realistically be able to work the area. It was, therefore (thanks to Hanna), under the auspices of the California Institute of Technology at Pasadena that he was actually to work, and it was to them that all of the fossil material he collected was to eventually go. Charles hints at the beginnings of this affiliation in a letter to Dr. Hanna:

> Dr. Chester Stock was here the other day and was delighted evidently with the results, but he does not want to see the material leave the state, he wants it at the univ. The man with him who seemed his associate is to go to Nevada to see Miss Annie Montague Alexander to try and raise the money. Dr. (William Diller) Matthew writes me that Dr. (John C.) Merriam is interested, so between the two, I hope to be employed.[19]

Hanna, who befriended Sternberg after they had communicated for some time by letter, had also visited Sternberg in San Diego to see some of his collection. He also was to take on the task of describing part of the invertebrate collection Charles had made from the Pliocene in the San Diego area. Charles was particularly looking for an "authority" to review the material, for he considered it necessary that he have the proper nomenclature on specimens being sold to others. Mrs. Kate Stephens of the San Diego Natural History Museum had graciously assisted Sternberg in some of the naming, but principally due to not having as large a collection for comparison, some of the namings were all but guesswork. Dallas Hanna was to later work out an arrangement with F.M. Anderson, an authority on the Cretaceous of the Pacific Coast to look over Charles' Cretaceous material, and later for L.G. Hertlein to describe his Pliocene material. Hanna was to indeed prove a valuable friend.

Hanna was most likely familiar with the McKittrick site because of the fact that he worked for the Pacific Oil Company, as well as for the California Academy of Sciences. The Pacific Oil Company owned land in the same area, north of the highway. It was Hanna who was to write letters of introduction to the Midway Royal Oil Company who either owned or leased the property where the work was to be done. Hanna also referred Charles to a personal friend by the name of Stevens who worked for the Associated Oil Company, and it was he who was to guide Charles to the bone beds once permission had been obtained for the dig.

As the property owned by the Pacific Oil Company was also composed of beds of asphaltum, Hanna thought that there would be fossils of value there, but Charles was to find that such was not to be the case. He only found limited and scattered evidence of fossil bone at this site. When this effort proved futile, the permission was obtained from Midway Royal through the intervention of Mr. Frank A. Garbutt, an officer of that company, for work to begin on their property. In his request Hanna was to state: "the living he (Charles) makes is very bounteous at best, so I think any favors we can extend him will not be amiss."[20]

During his stay in the area, which lasted until the end of October 1927, Sternberg was allowed to use a house right on the property upon which he worked. A Mr. Hall, then manager of the Midway Petroleum Company, not only allowed Sternberg access to their oil lease, but provided the housing too. The house that was offered was the former residence of a man named Bonner who presumably worked for the oil company. This housing was ideal since it served a dual purpose; not only did it provide housing for he and Anna, but they were able to convert the other half of the building into a laboratory where the fossils could be prepared prior to their transmittal down south. The fossils could be conditioned for travel here without having to suffer the indignity of being ill-prepared for travel over the rough roads of that day.

The area that was to finally be exposed to the light of modern day was in an oil seepage section where the tar or brea had accumulated to form the classic brea terraces now so famous. The rock itself is hardly distinguishable from a chunk of asphalt one might find along the edges of our better country roads today, and it has the same sticky consistency as the deathtraps at La Brea. Brea is as natural a trap for animals as the Venus Flytrap is for insects. Once caught, animals were doomed to struggle into the mire even further until exhausted, they fell prey to the enveloping menace. It was at this world famous locality that Sternberg was to work.

McKittrick is too small to properly label it a town, and perhaps it might be better characterized as a village or hamlet if that did not give one a vision of a more respectable and refined sort of encampment. The "town" today consists of a few commercial buildings only, and an odd assortment of houses that spot the area in an apparently unplanned way. The only two buildings of any prominence are the local shortstop grocery, and the attached post office.

The one-man post office was open and operative when we made a visit to the community, but the postmaster, although pleasant and helpful, was unaware of any of the history of the brea pits or their discoverers, and only knew that the pit itself was located outside of town on a small rise about one half mile south of the town just off Highway 33. The excavation site is marked with a historical marker so that the casual observer would be aware that this is not just any old pit they happen to be staring at, but a first-class, world-class pit. Considering the importance of the discoveries here, however, it seems anti-climatic to look at the historical marker with some fascination, and yet find that the local populace is basically ignorant of what had transpired here.

Sternberg was actually to work two main sites while at McKittrick. The first he was to label the "Sternberg Quarry," (perhaps in retaliation for Marsh's robbing of his famous Long Island Quarry) and it is located across the highway from the "University Quarry," where the previous work had been done by university crews. Dr. Stock asked him to work the second deposit because of the successful nature of his work in his own quarry.

Considering the isolation and barrenness of the area, one is certainly surprised to see that the region is still vividly bustling with the raucous greetings of oilmen, who are to be seen racing from one site to another in open pickups. Their bedraggled appearance is accentuated by their T-shirts which are smeared with the blear of black gold. The silk shirts described by Olaf Jenkins were not to be found. I couldn't help but compare the vision of these roughnecks with the one that I have of Sternberg and his sons romping around the Kansas fossil fields in earlier days, as the circumstances are very similar. These workers, like the Sternbergs, earned their money with back-breaking efficiency.

One suspects that this area has changed little over recent geological time. Over the last one hundred years there has been little evident change to the buildings, and its area residents appear little changed from what they must have been like when Sternberg worked in the area. Modern dress, haircuts, and the cars and trucks are the main detractors, if one dwells on the agelessness of the scene.

Charles was to mention that the beds of McKittrick are very different from that of La Brea. He was to remark to Hanna that there was no place where he was to find any deposit where there had been "liquid moving thick petroleum as in the La Brea beds." [21] He claimed that the bones had been deposited before the oil had "flowed out and impregnated the clays in which they were buried."

The brea pit excavation was under the direct supervision of Dr. Chester Stock, who was connected both with the Carnegie Institute and the California Institute of Technology. Stock was a paleontologist of some note, and was well acquainted with the general nature of the California oil fields. This background made him an ideal expedition leader, and he and Sternberg were to become good friends. It was Charles Sternberg, however, who was to provide the real collecting expertise at this fossil site and who would receive the credit for the successful excavation of its wonders. Later, Stock was to meet and spend some time with Charles' son George also, and both had a healthy respect for the other.

Stock's further contribution to the excavation project was such that he was to foster within Sternberg a real sense of the worth of this particular fossil site, and in so doing it prompted Sternberg to exert himself even more diligently than usual. This led to the uncovering of a substantial number of different species of fossils. The obituary for Sternberg in the *New York Times* of July 23, 1943 stated that Sternberg made "many of his best discoveries in the field of fossil hunting...in the tar pits of California" and that he was to unearth from these tar pits "eighty-five distinct species." This tribute gives one some sense as to the magnitude of this site, and the reason it was to become a fossil site of world-wide stature.

In an article in the Carnegie Institute's News Service Bulletin of 1938[22], Chester Stock was to write in retrospect about the importance of the McKittrick excavations. He was to state that there had been discovered to that date a total of 43 different fossil mammals. The larger forms included bison, camel, peccary, the western horse, antelope, pronghorn, elk and deer. Most importantly they were to discover American counterparts of the large South American ground sloths which greatly added to the knowledge of California's role in sloth development. The more common and expected forms of sabre-toothed tiger, grim wolves and short-faced bears were also found in the deposits.

Sternberg's collection list was even more extensive, for he adds skunk, badger, fox, jack rabbit, rats, mice, mastodon, coyote, lion, wolf, musk ox, vulture, turtle, and water beetles, although this is only a very small list of the animals found. The lasting value to this site, however, is due to the fact that many of the skeletons uncovered were still articulated, something which was not the case with the La Brea finds. The value of the site is only diminished by its age which was more modern than that at La Brea, and therefore of lesser value in the eyes of the many who view a site's significance by how old it is. To do so in the case of McKittrick would be an unfortunate mistake.

One of the more interesting factors about the McKittrick site, especially in comparison to its more famous counterpart at La Brea, was highlighted by Stock:

> At McKittrick, however, the carnivores are relatively not so abundant [as at La Brea] and the dog family which is present in largest numbers is represented principally by coyotes. Another significant fact is the closer resemblance the McKittrick mammals make to the living assemblage of the California area than is the case with Rancho La Brea and which may be interpreted as an indication that the former are nearer in geological time to the Recent.[23]

Among the various publications Stock was to generate based upon the fossils Sternberg found at the McKittrick site were one on a new genus of llama, and a new form of peccary then unknown in the California Pleistocene. In his "generic" paper entitled, *Tanupolama, A New Genus of Llama From the Pleistocene of California*[24], Stock attributes the mammalian remains found to be similar to the living llamas, and certainly of the family Camelidae. This important fossil find was described based upon a fragmentary mandible with some of the lower dentition still in place, and three skulls in varying states of completion. Regrettably the skulls are such that none provides a true impression of the braincase, even though they were of ample proportion. Although some of the identification points are not of perfect clarity Stock felt he could comfortably identify the fossils as of a new genus. In closing his preliminary report Stock was to note the similarity of *Tanupolama* to *Camelus americanus* Wortman, particularly in reference to the "rather weak incisiform canine," and he was to note that the close similarity in size and general shape were to make it generally resembling *Tanupolama stevensi*, as well as the Nebraska and California forms.

Stock's publication for the Carnegie Institution relating to the fragmentary and presumed fossil peccary is in the same series, and again names Sternberg as the collector. The fossils themselves consisted of a portion of a mandibular ramus with both permanent and milk teeth in place, and a metapodial. Stock suggests the evidence leans towards the genus *Platygonus* and notes that the fossil is somewhat comparable to a peccary metapodial found at Rancho La Brea, except that the McKittrick metapodial is distinctively larger in overall size, and when it is viewed from a lateral view it is appreciably narrower. It is, however, one of the few evidences of related animals from the two different sites.[25]

Charles Sternberg himself has written little about the area, but we do get a flavor for the site's significance from a newspaper account from the *Hays State Teacher's College Leader* dated Feb. 15, 1928. The article recounts an address made by the elder Sternberg to his son's geology class regarding the McKittrick site:

> A pool measuring approximately 60 by 50 feet yielded a vast number of specimens to Sternberg's shovel...'The thing that surprised me,' said the speaker, 'was the fact that many geologists had worked all around the pool and yet had not discovered it.'

Charles describes the Sternberg Quarry fossils as being in material of a very fine clay. The sticky asphaltum (brea) only extended down about two feet in this particular site, while the clay section "filled with water" lay below this and was therefore not penetrated by the tar. Throughout this site, however, there were pockets where the asphaltum passed through the "bone-drift above and extended down through the clay." Charles indicates that these pockets were often round and that in one particular spot the asphaltum penetrated to a depth of 17 feet below the level of the highway. The largest of these pockets was only several feet in diameter, but the average was far less. It is amazing, considering this fact, to realize that in one small pocket alone he recounts finding 60 jack rabbit skeletons, interspersed with myriad jumping rats and birds. These pockets were more recent, and the pockets bearing the remains were seepages through the clay layer that he felt was caused by the "mass on top melting in hot weather" and traveling down through the older layer by the very fact of its heating. This pocketing made it difficult to separate out the recent from the older fossils, which is a perennial problem for the paleontologist.

Sternberg was to explain the articulation of the skeletons in his quarry, as opposed to the disarticulation of the La Brea fossils, as being due to the bones being carried into their final resting place by water. The very nature of the death poses of some of his finds made it evident they had been borne to the site by water, rather than being mired in the La Brea manner. This intermixing of fossils was to Sternberg unique for he stated:

> I found horses and camels, llamas, antelope, bison and saber-toothed tigers, wolves, lions and bears, lying together--the first time after an experience of 60 years in the fossil fields, that I have found carnivores and herbivores mingled together in death.[26]

Amazing to anyone familiar with the rigors of paleontology, both from a physical and mental point of view, is the fact that Charles was able to function so credibly and reliably at his age on this kind of an expedition. While it is true that the site was close to habitation (and therefore not too rigorous to reach) it is also true that the very nature of the asphalt-like rock made it difficult to work. Because of its sticky composition, and because it tended to cause further difficulty in its absorbing heat, it was to prove a formidable opponent that caused much discomfort.

Because this was such a major fossil site it put further stress on Sternberg as he was the leader of the hunters and had the responsibility of seeing that the others followed directions precisely. The mental duress that such work could cause cannot be understated. Work in paleontology is much more difficult in respect to figuring out where to dig, what to expect in terms of kinds of fossils, and the like than is ever the case with the physical exertions. While the physical cannot be, and should not be overlooked, one still needs to say that it is the major mental effort that truly produces the greater fossil finds and site discoveries. Sternberg knew his trade well and was an ideal leader because he was good at both aspects of the job.

Sternberg was not just to limit his California collecting to this one region, but was to collect a little here and a little there in a number of different areas. At some of these sites he collected vast amounts of fossils. Some of the areas that he was known to collect from were the whole of the region around Coalinga for invertebrate fossils, with specific sites being eight miles north of the town, three miles northeast and thirteen miles northwest of Coalinga, and in other sites as far down as the bottom of the state. In 1925 he had collected around the town of Barstow in the Mojave Desert, collecting deer, horse, camel and wolf fossils.[27] He also was to collect in the Kettleman Hills not far from Coalinga, further south near the ocean at San Clemente, and then down towards San Diego. He traversed the San Diego region fully as well, collecting at Pacific Beach, Balboa Park itself, and then at the more famous Tecolote Canyon. He also was to collect in the

Coyote Mountains in Imperial County near San Pedro, as well as collecting out of the Eocene at Rose and Alverson Canyons.

Like most good fossil hunters Sternberg always unconsciously annexed scenery and fossil habitat as he traveled about. Every time he looked at an embankment or an escarpment he viewed it in light of whether it would be potentially fossiliferous or not. He knew the colorations to look for, the type of rock that holds the best fossil material, and he knew what to expect since he was well read on existing fossil sites and areas of potential as suggested by others. He also had the added benefit, as the world's then oldest collector, of being able to query a myriad of bone hunters and paleontologists personally, and who as often looked to him for the same guidance. Each helped the other, and he was frequently of significant help to others. It was generally through word of mouth that he was given jobs, for people recognize quality work, no matter what the profession and respond.

While a resident of California Sternberg did not simply restrict himself to California soil as a collector, but often left to perform duties out of state in much the same way as he had from his former base in Lawrence. The Baja peninsula was of particular interest to him since it was less well explored for its fossil treasures, and naturally, because he hadn't been there before.

His first trip to Baja was at the inspiration of a Canadian friend named Frank Moody, who was the Superintendent of the Pap Burns Oil Well, and Charles was prompted by Moody's indicating there were great fossils on the lease where his company did business. Moody had been to the site near Santa Catarina and had brought back an Ammonite to show Charles and this provided the necessary encouragement for Charles wanting to collect there. Charles was to lovingly refer to him as a "loud scotchman" in a letter to Hanna, and told Hanna that he should "pat him on the back and agree to publish him as the discoverer of this locality"[28] if the institution ever hoped to collect there. Whether or not this ammonite was the same *Pachydiscus catarinae* as later discovered by Sternberg or not is really not clear. It was, however, at the insistence of the Mayor of

Ensenada at the time, David Goldbaum, that Sternberg was prompted to become serious about the trip. Goldbaum wrote a letter to the Governor of the Baja Province (headquartered at Mexicali), A.L. Rodriguez, and asked for permission for Sternberg to enter the field there, and the permission was granted. They were not to actually enter the fossil field itself until they reached the vicinity below Santa Rosalia, in the middle section of the Baja. That is, and continues to be, fairly primitive territory by most world standards, and was hard to scout since the roads were so bad.

Charles was accompanied on the Baja trip by his friend Dr. Peter Alpeter of Bonita, who was to assist him. Alpeter is something of a mystery since I have been unable to determine anything about him either through Bonita sources or the San Diego Historical Society. It is only known from Charles' writings that he was a friend and that he was from Bonita, California.

They left on their trip May 28, 1928 (when Charles was only 78), and they left well-provisioned in a newly purchased Dodge car which Alpeter provided and drove. Their trip took them along the coastal route from Ensenada to Santo Thomas to El Rosario and through the San Miguel Mountains into the area they wanted to work near Santa Catarina. Through the kind assistance of Manuel Santillian, a geologist headquartered at El Rosario, they were able to obtain special insight as to what to expect in their travels, since Santillian was working on a topographical/geological map of the area's terrain. His kindness extended even further for he also provided them with a man to show them the way over the mountain roads, and he provided photographs of the flora to assist them as well.

Hospitality did not end with Santillian while they were in Baja, since they were also royally entertained by the son of the Superintendent of a San Diego based company, the Southwestern Onyx and Marble Company. This company had a quarry location at Onyx on the Baja. The son, Kenneth Brown, was the only American at the plant at El Marmol ("The Marble" literally, but loosely translated as "The Onyx"). This was the largest industrial complex for miles around as there was just no other industry in this remote site.

Max Miller, part of an expedition to the same vicinity for the San Diego of Natural History some 15 years later, was also to comment on how hospitable Brown was, and how there was still nothing of consequence in the area except for the Onyx itself.[29] Miller also comments upon how important Brown himself was to the local populace since he served as administrator, village doctor, civil clerk for marriages, and so on. The work crew, numbering around one hundred or so were all Mexicans, and all worked very hard for their living.

El Marmol played such an important part in the lives of those in the area that Miller said that most of the families within a radius of some 300 miles were dependent in some manner on the plant. This was a sizeable operation (in fact at the time the largest onyx quarry on earth), and Sternberg was impressed with the work that was done there in carving out huge slabs of this great stone; huge stones that were to find their way to the city of San Diego for processing. This Mexican marble, or onyx, was a form of calcite rock and has often been used in decorative jewelry.

Once they finally reached the site for which they were searching near Santa Catarina, they began in earnest to investigate the wonderful terrain. A recent travel guide describes the famous fossil site now for those brave enough to still travel over those rocky roads to get there. The guide says that the road takes one of its numerous forks at a point some 220 miles or so below Ensenada, near the small villages of El Aguila and Guayaquil. If one takes the right hand fork they find a water-damaged and rutted road leading to Santa Catarina Landing. At about 12 miles along this road one passes the fossil locality, before the road wends its way back to Arroyo de Santa Catarina. It was at this site that Sternberg and Alpeter were to spend most of their time. In the course of this investigation they were to discover a new ammonite, a large-sized ammonite *Pachydiscus catarinae*, as well as wonderful examples of the worm-whorled *Hamites duocostatus* and *Nautilus sternbergi*.

Sternberg indicated that they stayed near the ocean, having been given a room to stay in by the head Mexican supervisor at Brown's onyx company. They hired the man's son to assist them, as well as a pack mule for transporting the fossils. He described the fossil site where the ammonites were found as a "narrow gulch," and said that the level land to either side of the gulch housed a bed of shale in which they found the cache of ammonites. The fossils were found in varying conditions, but generally the best preserved were found within sandy concretions which could easily be popped open with a large pick.

Their cache for the season from this locality was to total a substantial number of the larger variety ammonites. He notes that some weighed over 100 pounds, and that 35 of the larger ones were between 17 and 20 inches in diameter. The Mexican's son was to pack the fossils by mule down to their car, which they had been forced to park about a mile away. They were then hauled by car to the ocean where they could be transported by Onyx Company boat to San Diego.

Another Sternberg collection site in the Baja, at San Antonio del Mar, was also to prove a good collecting site for invertebrate fossils. F.M. Anderson in his *Upper Cretaceous of the Pacific Coast*[30] was to indicate that Sternberg's fossil trove was to include the following: *Gyrodes conradiana, Tessarolax incrustata, Turritella peninsularis, Dentalium (Entalis) whiteavesi, Spondylus* cf. *S. rugosus, Gryphaea* sp., *Ostrea* sp., *Parallelodon brewerianus, Parallelodon vancouverensis, Crassatella tuscana,* and *Inoceramus pacificus.*

Charles was to mention other sites in the Baja, but they are little known comparatively and cannot be determined precisely. He mentioned working across the Bahia de Todos Santos from Ensenada at a place called St. Quintana (which is not on my maps) and that he collected Cretaceous material there, including ammonites. Perhaps he has confused this town with San Quintin which is a sizeable town south of Ensenada, as he was known to collect there as well. About five miles from this site he found a location that sported a wide variety of baculites. In a letter to G. Dallas Hanna he remarked upon the finding of forty different species from the Cretaceous on this trip, with most of the fossils being fossil shells. He did, however, also accumulate a large representative collection of the aberrant Cretaceous rudistoid (pelecypod) *Coralliochama orcutti* White, that is striking because of its unusually elongated shell.

On a second trip to the same area in July 1925 Charles noted collecting fifty various species, and mentioned that Orcutt, who had described the *Coralliochama* from this site initially, had erred in indicating that this was the lone species to be found at the site. Charles found a variety of small specimens of other species. It was clearly another case where he found fossils where others couldn't. Charles was to note that in his opinion the only difference between this particular site and the beds of Rose Canyon (Eocene) were that the Baja site lacked *Amauropsis alveatus* and *Cardium cooperi*.

According to Hanna, F.M. Anderson was particularly impressed with Charles' Cretaceous collection from Lower California because of its "unusual character, and its obvious value for a popular exhibit,"[31] and as such the academy wanted to purchase it for their work. They did, however, have two main regrets: first, that there were so few species actually represented in the collection, and secondly, that the collection lacked solid geologic data relating to the stratigraphic occurrence of the fossils. This is perhaps indicative of the fact that Charles, while an excellent collector and an erstwhile student of terrain, still fell short in the area of geologic determination. Because of this failing, however, Charles was to recommend that the academy fund a month's trip there to better develop these deficiencies, but alas, McKittrick was to take precedence and the work had to be taken up by others.

NOTES

1. Henry Fairfield Osborn, *Cope: Master Naturalist*, Princeton Univ Press, Princeton, NJ, 1931, Pg. 263

2. SDNHM Ltr, Aug 10, 1920 (To Dr. Harry M. Wegeforth from San Diego), Materials Courtesy of San Diego Nat. Hist. Mus. Archives

3. SDNHM Ltr, Sept 11, 1920 (To J.W. Sefton, Jr. from W.O. Bourne, from San Diego), Materials Courtesy of San Diego Nat. Hist. Mus. Archives

4. Personal Communication

5. Frederick B. Schram, "People and Rocks; Geologists at the Museum," *Environment Southwest,* San Diego Nat His Mus, Autumn 1984, #587, Pg. 16

6. San Diego Museum Bulletin, Nov 1934

7. Paper Unknown, Partial Date Readable; July 6, 19--

8. *San Diego Union,* Jan 17, 1937

9. *The Natural History Museum Bulletin,* #130, Apr 1938, San Diego Nat His Mus

10. USNM Ltr, May 6, 1935 (To C.W. Gilmore from George F. Sternberg, from Hays, KS)

11. Senckenberg Museum Ltr, Mar 8, 1922

12. SDNHM Ltr, Oct 15, 1923 (To Fred Baker at Point Loma, CA from Clinton Abbott), Materials Courtesy of San Diego Nat. Hist. Mus. Archives

13. SDNHM Ltr, Nov 11, 1922 (To J.W. Sefton, Jr., V/P San Diego Savings Bank from Clinton Abbott), Materials Courtesy of San Diego Nat. Hist. Mus. Archives

14. *San Diego Union,* Aug 12, 1939

15. *Kansas State Teacher's College Leader,* Feb 2, 1939

16. n.a., "Geologic Maps of California 1839-1989," *California Geology,* Sacramento, Jan 1989, Pg. 21

17. Olaf P. Jenkins, *Early Days: Memoirs,* Ballena Press, Ramona, CA, 1975, Pg. 57

18. Ralph Arnold and Robert Anderson, *Geology and Oil Resources of the Coalinga District, California,* United States Geological Survey, Bull #398, 1910, Pg. 79

19. CAS Ltr, Oct 5, 1925 (To G. Dallas Hanna from McKittrick), With Permission of California Academy of Sciences

20. CAS Ltr, Sept 3, 1925 (To Frank A. Garbutt of Midway Royal Oil Co., Los Angeles, from G. Dallas Hanna), With Permission of California Academy of Sciences

21. CAS Ltr, Sept 13, 1925 (To G. Dallas Hanna from McKittrick), With Permission of California Academy of Sciences

22. Chester Stock, "Recent Excavations in California," Part II, *Carnegie Institution of Washington News Bulletin*, Vol IV, #32, 1938, Pgs. 262-263

23. Ibid, Pg. 263

24. Chester Stock, "Tanupolama; A New Genus of Llama From the Pleistocene of California," *Carnegie Institution of Wash Pub* #393, 1928, Pgs. 29-37

25. Chester Stock, "A Peccary From the McKittrick Pleistocene, California," *Carnegie Institution of Wash Pub* #393, Washington, Sept 1928, Pgs. 25-27

26. Charles H. Sternberg, "Extinct Animals of California," *Scientific American*, Sept 1928, Pg. 225

27. CAS Ltr, Sept 13, 1925 (To G. Dallas Hanna from McKittrick), With Permission of California Academy of Sciences

28. CAS Ltr, Dec 3, 1925 (To G. Dallas Hanna from McKittrick), With Permission of California Academy of Sciences

29. Max Miller, *Land Where Time Stands Still*, Dodd, Mead & Co., NY, 1943

30. F.M. Anderson, *Upper Cretaceous of the Pacific Coast*, Geo Soc of Amer, Memoir #71, NY, 1958, Pgs. 64-65

31. CAS Ltr, Jul 18, 1928 (To Charles H. Sternberg at San Diego from G. Dallas Hanna), With Permission of California Academy of Sciences

CHAPTER NINETEEN
STERNBERGS IN THE SAN JUAN BASIN

In May 1921 Charles moved from his San Diego residence to Los Angeles, and then after briefly settling in he headed towards the Cretaceous beds of San Juan County, New Mexico. This was a substantial trip for Charles since he was now a resident of California, his destination was some 800 miles away, and because he was now a mature 71 years of age. He did, however, have the advantage of traveling there in his one-ton Ford truck. The truck was driven by John Bender, who was apparently just one of a number of temporary contractual drivers that Charles was to employ over the years. Bender seems not to have been anything other than a driver, and campmate, as Charles does not record him in the customary manner as participating at all in collecting activities.

Charles had anticipated this trip for sometime, and had high hopes for a success that would place him among the first major contributors to paleontology in that region. Barnum Brown had been there for a shortened trip a few years prior, and C. Max Bauer, J.B. Reeside and F.R. Clark of the U.S. Geological Survey had made a fairly representative collection of turtles and some dinosaur material more recently, but for the most part the territory was largely still unknown. The site had been suggested to him by Dr. Edgar L. Hewett of the Museum of Man in Balboa Park.

The trip into New Mexico brought Charles into contact with the Navaho Indians, but these people were very different from the Indians he had encountered in early Kansas and Oregon. The Navaho were a friendly tribe, and obviously the 40-plus years that had passed had done much to mollify the Indians, and so they were to prove a real enigma to Charles. He was quick to label them as a "self-respecting race of shepherds and gardeners"[1] and it is obvious from his writing that he liked the people.

The expedition of two traveled from Los Angeles through the San Bernardino hills to Needles, crossed the Colorado River by ferry to Oatman, and into Kingman, Arizona. Charles indicated in his autobiography that truck trouble prompted the necessity of repairs outside of Williams, west of Flagstaff. In staying true to his nature, Charles was to send Bender on to Williams to take care of the mechanical difficulties while he pitched camp near the road, and went into the rocks in search of fossils. His efforts were basically unrewarded, but he was able to ascertain that the rocks were of the Carboniferous Period, since they contained large quantities of *Productus* and other period shells.

Traveling on through Flagstaff and heading east they came to Winslow and then Holbrook, where nearby he recounts having taken time out to visit the famous Petrified Forest of Arizona. He was to note his displeasure again over the "vandal hand of man" which was so evident in its destruction of these natural wonders. Among his remarks was a suggestion that a custodian should be hired to oversee this wonder of the world, as he felt that would help to stop the wanton destruction of these fine examples of prehistoric flora. He also desired for the major museums of the country to each take a prime example of one of these petrified giants, so that weathering by the elements and further destruction by man would stop their obliteration. Not bad ideas.

After a series of small mishaps they finally made camp in the area of Butler's Pass near the eastern border of Arizona and Charles comments further upon the obvious destruction done by man. He indicated finding a vast quantity of ancient pottery sherds, and also that he found where "an ancient graveyard had been plowed through in grading a road, and human bones and potsherds of this beautiful pottery had been thrown outside of the road by the grader."[2]

He was further frustrated while at Butler's Pass because word came to him that his wife Anna was quite ill. After waiting there three days for further word he finally received a note from her indicating she would have to undergo an operation, and so he set forth immediately by train back to Glendale to be with her. Anna had her operation on the 22nd and was pronounced healthy enough for Charles to return to his work about three weeks later. So on June 14th he headed back to New Mexico and the waiting John Bender. In his autobiography he acknowledged that George's first wife, Mabel, was to come to their residence and help to care for Anna, and that they both were pleased at the comfort she provided.

Upon Charles' arrival at Thoreau they set out towards Pueblo Bonito to the north, and made camp finally at Chaco River Canyon. This whole area in the northwestern corner of New Mexico is a fossil hunter's delight, and Charles was to take full advantage of his time there. Pueblo Bonito itself was actually an archaeological site being worked by the United States National Museum, under the leadership of Neil M. Judd. The site, excavation of a great pueblo of an ancient people, was of interest to Charles and he seems to have taken time out to discuss the excavation with Dr. Judd.

Historically Pueblo Bonito was important in a number of ways. Located in Chaco Canyon, it was clearly a substantial undertaking for those who built it, and for those who would try to unearth its secrets. The ruins cover an area of about three acres, and it has been likened to a gigantic apartment building. The ruins show at least 800 different rooms, or apartments, as part of the complex (which in some cases is multi-storied), and it is said to have housed over 1500 individuals. The complex was apparently abandoned about 900 years ago, for reasons unknown. Its ancient age and immense size make it of significant worth among ancient pueblo sites.[3]

Sternberg indicates that shortly after his arrival he was to find a wall of sandstone nearby that contained evidence of a sea-floor, for there in the wall were the impressions of an early seaweed called *Halemenites major.* At the top of the sandwall he was to find abundant evidence of a type of *Inoceramus* that he claims had apparently never been worked. As he describes them as being up to eight inches in length it is certainly clear that they were not of the same type as their large Kansas relatives, as they measured in feet and not inches. This was an important discovery for Charles since it spurred him on to early success in the area, as well as giving him an idea of just exactly what he could expect in terms of its fossil treasures. It was not, afterall, the first time that Charles had been in a Cretaceous environment!

At one of the nearby residences Charles was to hire a young Indian boy named Dan Padilla, along with his team of ponies, and then later another Navaho youngster by the name of Ned Shouver. It is evident from Charles' writing that he was impressed with these young Indian fellows, and with their hard work, even though at times there was a problem in communication. He did mention, however, that he did the cooking himself "as a matter of self-protection."[4] One can assume from this that their abilities in this area were limited, but he did make sure they had the honor of taking care of the daily dishes.

In this first year of what was to prove to be a three-year commitment Charles was to end up spending a full five months in the badlands. The look of the terrain was again a very familiar kind of land for Charles and he seems to have been very comfortable there, even at his advanced age. He was to describe these badlands as being somewhat typical, and he compared them at one point to the Red Deer area of Canada. He was to comment on their stark beauty as well, emphasizing the marvelous fluted columns and the sculpted statuary of the outlands.

With his two young helpers he began to canvass the countryside. Since Bender was not mentioned further (in any of a number of different narratives) it appears he must have gone back to his home, or moved on to other business. In either case, Charles' main help was certainly these two novice collectors who had never had any previous training in fossil collecting. He did note that both boys became proficient fossil hunters, and that Padilla was to find the best specimen of the season; a fossil turtle called *Baena* while young Ned was to find three of the shells of the turtle *Adocus*. Since Charles was to add the postscript "as far as finding material was concerned"[5] it seems clear that their abilities did not translate to fossil collection or preparation as well.

Before they were through they were to work through what would prove to be the best sites in the badlands: Ojo Alamo, the Kimbeto (Kimbetoh), Escavada and Hunter's Wash, Meyer's Creek and the Kirtland shales. These are the same areas that have continued in fossil prominence in New Mexico to this day.

It was thanks to the suggestion of Dr. Reeside that Charles was to have a really successful trip that first year, for it was he who was to steer Charles towards the site around Ojo Alamo where Charles would find a great number of the many different fossil turtles that became such a hit with Dr. Gilmore of the U.S. National Museum. This additionally spurred Gilmore to collect in the area himself in a later season.

The main finds in that first season were isolated dinosaur material, various fossil turtles, a fossil crocodile skull, and two substantial finds of Ceratopsian skulls. It was the two huge skulls that were found that made the expedition in Charles' eyes, and for good reason. The finding of the best of these skulls was interesting because it was really so casual and in a sense anticlimactic.

In the early part of his second season's searching frenzy, where Charles would just canvass an area for material before actually doing any collecting, he would locate the material only, and then mark it for further examination by placing a small hillock of stones nearby so that it wouldn't be lost to sight again. He notes that he "discovered only the small edge of the *Pentaceratops* skull projecting from the face of the clay bank, and did not think it was worth while going back to investigate further."[6] In other words, he didn't even bother to mark the find because he thought it too limited to bother with at all. In the meantime he was to discover what proved to be another Ceratopsian which was related but different and hence his time was taken up with that specimen. For whatever reason he decided later to go back and look again at the first find, even though it had been found in the first week of searching, and he was surprised to find upon more careful investigation that it was very complete, and it proved to be the largest catch of his second season.

Charles, in an unpublished fragment found among his things at the San Diego Museum of Natural History was to describe the first dinosaur find as follows:

> We drove to Kimbeto [Kimbetoh] east of Pueblo Bonito about 10 miles. Here was a trading post and across the Kimbeto wash, Mr. Tyler cared for a herd of thoroughbred sheep, owned by the government. In its efforts to induce the Navaho's to improve their herds by buying fine bucks. After exploring Escavada wash, and the head of Meyer's Creek, I was about to go to new fields when I remembered some bones I had noticed near Tyler's ranch. So we went to work to uncover them, and found to my delight a new horned dinosaur skull...This skull was 7 and a half feet long. It is the first horned dinosaur found west of the Rocky Mountains.[7]

This find was to be eventually sent to the Upsala University in Sweden where it was to be named by Dr. Wiman as *Pentaceratops fenestratus*. The name of this Ceratopsian was apparently prompted by the fact that it differed from other Ceratopsians in having a single fenestra (aperture or hole) in the squamosal bones. The find, however, must have been sold or received after the second find (in the following year) in order for Professors Osborn and Matthew to have named the new dinosaur *Pentaceratops*. In addition to this valuable find Charles was to discover and sell to Dr. Wiman the major portion of a duck-billed dinosaur skeleton named by Wiman as *Parasaurolophus tubicans*.

In the following year Charles was to uncover the second Ceratopsian. The significance of these two specimens cannot be understated, for they were to establish a new species. Professor Osborn of the American Museum was to describe the new form in October of 1923, giving it the name *Pentaceratops sternbergii* in honor of its finder. The name itself had apparently been suggested by William Diller Matthew due to its appearance of having five horns. This has since been proven to be incorrect in that the animal is actually only three-horned, and the two "lateral horns projecting downwards and backwards below the orbits"[8] as described by Osborn have been found to be false horns, or cheek bones (found in other Ceratopsians as well although not normally as prominent) that appears to have offered some measure of protection to the animal's neck from the bite of other predators.

Although such is not recorded by Sternberg, Osborn notes that the site itself was located nine miles northeast of Tsaya, New Mexico, and he indicates that he was to send the paleontologist Charles C. Mook, and Peter Kaisen who had been Brown's assistant in the Red Deer, to assist Charles. One wonders whether the small feud between Kaisen and Sternberg that had been initiated in Canada had been settled sufficiently by this point in order to have made this a pleasant experience for the party. How much help they provided remains to be seen since Charles had managed to excavate and take up the much larger and complete skeleton of the *Pentaceratops fenestratus* with only the help of his Navaho Indian boys and one of their fathers. Perhaps Mook and Kaisen provided the most help in surveying the site and in attempting to age the deposit.

Pentaceratops skeletons are still known only from the San Juan Basin, and as this animal has not been discovered elsewhere it makes one wonder how specialized it was. Osborn described the deposit age in which Charles' *P. Sternbergii* find was made as being "intermediate between the Judith and Belly River and the Lance division of the Upper Cretaceous." The formation is labeled the "Fruitland." The specimen was given AMNH #6325, and has been described as having an elongated frill and skull, as having a moderately sculptured crest, and it has the large frill openings characteristic of the highly adorned *Chasmosaurus*.

In 1922 Charles was to make his first real strike of fossil turtles in the San Juan. This thrilled him since anyone reading his works quickly senses the fondness he had for fossil turtles. Charles, in his inimitable way, was to describe the environment that existed when the turtles were alive as being "warm moist country covered with luxurious jungles of vegetation." The flora he indicated was "covered with palms and magnolias, figs dropped their ripe fruit, roses bloomed while evergreens, including the redwood, graced the forest and the gentle winds made music among their rank foliage."[9]

In 1923 son Levi was to take a year's leave of absence from the Royal Ontario Museum and was to go to work with Charles in the San Juan Basin. During that season Levi was to collect the second known specimen of *Parasaurolophus tubicans* for the Toronto University.

After this date the San Juan remained untouched by Sternberg hands until 1929 when George F. Sternberg was to travel there on an expedition with the National Museum. The party, which consisted of Charles W. Gilmore and Norman H. Boss of the museum, and George and his son Charles W. Sternberg, met at Kimbetoh in May by prearrangement. Gilmore notes that the finds of the expedition were generally unsatisfactory due to the lack of articulated specimens found in the region, which he seemed to attribute to the specimens being scattered well before burial, most likely as a result of flooding and predators.

Gilmore explained that they found a large complete fossil turtle, *Neurankylus*, and isolated dinosaur bones at the first site after two weeks of work. The limited success therefore caused them to move their base to Ojo Alamo to the north, where they met with similar results so they moved on to the Hunter's Store area, in the Kirtland. Here they met with some success finally. They found additional dinosaur material consisting of an articulated caudal vertebrae series from a hadrosaurian and a portion of the skull of a horned dinosaur similar to *Chasmosaurus*. They also found an abundance of turtle specimens including *Baena, Adocus, Plastomenus, Aspiderites, Amyda* and the same *Neurankylus* as at the other site. Gilmore, although dissatisfied with the overall finds, was pleased with the information obtained about these specimens and felt that their value was still well worth the time put in.[10]

Interestingly enough, George was to find what was described as a "nearly complete right squamosal" of a *Pentaceratops sternbergii* (USNM #12002) in the Kirtland formation, whereas Charles' skull find came from the Fruitland.[11]

It appears, therefore, that the first three representatives ever found of *Pentaceratops* were found by Sternbergs, even though they were years apart. Obviously the San Juan Basin was kind to the Sternberg family, and their mark will indelibly be a part of the paleontological history of that region.

NOTES

1. Charles H. Sternberg, *Hunting Dinosaurs in the Bad Lands of the Red Deer River, Alberta Canada*, NeWest Press, Edmonton, 1985, Pg. 188

2. CHS, *Red* (NeWest), Pg. 187

3. Frank C. Hibben, *Digging Up America*, Hill & Wang, NY, 1960, Pgs 156-157

4. CHS, *Red* (NeWest), Pg. 200

5. Ibid, Pg. 201

6. Ibid, Pg. 201

7. SDNHM Fragment (Trip Recap), n.d., Materials Courtesy of San Diego Nat. Hist. Mus. Archives

8. Dr. David Norman, *The Illustrated Encyclopedia of Dinosaurs*, Crescent Books, NY, 1985, Pg. 143

9. SDNHM Fragment (Trip Recap), n.d., Materials Courtesy of San Diego Nat. Hist. Mus. Archives

10. Charles W. Gilmore, "Fossil Hunting in New Mexico," *Explorations and Field-Work of the Smithsonian Institution in 1929,* Washington, 1930, Pgs. 17-22

11. Ibid, Pg. 163

CHAPTER TWENTY
CHRISTIANITY

To say that Charles Sternberg was a vibrant and alive Christian could hardly be understating the case. He was in Christian parlance a man with missionary zeal; a prime example of what man can accomplish in a lifetime, in the name of God and in thought of other men, rather than in consideration of personal gain.

Sternberg's Christianity presumably began easily for him, for it didn't have to be searched out in the same sense that it would be for a man brought up in a household alien to God, and Christian principles. Since his father and his grandfather on his mother's side were both clergy, and both teachers of theology, it is clear that he would have received biblical instruction at an early age. This teaching would have been enhanced by the fact that his schooling at Hartwick had a curriculum geared towards biblical principles as well.

If there were a measurable litmus test for the authenticity of one's faith it would probably attempt to measure how an individual would react, when they were placed in a position where openly displaying their faith would cause them some real or imagined personal discomfort. Many Christians when faced with such choices take the easy route and simply remain mute in the face of defiance. Charles was certainly not this kind of Christian, but was vocal and forceful about his beliefs. We find that he even went so far as to make special effort to attend church whenever reasonably possible while out in the field. His trips into Seymour, Texas, were good examples of this.

Even in advanced age Charles continued to be stout-hearted in things biblical. At age 87 he was still teaching an adult Bible class at the First Lutheran Church in San Diego each Sunday.

Judging from the biblical thoughts and paraphrasings of biblical passages that pervade his writings, it seems eminently clear that Charles had a solid working knowledge of the Bible. And he went beyond simply knowing it, but put into practice his beliefs. Since this is what separates the pedagogue from the true-charactered Christian we can safely say that Charles Sternberg was a practicing Christian, and would have been considered so by most of the fellowships of the major Christian denominations of his day. Those who chanced to meet him on the byways of Lawrence, San Diego, or the Permian outback would have fast become aware of his love of God. His life simply exuded Christianity.

While there have been those who have labeled Sternberg as a religious fanatic, principally due to his strong stance on "religious" issues, these have been a decided minority. For all his fervor most people could still tell that his faith was genuine and that he had a clear concern for others too. While his views were most certainly strong, and his exposition of them equally hardy, he never yet gave one the impression that he was intolerant or uncaring. Nor did he try to foist his convictions upon others when they made it clear their interest lay elsewhere. He seemed to have developed a firm balance between wanting to share his faith with others, and yet respecting their right to reject his religious convictions. There can't be enough emphasis placed here, as it is here that the fanatic and the zealot part company. Charles was clearly a caring zealot.

As with any other man, however, Sternberg was not perfect and was not without his detractors. Al Romer, the famous Permian paleontologist, was said to have referred to Charles as "that sanctimonious S.O.B."[1] Whether Romer took this view because of his personal dealings with Sternberg or for some other reason it is not clear. In my opinion it perhaps says more about Romer than it does about Sternberg.

Examples of his "practicing what he preached" are not difficult to find, and again they pervade his writings. His tolerance towards Othniel C. Marsh, and in a lesser degree of Edward Drinker Cope, in light of their actions is one notable example. His very obvious omissions of reference to Marsh, who did both Sternberg and Cope major wrongs, was certainly commendable, particularly in view of Charles' very vocal love for Cope. One might have expected a lashing of Marsh would have followed, but it didn't.

Other examples of his practiced-Christianity abound. One could cite his lack of personal jealousy over the successful careers of his fossil hunting sons, and the evident joy that he derived from watching them find fossils even greater than those he himself had found. How often do we find hurtful conflict between father and son when the father either smothers the son or the son overshadows the father? The times when the father-son relationships work well in these same kinds of situations are rare, but it worked for the Sternbergs.

One could also mention Charles' always giving the credit for found fossil to his great God, rather than taking upon himself the "glory" for these fabulous finds. This is a major hurdle for most Christians, but one that Sternberg was to overcome with apparent ease. The Frank Sinatra syndrome ("I'll do it my way") was very prevalent even in that day, but Sternberg could rise above that to the point of giving the credit to God.

The fact that Sternberg's book of poems, *A Story of the Past or The Romance of Science*, is an interblending of both his science and religion emphasizes the weight that each was to play in his total lifestyle. It therefore seems to be a wise place to look further for Sternberg's views on these two substantial subjects. He could not personally separate one from the other since both were of major importance to him. That he was able to take these two subjects, so often thought of as being diametrically opposed, and was able to blend them into a natural harmony makes it a matter of great interest. The science (or theory) of evolution, and the biblical perspective of God and His creation are not reconcilable in the minds of many, and it is that very fact that spurs interest here. If he had been a lacklustre Christian the matter would not be worth a second thought, but because he was indeed both consistently and vibrantly a conservative Christian and a scientist, it makes this a subject worth pursuing.

Historically this opposition between the "Church" and "Science" has not always been. In fact, there was a time when the clerics were the best and brightest of the scientific men, in part because they were among the best educated. There have been over the years some notable scientists who professed their Christianity quite freely, and seemed not to feel any need to sacrifice one for the other. But then again one must admit that this has been less evident since Darwin's theories began to take hold. One might mention perhaps Asa Gray, Louis Agassiz, Benjamin Silliman, Joseph LeConte, Thomas Condon, James Dwight Dana, Clarence King and Sir Isaac Newton as prominent examples, and there are certainly many others of note.

However, with the age of specialization came a change here too and the priesthood changed with it. With the Reformation and the resulting Protestant split came further necessity for ministers ministering to their flocks, rather than pursuing a scientific bent. (The Abbe Henri Breuil and Pierre Teilhard de Chardin perhaps being somewhat anomalies in this regard).

Thomas Condon, the great and substantially self-taught geologist of early Oregon, is characterized by his biographer Robert D. Clark as displaying the "determination to accept the findings of geology and Darwin, without giving up his religious beliefs" and who "saved the faith of many and gave both evolution and the methods of science a fair start in Oregon."[2] The question then arises, I suppose, is whether or not either his science or his religion had to be compromised in the process. As a Christian myself I would have to suggest the answer is "yes."

It should be interjected here that Condon and some of these other men were decidedly more liberal in their Christianity and therefore could perhaps more easily reconcile their brand of "Christianity" with prevailing scientific thought than could the more conservative Christians of Sternberg's character. Yet, in spite of the fact that there were and are prominent Christian men of science, there developed a division between the Church and science. To some it continues to be a source of vivid conflict. There is, however, some question as to how necessary or genuine that conflict is, since a great deal of the trouble seems to be mainly one of perspective.

Andrew D. White, in his epochal but highly biased work *A History of the Warfare of Science with Theology in Christendom*, states that he views the conflict as not being so much between science and religion as it is between what he calls "science and dogmatic theology,"[3] although he could just as easily have said "between dogmatic science and theology." His point, however, is well taken, for it has historically been the staunch and unyielding that has proven to breed the greater difficulty when compared to open and alterable dialogue based upon merit. Although this protective behavior goes well beyond merely guarding "truth" against assailment, which is wrong action in itself, it is not to say that right doesn't always remain right, as truth should never be compromised. It is in a sense, suggesting that one ought to stay away from the Pharisaical practice of letting the "letter of the law" be the law of the land versus the "spirit of the law."

Ronald B. Parker in his latest book, *The Tenth Muse; The Pursuit of Earth Science*, recognizes that a great part (but not all certainly) of the problem with science's internal structuring is the result of the closed mind. This equally applies to the religious in their attitudes. Parker describes the ostracization that occurs when a scientist takes a stand against the norm. Such stances, as we all know, can prove to be political suicide, such as can happen today when one uses language that is not judged to be politically correct. Parker's example comes from the science of crystallography and relates to the discovery of an icosahedronal crystal that bucked the established rules. Parker describes the dilemma well:

> As I mentioned previously, no exceptions to this fitting of nature into the concept of symmetry groups had ever been found. Perhaps it is more realistic to state that no exceptions had ever been reported in the scientific literature. A crystallographer who found

data that didn't fit the scheme would have just assumed that a mistake of some sort had been made. If the anomalous work was shown to a colleague or a superior, there would be strong objections and a questioning of the worker's competency. If an article describing the crystal that didn't fit the rules was submitted to a journal, it would surely be rejected. One can assume that an innocent and eager graduate student discovering a violation of the space-group dogma and telling a professor about it would suffer career termination. Almost all research scientists were once graduate students, so the pressure to make no waves is pervasive.[4]

Other examples are as easy to find, and in any of the related sciences. Loren Eisley, for instance, in speaking of the famous Foxhall jaw was to intimate that the jaw's modern appearance was to exclude it from active consideration as being of ancient man. He clearly defined part of the problem when he stated, "Only time will tell how many other ancient human relics have been discarded simply because they did not fit a preconceived evolutionary scheme."[5]

One cannot argue with the "experts" because they always have the benefit of citing this piece of evidence or that "with authority." Those who would argue against them are made to appear as blithering idiots even though their logical and less-biased thinking may have more validity. So it is that the man with the naturalist's eye, who looks with depth at the flora and fauna around him, often sees more than the jaundiced eye of the scientist.

This is similar to the meteorologist, who works hard at trying to predict when it will rain, only to be outwitted by the layman who without benefit of all the trappings of the meteorologist walks outside and predicts correctly that it will rain by looking up at the clouds overhead. God gives men observational powers and he wants him to use them.

It is clear that one must conform, however, in the scientific world as well as in the religious world. It is only the ablest and strongest who are able to surmount the difficulties that closed structures offer, and who can successfully buck the establishment and get it to change. John McPhee in the latest book of his geologic trilogy, *Rising From the Plains*, sums up the abrasion well in a story related by his central "character," David Love, a present-day supervisor with the Laramie branch of the United States Geological Survey:

> We passed St. Matthew's Episcopal Cathedral, which also--as Love has reason to regret--contained in its walls brachiopods, crinoids, and algal buttons. He once taught Sunday school there. He took the kids outside and showed them the fossils in the church walls. He described the environment in which the creatures had lived. He mentioned the age of the rock. He explained how things evolve and the fit prosper. Here endeth his career in sedimentary theology.[6]

While the churchman's attitude was somewhat intolerant perhaps in this instance, it is also certainly clear that Love lacked tact. One just does not suggest in a church setting that evolution is the law of the land, and one must additionally wonder how it ever came that Love was teaching a Sunday school class when he was not in agreement with the teachings of that church. If men of sound and sane thought would view and study the issues more objectively they just might find that each has grossly misunderstood much of what the other is saying and some reconciliation might actually occur. This is not to say, however, that outright error should be tolerated under any circumstances. Sternberg was one of those rare individuals who seemed to be able to transcend this breech and who found reconcilement in his understanding of the creative process.

Since the age of the earth is often considered a focal point for determining where one stands in relation to religion and science (although this is perhaps an unfair assessment) it is helpful to view Martin J.S. Rudwick's intriguing observation about the state of things in the seventeenth century relating to this subject:

> In assessing the seventeenth century belief that the Earth was only a few thousand years old, historians of science have perhaps been less ready to grant to this limited view of time the historical sympathy and understanding that is now accorded to the limited space of Ptolemaic--and even Copernican--cosmology. Yet the cases are closely parallel...It was not simply the force of mental conservatism, still less the threat of ecclesiastical sanctions, that made almost all the thinking men agree that the Earth was of quite

recent origin--though indeed very ancient by the standards of human life. Even Thomas Hobbes, for example, who was certainly heterodox in many of his opinions and perhaps even an atheist, could take this particular belief for granted as easily as any more orthodox scholar.[7]

The real schism that doomed church and science to conflict is generally traced back to the denunciation and imprisonment of Galileo by the Roman Catholic Church for so-called heresies. This was because his view of the heavens was contrary to their dogma. This unfortunate incident was perhaps inevitable, since it was a highly visible case that pitted human will and doctrine against doctrine. Was the Church really any more malicious or hurtful in protecting itself than we would find with our modern governments or organizations? I think not. (Although I would suggest the Church certainly erred in its handling of matters).

We clearly have a modern parallel in the National Rifle Association's espousal of their right to bear arms, even when those arms are semi-automatic. While semi-automatic weapons are clearly outside the purview of the sporting rifle class, and are not for hunting game, yet the NRA feels they must be protectful or lose what rights they have in other areas. In the case of Galileo, however, perhaps because it was the volatility of religion that was involved, the matter quickly flared up, sparks ignited elsewhere and arsonists and fire-fighters have been busy ever since. When Darwin came along he unwittingly added the oil and things really went up in a raging, obscuring inferno. Only recently have we seen a limited abridgement of hostilities in some quarters, although to call the battle lost or won seems inappropriate when the combatants remain so adamant and entrenched generally.

Sternberg's evolutionary viewpoint would have been born in early Bible teaching, and nurtured by his father's strong beliefs. It would have been further honed by the substantial friendships and discussions with men such as Benjamin Mudge, Samuel Williston and most particularly Edward Drinker Cope. For better, or for worse, they were to help shape Sternberg's views on this, perhaps the most controversial subject for a Christian and a scientist.

Son Charles Mortram Sternberg was to recount that his father was to write an article (or address) at the age of nine while a student at Watertown. This article "proved" that the biblical concept of a universal flood was correct, since Charles had found near Cooperstown rock containing marine shells. Charles obviously was interested in fossils, evolutionary theory and the biblical flood at a very early age. Regrettably this paper apparently no longer exists.

The geological knowledge of Charles' father Levi was well beyond that of the average man of his day. This would appear evident to anyone reading an article he published in 1868 in a Lutheran publication, the *Evangelical Quarterly Review*, entitled *Geology and Moses*. In reading this article it becomes evident that Levi had a sound understanding of what the prominent geologists of the day were saying. He also was obviously well versed in the fossil sites around the areas where he had lived and elsewhere because he illustrates his arguments by making reference to them. Clearly he was an educated man who was aware of his surroundings, and with such being the case he most certainly was a major influence on his son.

Levi Sternberg, however, strangely seemed to buy into a form of uniformitarian doctrine from what one can derive from his article, noticeably so, as he portrayed his views of geology as a science:

> Its leading principles, however, are not unsupported by hypotheses, but are established upon the solid basis of rigid induction. Carefully observed facts in nature, are the materials of its reasonings; analogy is its process. Causes now in operation are assumed to have produced similar effects in the pre-historic period. The dynamics of the past, though often, perhaps, more intense in their operation, were of the same nature as those of the present. The same general laws of nature have been in force since the heavens were created.[8]

It is equally clear that Levi Sternberg had a firm commitment to a traditional view of creation and did not feel that there was validity to the evolutionary doctrines. This is clearly evident as Charles was to note in an article of 1903 in the *American Inventor*:

> So carried away was I [Charles] with this idea of progress, my venerated father did all in his power to dissuade me from my purpose. He told me that it was impracticable; if I were wealthy I might find recreation and pleasure in the work, but I could not possibly earn a living by it. Besides I might lose my birthright, and be drifted away on the hopeless seas of infidelity, after the study of 'Natural Selection' and 'The Survival of the Fittest.'[9]

Sternberg plainly did not take his father's advice as it related to entering the fossil profession, managed to make a fair living at the vocation in most years, and was certainly successful by earthly standards. What is not so clear, however, is how much effect the doctrine of evolution had on him specifically.

In 1885 Charles was to write an article for the *Kansas City Review of Science and Industry* entitled "*The Flora of the Dakota Group.*" In this relatively short piece we get one of his longest comments on his view of the evolutionary theory, albeit an early one:

> This Dakota flora is a wonderful disproof of the Theory of Natural Selection. Here we find at the base of the Cretaceous, millions upon millions of years old, a flora as perfect as any of the present day. There has been no improvement during all these ages. Some of the species are called new, more on account of the position in which they are found, than from any dissimilarity between them and those of recent species.
>
> What has nature been doing in the vegetable kingdom during these countless centuries? Is the line of development confined only to animals? And more wonderful still, these perfect plants appear for the first time in the earth's history. Like the hero of old that came into the world full-grown and ready armed, so the grand flora of the Dakota appeared with no intermediate species between it and the coal plants of the carboniferous. Let the scientists of the Darwinian school explain this fact and make it conform to the theory of Evolution, if they can.[10]

There is no doubt, judging from this statement, that Charles had no firm belief in the evolutionary process at this early date, and it appears that he was strongly against it. It can positively be said that he felt natural selection was not the causative agent for the evolutionary process, if he believed in evolution to any extent at all. Charles' views are less pronounced in later writings, and we are therefore able to assess little about any changes to his views as he got older.

And yet Sternberg, it appears, was to come to believe in some form of evolution (or perhaps adaptation is a better word) if one reads all of his writings. The components of that evolutionary viewpoint are not quite so easily defined, but we will try to do so. Let's begin in his autobiographical writings.

In *The Life of a Fossil Hunter* Charles lightly hints at a pro-evolutionary viewpoint, or certainly a non-orthodox Christian viewpoint for that age, when he says..."but the glorious picture is only for him who gathers the remains of these forests, and by the power of his imagination puts life into them: for it is some five million years, according to the great Dana of my childhood days..."[11] Since time is so central to the evolutionary concept, and because it was more common to find the Christian backing a 5,000 year old earth, it should not seem unfair to label his position in this respect as non-orthodox.

Then Charles goes one step further to state that he decided to make collecting fossils his life's work, "...I would make it my business to collect facts from the crust of the earth; that thus men might learn more of 'the introduction and succession of life on our earth.'"[12]

The word "succession" is the key word here, and the obvious implication is that life as Charles saw it was perhaps not introduced in the literal six day format so often espoused by some Creationist groups, but was a series of creative efforts or periods. One should be careful here, however, to state that this view does not in itself violate a literal reading of the Bible as is often argued, and is therefore an acceptable position for a conservative Christian to take. While the original language may suggest literal days, there are competent and conservative theologians who have argued the point effectively that "periods" is a justifiable translation.[13]

These comments then are enough to lead one to believe that Sternberg felt the earth was substantially older than the 5,000 years suggested by the oft-maligned Bishop Ussher and his adherents, and that life was some sort of successionary process (whatever that might mean). This in itself, however, does not prove that Charles was a standard or ardent evolutionist, nor even that he was an evolutionist, but it does at least allow for the sufficiency of time for the evolutionary process to be considered as an option in his thinking.

Sternberg is much more exact in another passage where he describes the finding of a fossil camel bone in the John Day of Oregon. This appears to leave little doubt that he believed in some sort of "evolutionary" process:

> ...I was constantly looking for a camel in the older beds, and I cannot express my delight when one day, as I was exploring the John Day beds, I came across a skeleton which had been weathered out and lay in bold relief on the face of a slope. I knew before I picked up the cannon-bone that my belief was verified, and when I took up the two bones separately, the fact was proved beyond a doubt that in this ancestor of the living form the metacarpals of the fore foot and the metatarsals of hind foot were respectively distinct...Professor Cope named it in my honor *Paratylopus sternbergi*...
>
> I arrived at this conclusion with regard to the cannon-bone of the ancient camel as Darwin, Marsh and Huxley arrived at the conclusion that the ancient horse had three toes. They recognized that the splint bones of the horse represented the side toes of rhinoceroses, one on each side of the middle metacarpals and metatarsals respectively, and they decided that they were the remnants of side toes in the ancestor of the horse. And later we found a three-toed horse.[14]

There can be little question, it would seem, that Sternberg is espousing a pro-evolutionary viewpoint to some degree, and that he was clearly in accord with Marsh's horse pedigree that is so often disputed by Christian authors. With this in mind it would perhaps be helpful to consider the related and equally controversial subject of the universal biblical flood as it might contribute to a better understanding of Sternberg's Christian views.

From all evidence Sternberg seems to have firmly believed in the traditional and global biblical flood, and yet some would consider the two views (i.e. the flood and evolution) as being totally incompatible. On one occasion, for instance,

Charles was to make an offhand remark as he was digging in his newly discovered "Sternberg" Quarry at Long Island in 1882. He was responding to the owner of the quarry land, a Mr. Overton, who had yelled to Charles inquiring as to exactly what he was doing. Charles yelled back that he was "digging up antediluvian relics."[15]

There are other writings where Sternberg alludes to the biblical flood, and others yet where it appears that his allusion to a flood is perhaps the same as the Noachic flood, but these references might be considered more speculative. Since the biblical flood is such a central issue for Christians and scientists it seems logical that one's evolutionary viewpoint is molded in great part by one's views about this event.

Charles Coulston Gillispie in his well-researched book *Genesis and Geology* denotes the importance of the biblical flood by explaining Lyell's predicament:

> Lyell was, of course, perfectly aware that the flood was his chief enemy, because to many minds the diluvial theory alone seemed capable of affording an explanation of natural phenomena in accordance with scriptural history. And being chary of disturbing religious convictions unduly, he impugned the deluge explicitly in only one passage, and that one rather in the nature of a digression.[16]

If we are to look for a good source from which to make some observations as to what Sternberg's evolutionary viewpoint might have been we would most certainly want to consult the writings of Edward Drinker Cope. Cope was to publish an integrated series of essays under the book title, *Origin of the Fittest*. This book, by its very title asks to be considered in relation to Darwin's revolutionary work *The Origin of Species*. It is an analogue of that, as well as Wallace's phrase "survival of the fittest." Cope was to publish this series in 1887, at the height of the debate about evolution, and only 28 years after Darwin's work was first published.

Cope begins in chapter one of his book by describing the scope of evolution as he saw it:

> It attempts nothing less than a history of the process of creation of the universe, so far as we can behold it; and is, therefore, an attempt to formulate the plans and thoughts of the Author of that universe. Hence, it is not surprising that it excites the interest of the best of men, especially as it is one of the results of the efforts of a class of these, crowning many centuries of labor and thought.[17]

That Cope was to place great emphasis on what he considered the effects of evolution is without question. The fact that the book's print is small and densely packed, and still runs to over 450 pages, shows how important Cope considered the subject. Since Cope was to have such a profound early effect upon Sternberg, and because they had occasion to share similar views on other matters we can only hope to learn something about Sternberg in the process.

Cope goes further in the first chapter to state that the hypothesis of evolution has been placed on the "basis of ascertained fact,"[18] and this due to what he considered the frequent observations of changes from type to type. Cope, at least, viewed evolution as established fact and not theory. However, it is also necessary to state here that there are evolutionists and there are evolutionists, and that the Thomas Henry Huxleys and the Joseph LeContes do not mix as well as water with water. Cope's view of the evolutionary process was not standard Darwinism by any means.

Cope's views were quite radical in their own right, both for a "Christian" and a scientist. He agreed apparently with Darwin that man had descended from an ape, or an ape-like creature,[19] but Cope could not divorce God from the internal workings of the creative process. Cope calls it the "hypothesis of development," where that key word "development" seems to be a synonym for the word evolution. In fact, one often finds the theory itself referred to as the "developmental theory."

Cope felt that his view of things was perfectly in keeping with his "advanced" Christianity, for he states:

> That they [following pages] will coincide with the spirit of the most advanced Christianity, I have no doubt; and that they will

add an appeal through reason to that direct influence of the Divine Spirit which should control the motives of human action, seems an unavoidable conclusion.[20]

Cope's advanced Christianity might be better described as a liberal Christianity, or even more likely as not being Christian at all. Clearly anyone acquainted with Christianity recognizes that it is based on long established principles which are held to be inviolate, and unchanging. Since the Scriptures are held to be sacred they are also held to be beyond "advancement," as Cope would call it.

With what we have already read it is not hard to believe that whatever Charles' beliefs were in this area they certainly were not taken from Cope's beliefs because Cope was simply too far left in his views. While Cope had obvious effect on Charles directly in other areas, they too had differences, and this appears to be one important area where they were likely at odds with one another.

Joseph LeConte, a scientist contemporary with Sternberg and an avowed Christian, is perhaps another source to consider, since due to his prominence, it is likely that Sternberg would have been aware of his writings on evolution and Christianity. The only difficulty in approaching LeConte is in knowing at what point he was writing, for he began as a liberal Christian with a conservative position on evolution before becoming an avowed evolutionist and an even more liberal Christian. In an inaugural address before the state house in Columbia, South Carolina, LeConte was to acknowledge that the perceived "collision" between religion and science was in reality "only apparent, and the evil therefore transitory, while the service is real and lasting."[21] He foresaw a time when geology would be seen as the "hand-maid of religion among the sciences." In all reality it seems we can safely say that to date LeConte has not yet proved to be a prophet.

Without hard factual evidence it is hard to say that Sternberg definitely shared either Cope's or LeConte's views about evolution (in any sense of the word). It should be said as well that his Christianity was considerably more orthodox than was Copes or LeContes, as Cope was to drift off in the late part of his life into a Unitarian view that would not today be considered Christian by most, and LeConte was to gradually become more liberal in his Christian outlook. This very fact might lead us to believe that Sternberg would not have held a position similar to theirs. Until further proof should come to light speculation as to how Sternberg was directly affected by the views of others might prove of little value, although it still seems to be a necessary exercize.

A recent book by Roger Lewin, entitled *Bones of Contention* takes a good, long look at the conflict between religion and science, albeit indirectly. Lewin's thesis is that each scientist is the product of a number of biases, and that no matter how meticulous he feels he has been he still is provably biased to the point that his science is directly (and often measurably) affected. Lewin cites a score of examples where well-meaning scientists have made every attempt to be open-minded and yet we find their works unconsciously tainted with their preconceptions. One of the longest ranging of his examples is the Richard Leakey KBS tuff controversy perpetuated by his adherents and detractors.[22]

In this instance he indicates that the actions of the principals, and often their ending data, was clearly tainted by their desire for the story to end a certain way. Lewin did not, and I certainly don't either, intend to castigate these fine scientists for their oversights since we all do the same thing in one form or another. I have had the same problem in writing this chapter as I have often found myself as an evangelical Christian wanting to put certain words into Sternberg's mouth that simply weren't there. The best advice I received from anyone on this matter was from my former pastor, David Brooks, who said if you're writing a biography you should tell what you know about Sternberg's beliefs in this area, even when you may not believe in them yourself.

Samuel W. Williston, a friend of Sternbergs and a fellow scientist, also held evolutionary views that were somewhat out of the mainstream, although he never professed to be a Christian. As a young scientist he did not believe in evolution, but then later was won over to that point of view. In an article in the *New York Times* he was to respond to an interviewer's question as to whether or not he was a believer in "Darwin's evolution doctrine:"

> Yes; still, I confess there are some very great scientists, who, though now apparently in the minority, antagonize that doctrine with many very cogent reasons. There was the late Agassiz, whom I regarded as even above Humboldt, and second only, perhaps to Cuvier, as a naturalist. He had a very fine, well-studied theory of his own, and even shortly before his death, while admitting that the evolutionary doctrine was gaining ground under such leaders as Huxley, Tyndall, Darwin, and others, he still clung to his own matured views, and his son, who is following his footpath as a naturalist, only in a limited and more specific field, clings to his father's views.[23]

While his friend Williston does not mention him by name here (undoubtedly because Sternberg did not have the credentials of the others noted) it is certain that Charles would have fallen into the general camp of those described. Clearly too Williston had a respect for these scientists even though he did not agree with them on this issue. It is a shame that scientists today can't seem to take that same respectful attitude toward one another.

One suspects that Sternberg held perhaps a similar view in part--that is, finding much of merit with the doctrine, as a result of what he had personally observed in the earthen graves of his finds, and yet wanting to hold to the traditional teaching of the Church due both to his own reading of the Bible and the writings of others who were grounded in the Scriptures. Indeed, this is the same dilemma one is faced with today. Some of the questions are as still as difficult to answer as they were then.

Perhaps this is a good place to interject that no reliable theologian would argue that Genesis is intended to be a geological treatise. It is cursory in both its language and its intention, purposefully. It is not meant to give mankind anything other than a very brief overview of things and hence much is left out. This is not, however, to say that what it does say is not to be taken literally, and I personally believe that the Scriptures are inviolate and without error, as I am sure did

Sternberg. Even within the conservative church there remains a great deal of difference in interpretation because Scripture in this area leaves room for such. The fact of man's salvation does not rest on this view or that view of the age of the earth, or indeed, whether the six days of creation are literal 24-hour days as we know them, or creative periods. Neither necessarily violates a literal view of Scripture. The same thing, however, cannot be said about evolution, if the user of that word means by it that there is a morphological change that occurs that makes a former dinosaur a bird or an ape a man. Changes, however, that occur where a rhinoceros undergoes some change to become a slightly different type of rhinoceros is perfectly in keeping with Scripture because the "type" of the animal is not changed. These, at least in the fullest sense of the word, are evolutions as well, although because of the evolution of the word evolution perhaps using "adaptation" is a wiser term to use. Some choose to use the terms micro and macro-evolution to explain their understanding of the distinctions of the term.

Christians certainly hold to Adam being the first man and all individuals being his descendants. Since we now have on the earth white, brown, black and yellow men there has obviously been some kind of "evolution," micro-evolution or adaptation that has gone on. One's choice of a term often carries meaning with it that one doesn't necessarily intend. The word "creationist" carries also a great deal of meaning that can differ for the same kinds of reasons.

It would appear from the limited amount of information available on Charles' brother Levi that he perhaps did not believe in evolution, or was somewhat reluctant to use the term. In one of his few popular articles, entitled *When Dinosaurs Ruled the Land,* he seems to purposely avoid use of the word "evolution," but instead refers to something he labels as "developing." He says in part:

> It was apparently during the Triassic period that a small animal developed that was to become the dominant form of animal life upon our earth...

and

Many new species of animal developed, and are developing at the present time. This great book took many millions of years to write.[24]

Since Levi was in his own way to remain the closest to his father (in large part owing to his being the youngest son) he was perhaps less evolution-oriented than his older brothers. It certainly would seem odd, in any event, that he would write such a lengthy piece without once mentioning the term "evolution," unless it was intentional. At the same time, however, it is evident that Levi believed in an ancient earth, as apparently did his brother Charles.

Some of the most overt displays of Charles Sternberg's Christian beliefs can be found in his book of poems, *A Story of the Past or The Romance of Science*, published in 1911. The poem's various titles essentially relate to either geologic or Christian subjects, or a blending of both. While the poems will hardly cause a ripple in literary circles, they are important because of the picture they give us of their author.

The poems do contain a clear and simple statement of Sternberg's feelings on many important matters, and show also that he gave thought to these same matters. He was not a man who took his religion or his view of the world lightly. In his poem entitled "A Story of the Past" he was to write:

> Who made men's minds, if not our God?
> Who gives him power to think?
> Are we but creatures of the sod?
> Naught but a simple link.
>
> In the great chain of life He's riven,
> As helpless as the clay?
> Are we but like the cattle driven,
> Live only our short day?
>
> O no! though rulers of our fate,
> Our God will use us still
> To bring about His plans so great,
> Still subject to His will.
>
> And I believe that God chose me,

> Unworthy though I am,
> To add my mite to Life's Great Tree,
> To carry out His plan;
>
> To show to man His wondrous work
> Through all the scenes of time,
> Since life in humble beings lurk:
> Rich facts for us to mine.
>
> I here profess my strong belief
> In my revealed Lord;
> I've found Him in the rocky leaf,
> And His inspired word.
>
> For forty years I've lived with God,
> Oft from the haunts of men.
> I've thought upon His wondrous word
> And scenes beyond our ken.[25]

These lines are written by a man who places great credence in the Bible as being the Word of God, as well as placing great emphasis on the need to learn as much as possible about life that formerly inhabited the earth. The phrases in juxtaposition to each other, "rocky leaf" and "inspired word," are fitting testaments to the way Sternberg was to lead his life. He believed it, and he led it as best he was able.

In a second poem, "A Song of Love," he delineates his Christianity further by evincing a fairly orthodox Christianity. In that poem he states his doctrines: Jesus being Lord, Christ's blood being redemptive, through belief in Christ one is saved, Christ rose from the dead, sin is rampant in the world, and men are as sheep who need to be cleansed. Although there is nothing new in the doctrinal points he so vividly illustrates (thank goodness!), one cannot but help appreciate the vehemence with which he approaches the subject. Sternberg was a man "on fire for the Lord."

In another poem, "One Hundred and Seventh Psalm," he takes the 107th Psalm, which expounds upon thankfulness to God for His manifold blessings, and encapsulates it in one of the stanzas:

> Children of men, praise ye the Lord
> For all his goodness given,
> On earth, His blessings He'll accord,
> With promise of His Heaven.

And in another:

> Ah, yes! His mercies will endure
> While countless ages roll,
> Oh, sin-sick soul, He will cure
> The weary burdened soul.

In another, "And They Shall See His Face," he uses text derived from a study of the Book of Revelations, chapter 22, verse 4. This biblical verse, found in the very last chapter of the Bible, is a prophetic verse and speaks about the way in which God's followers will be after the massive destruction of Armageddon takes place and the Millennium is ushered in:

> And there shall no longer be
> any curse; and the throne of God
> and of the Lamb shall be in it, and
> His bond-servants shall serve Him;
>
> and they shall see His face,
> and His name *shall be* on their
> foreheads.
>
> And there shall no longer be
> *any* night; and they shall not have
> need of the light of a lamp nor the
> light of the sun, because the Lord
> God shall illumine them; and they
> shall reign forever and ever.

This poem, which mirrors the verses quoted, clearly indicates Sternberg's surrender to the doctrine of Christ's return to take His followers with Him, and makes it very evident that Sternberg's Christianity was a traditional, orthodox Protestant view:

> To him that overcometh
> A crown of life shall wear,
> Forever ever dwelleth

> All free from pain and care,
>
> A new name in his forehead
> Blest by the God of grace,
> For hath not Jesus promised
> That "they shall see His face!"

And then:

> Yes, see His face. Oh, rapture!
> How glorious is the thought,
> To be with Christ our Saviour,
> Whose precious blood has bought
> And made us Kings and Princes,
> To reign with Him for aye,
> To sound abroad His praises
> Through an eternal day.

Yes, there is certainly no question about what "brand" of Christianity that Sternberg was to avow to, and no question as to how well he was to adhere to those same tenets. The only question left unanswered then is how did this orthodox belief correspond with the prevalent world attitudes on evolution and the descent of man. And, if we can answer that question, will it somehow bring us closer to an understanding of the way that science and religion ought to work together?

It can be said that in some ways science and religion butt heads more because of their similarities than their differences. Some of these "similarities" can be found in a quote from the science/creationist historian Dorothy Nelkin:

> Their [creationists] main objection to evolution theory is that it 'incorporates all the attributes of a religion,' it is 'a doctrine of origin' that replaces God with eternal matter and creation by random mutation; it is a doctrine of salvation not through faith but through foresight and the manipulation of nature. Thus, claim the creationists, it violates traditional religious assumptions and endorses its own system of ethics.[26]

Although we have said much in this chapter about what the "might have beens," it is evident that we have had less to say about the hard facts of some of these important issues. It is difficult, except by reading through the lines, to make any definite statements as to just exactly what Sternberg believed since he was not to firmly state his position on some of these matters, and when he has spoken firmly it has generally been from early writings which seem to be somewhat compromised by later writings.

The best view (in my opinion) of Sternberg's evident views on evolution and geology come from an early work. In an article of 1885 he hints at much, and particularly suggests at a fairly substantial change from the views he expressed in an article published earlier in the year, *Flora of the Dakota Group*. Since both articles were published in 1885 it would seem that neither should conflict with the views of the other, and yet there are some differences.

Charles begins the earlier article with reference to geology being a world study, and indicates that this study began when the earth had cooled enough "to allow the atmosphere to disgorge its contents on the steaming earth." This is a geology not at odds with the Bible, but at the same time, not necessarily in accord with standard biblical teaching of the time. This was a geology based on science. He goes on then to blend this view of geology and religion:

> The study of geology is the grandest of the natural sciences, not only because it takes into consideration the largest object with which we have to do, but it includes vast lapses of time, so long in fact that it does not prove that time is eternal, it shows at least that it is very old. The text-books of the geological student are the solid rocks in which the Creator has written in never fading characters the history of each succeeding period, its life, climate, depth of sea, etc.[27]

Clearly again we have reference to his belief in an old earth as well as reference to a succession of life forms, but still we cannot fail to recognize that Sternberg's view was not an atheistic process but a process whereby God was always the controlling factor. And yet this is too not a theistic evolution dogma we find so common in some circles today.

He delineates slightly on the successionary process when he further says:

> But there is no wider field for the imagination than to people the old seas and lands with animals restored from their buried relics,

to clothe the ancient lands with verdure, and map out the continents, study their physical geography, the depth of the sea, climate, and above all to study the introduction and succession of life on the globe and other manifold changes brought about in the animal creation, their development from lowly forms, and their progression through successive changes, until the present state of perfection has been brought about.[28]

One cannot read this passage without recognizing that Sternberg must have believed in some form of evolution. It is equally clear from this passage, however, based his use of the word "creation" that he held to God's controlling hand, and therefore one can assume that he also held somehow to the fixity of species.

Some of the more well-schooled Christian earth scientists of today would probably balk at even being called "creationist" because that term brings with it baggage similar to the term "evolutionist." This primarily because the term has come to be somewhat synonymous with the "young earth" concept, and its espousal by the Creationist Science Institute of San Diego and like adherents. And yet the term in its proper use should mean something quite different, i.e. someone who recognizes God as the creative force of all created things and beings.

One of the finest modern authors on geologic science from the viewpoint of a practicing Christian is Davis A. Young, an Associate Professor of Geology at Calvin College. In his fine book, *Christianity and the Age of the Earth*, Davis clearly takes aim at those who would too narrowly define the term:

> On the other hand, what is more likely to undermine Christian faith is the dogmatic and persistent effort of creationists to present their theory before the public, Christian and non-Christian, as in accord with Scripture and nature, especially when the evidence to the contrary has been presented again and again by competent Christian scientists. It is sad that so much Christian energy has to be wasted in proposing and refuting the false theory of catastrophic Flood geology. But Christians need to know the truth and to be warned of error. Creationist articles constantly make sweeping statements to the effect that Christianity is now on surer ground because of the discoveries and theories that creationists have concocted...'Proving' the Bible or Christianity with a spurious

scientific hypothesis, can only be injurious to the cause of Christ.[29]

The historian Dorothy Nelkin is of the opinion that the "creationists" will continue to be of influence, for she says that "faith in science persists when it fills a social need," and that at present science to many people "has lost its credibility." She sees the revival of creationism as a revival of fundamentalism, and that it fills a social void. She says that creationists, "using representations that are well adapted for the twentieth century...offer intellectual plausibility as well as salvation and the authority of science as well as the certainty of Scripture."[30]

While all of these changes, be they ever so slight, are helping to perhaps bring the views of the general populace closer together it probably has not done much to bring accord between the staunch religionist or scientist. Sternberg was perhaps able to bridge some gaps at a time when others could not easily do so, and this was apparently because of his simple approach to both science and Scripture. Sternberg's non-threatening way of dealing with situations (even the most adverse) should certainly be an apt lesson for us today.

Perhaps Charles' clearest exposition on evolutionary matters overall can be found in his curiously titled *The First Great Roof*, which explains his desire to be a collector of facts, rather than an interpreter of those same facts. He wanted to leave to others the agony of interpretation:

> I have not had the long training necessary to even attempt to make a technical description according to rules of paleontology. My greatest glory has been that I have for nearly 30 years been a private soldier for science in the fossil fields of North America, gathering facts, new species or old, the direct handiwork of the great Creator of all things; and consequently I know but little of the great literature which my labor and that of others has made possible. I have consequently left others to wrangle over the fragmentary relics of creative mind. I let one build up a genus or species for another to tear to pieces, spending a lifetime in proving that this theory was right and that was wrong. I find men's theories come and go like the leaves of the forest, but the facts collectors gather, last for ever, as they are truths, and are immortal. I write, then, from the standpoint of a man who from

childhood has loved nature; and as a soldier in the great fossil camps, away from men and their works my mind has been little prejudiced in favor of this theory of evolution or that, and my habits and mode of thought are those of the camp, the battlefield, and not of the study and the laboratory; my ideas then are original and they belong to me, and not to some other man.[31]

If indeed Sternberg was "little prejudiced" he still seems to have come to a point of some belief in "evolution," at least if we use the term in its more precise meaning. And yet we know as well that he remained a staunch and vocal orthodox Christian right up to his last breath. Considering these facts it leads one to say then that Charles probably believed in a micro-form of evolution, one that did not involve any change in species (or type to type change). His "men's theories [that] come and go like the leaves of the forest," is obviously in juxtaposition to his own view which relates to God's permanence and control.

After speculating on much I recently was able to confirm some of what I had thought to be the case. In an interview with Charles Mortram Sternberg (at age 93) he was asked by the interviewer whether or not his father believed in evolution. He responded, "my father accepted evolution. It is possible for people to believe in evolution and in the Bible --- there is evolution in everything."[32]

It is interesting to me, however, that Charles' son uses the phrase "accepted evolution," as it seems to imply something akin to force-feeding. It, at least to me, does not have the tone of a free and unencumbered acceptance. I suspect that son Charlie might have held a different understanding of the term evolution than did his father, something perhaps akin to the differences Phillip E. Johnson, Davis A. Young and others cite with the word "creationism."

Ray Martin, son to Charlie Sternberg, when asked of his father's position on evolutionary theory responded similarly: "I have no idea how grandfather or my uncles felt about evolution. My father did accept the fact of evolution but probably knew little of the mechanics of it and found no conflict with the Bible."[33]

Of course this too comes back to the real crux of the matter, which is, what does one mean by the term "evolution?" Phillip E. Johnson perhaps best describes the conflict:

> The concept of creation in itself does not imply opposition to evolution, if evolution means only a gradual process by which one kind of living creature changes into something different. A Creator might well have employed such a gradual process as a means of creation. 'Evolution' contradicts 'creation' only when it is explicitly or tacitly defined as *fully naturalistic evolution* -- meaning evolution that is not directed by any purposeful intelligence.[34]

Sternberg's sheer vitality and verve both in his religious outpouring and his scientific activity are noteworthy. If he could hold both to be true and valid, and capable of being balanced, why cannot we? There seems to be no blatant error in his reasoning that I can uncover, so why cannot we today too learn from this? I think that perhaps the main difficulty lies in that God has been removed from the lives of many today and that they therefore have taken an anti-God attitude in all things, and have tried to explain all things mechanistically. To suggest that God has any part in any thing is to them anathema.

Regardless of Sternberg's views on evolution, which remain cloudy to us even after all that's been said, we still are left with the one thing that counts: Sternberg was a practicing Christian, a vibrant Christian, and a man concerned about others. It was his belief in Jesus Christ which sustained him, and which made him the man we have come to admire. His life was a testament to the God he served, not perfect, but built upon the foundation Christ demanded, "love thy neighbor as thyself."

As a final note it is perhaps appropriate to quote from Charles' Lutheran pastor in San Diego, Delmar L. Dryerson. In a 1938 tribute to Charles, on the occasion of Sternberg's birthday, Dryerson was to write:

> Your example of giving to the Lord of your time, talents and substance is a great inspiration to me as pastor as well as to the people. May God bless and grant many years in His service.[35]

NOTES

1. Robert Hook, Personal Communication

2. Robert D. Clark, *The Odyssey of Thomas Condon,* Oregon Historical Soc Press, Portland, 1989, Pg. 424

3. Andrew D. White, *A History of the Warfare of Science With Theology in Christendom,* George Braziller, NY, 1955, Pg. ix

4. Ronald B. Parker, *The Tenth Muse: The Pursuit of Earth Science,* Charles Scribner's Sons, NY, 1986, Pg. 40

5. Bolton Davidheiser, *Evolution and Christian Faith,* The Presbyterian & Reformed Pub Co, Philipsburg, NJ, 1969, Pg 162, Quoting from Loren Eisley's *The Immense Journey,* Random House, NY, 1957, Pg. 18

6. John McPhee, *Rising From the Plains,* Farrar, Straus, Giroux, NY, 1986, Pg. 72

7. Martin J.S. Rudwick, *The Meaning of Fossils,* Univ of Chicago Press, Chicago & London, 1976, Pgs. 68-69

8. Levi Sternberg, "Geology and Moses," *Evangelical Quarterly Review,* Aughinbaugh & Wible, Gettysburg, Vol #19, 1868, Pg. 138

9. Charles H. Sternberg, "The Life of a Fossil Hunter," *The American Inventor,* #10, 1903, Pg 311

10. Charles H. Sternberg, "The Flora of the Dakota Group," *Kansas City Review of Science & Industry,* 1885, Pg. 11

11. Charles H. Sternberg, *The Life of a Fossil Hunter,* Henry Holt & Co., NY, 1909, Pg. 17

12. Ibid, Pg. 17

13. See 1 Pet 3:8, Ps 90:4 and Is 55:9

14. Charles H. Sternberg, *The Life of a Fossil Hunter,* Henry Holt & Co., NY, 1909, Pg 188

15. Ibid, Pg. 128

16. Charles Coulston Gillispie, *Genesis and Geology,* Harper & Bros, NY, 1951, Pg. 128

17. Edward Drinker Cope, *Origin of the Fittest,* D. Appleton & Co., NY, 1887, Pg. 1

18. Ibid, Pg. 2

19. Ibid, Pg. 128

20. Ibid, Pg. 128

21. Lester D. Stephens, *Joseph LeConte: Gentle Prophet of Evolution,* Louisiana State Univ Press, Baton Rouge & London, 1982, Pg. 62

22. Roger Lewin, *Bones of Contention,* Simon & Schuster, NY, 1987

23. *New York Times,* (Wed), Jul 17, 1878

24. Levi Sternberg, "When Dinosaurs Ruled The Land," *Canadian Nature,* Vol #17, #1, (Jan-Feb), Pg. 15

25. Charles H. Sternberg, *A Story of the Past or The Romance of Science,* Sherman, French & Co, Boston, 1911

26. Dorothy Nelkin, *The Creation Controversy,* Beacon Press, Boston, 1982, Pg. 87

27. Charles H. Sternberg, "Practical Studies in Geology," *Kansas City Review of Science and Industry,* Vol VIII, 1885, Pg. 481

28. Ibid, Pg. 482

29. Davis A. Young, *Christianity and the Age of the Earth,* Zondervan Pub Corp, Grand Rapids, MI, 1982, Pg. 150

30. Dorothy Nelkin, Pg. 195

31. Charles H. Sternberg, "The First Great Roof," *Popular Science News,* Vol XXXIII, June 1899, Pg. 126

32. Interview with Charles Mortram Sternberg by Free-lance Broadcaster Laurie LaMaguer, Courtesy of Tyrrell Museum, Dept. of Paleo.

33. Personal Communication

34. Phillip E. Johnson, *Darwin On Trial*, Regnery Gateway, Wash DC, 1991, Pgs. 3-4

35. First Lutheran Church Ltr, (To CHS on his Birthday 1938 from Rev. Delmar L. Dryerson), Materials Courtesy of San Diego Nat. Hist. Mus. Archives

CHAPTER TWENTY-ONE
NUMBER ONE SON: THE EARLY YEARS

One of Charles Sternberg's main delights was to remind others of the accomplishments of his oldest son George. George is frequently mentioned in all of Charles' writings, and held a position of prominence, as the oldest often does, that the other sons really couldn't attain, even though they were also close to their father, and were equally talented.

George Fryer Sternberg was born of Charles and Anna Sternberg on Aug. 26, 1883 in Lawrence and was christened with the namesake of his grandfather on his mother's side, George Miller. George's uncle, George Miller Sternberg, was likewise named after the same man. As the oldest of the Sternberg children that were to live past early childhood (which were to eventually number at four), George, like most eldest children of that day and age, was to most closely follow in his father's footsteps. Since he was chronologically the oldest, (although Charles mentions the early loss of their first boy child in his memoirs he does not acknowledge the name of the child, nor the circumstances) he was therefore the earliest recipient of his father's teaching. He was truly a first-born in the fullest sense of that word, and always had a special place in his father's heart.

George was born in Kansas, and of all the various family members he was to maintain the most obvious love for Kansas, with the possible exception of his father. The fact that George was to come back to Kansas and end his days there says something about his love for this area, and this love was even to find voice in some of his writings.

George, like his dad, was to develop an early penchant for the fossil hunting profession and was to most closely follow his dad's course. He was the only one to remain an independent collector, since the other brothers, Charles Mortram and Levi, were both to become professional men. Their positions in the science had more stature (but not more importance) than the position of field collector. George himself always preferred that people refer to him as a field vertebrate collector or paleontologist, with emphasis on the word "field."

George's early efforts at fossil collection are highlighted for us in an early university publication called *Aerend:*

> At the age of nine years I took a trip into this region with my mother and brother to join my father who had his camp pitched at Elkader, Kansas. Little did I realize then that years later I would still be tramping over those old chalk rocks eager to find the slightest trace of a bone or bones which might reward me for my efforts. Perhaps that trip was the one which convinced my father that I should be trained to take up the work he had been following for years. At least I made a discovery which pleased him very much. He has often told me in the years that have passed since, that I was a natural-born fossil hunter.
>
> As I was the oldest son I was allowed to follow father about over the rough exposures while he searched for some exposed part of a buried skeleton. And one day while he was busily engaged in taking up a specimen he had found, I wandered off on my own account, perhaps in search of a shady nook to get out of the burning sun whose rays in these exposures made life almost unbearable at times. Perhaps I was in pursuit of one of the numerous cotton-tail rabbits which would scamper from a shady rock into one of the many holes. Or was I really looking for a fossil? I do not remember. But I have a very vivid impression of coming suddenly upon several large pieces of bone washed down the side of a steep exposure; and upon looking up, I saw other parts protruding from a soft ledge above.
>
> I remember running back to my father shouting, 'I have found a

fossil, I have found a fossil.' Father seemed little interested at first. And now after thirty years in the work, I realize how often we are called to view inexperienced collector's finds only to see a number of odd shaped rocks, a recent skeleton of a buffalo, or fragments of an extinct animal which are of no use to anyone, due to their fragmentary nature. In due time father went over with me to see what I had. I shall never forget the change of expression which came over his face when we arrived at the spot. Nor will I ever forget his first words, ' George, thank God, you have found a plesiosaur.'[1]

George was to go on to say that this was so thrilling to him that it had never been forgotten, and he felt sure that he could still go back to that very site (after some 30 years had passed) since it was so vividly etched in his mind. The passage also again stresses that bond between father and first-born son, and it is evident how lovingly George's father nurtured in him the love for fossils, and the love of the art of finding them.

George was to become his father's main assistant at the age of 12 according to Charles, or 14 according to George himself. Whichever is right is not really important, only the fact that it was very early in life that he began to work diligently with his father. This early proficiency was to serve him well in later years, and was to make him a much sought after collector; even more so perhaps than his own father who was certainly of great reputation in this regard.

One of Charles' early reminiscences of his son as a collector involved the late-taking-up of a giant fossil fish, *Portheus molossus,* in Logan County, Kansas. November is very late for fossil-hunting almost anywhere, but particularly in the midwest, and the situation was aggravated further by the fact that the weather had already definitely changed. Since the fossil fish upon which they were working was partially recovered they were afraid that it would not weather well and thus felt they must proceed or lose the fossil entirely. Charles was to recount his vivid recollection of George's part in the collection:

A violent windstorm was raging at the time, and to complete the slab, George had to bring water from a tank a hundred yards away. I can still see that boy running up with his pail of water, trying to carry it so that it would not be emptied by the raging, howling wind that was almost tearing his coat from his back, while I stood and shouted, 'Hurry up! The plaster's hardening!'[2]

The brothers and their father were to share a special camaraderie and one can't help but imagine the picture this fine family of four must have presented to those who would come upon them in the wilds of the badlands as they would be crouched in some out-of-the-way cleft meticulously etching away the matrix that surrounded some great prehistoric treasure. The family similarity would have been obvious to anyone who should happen upon them, for they all were lean and weather-worn men with that very distinctive Teutonic nose and the same languid eyes. One also is sure that they would have been found to be festive and jocular with one another, and sporting about since they had such a love for their profession, and one another. It is hard to find this kind of tight family unit around today, and that makes this family of even more interest.

There were occasions when the foursome would be joined by others. When George was to marry it was not uncommon for him to bring along his first wife Mabel Smith Sternberg and their baby, Charles. Charles H. Sternberg mentions this about his son George in his article entitled "In the Niobrara and Laramie Cretaceous" in the *Transactions of the Kansas Academy of Science* for 1909, Vol. XXIII, and one suspects that they welcomed the change of camplife because the visits of this type were so infrequent.

George was to marry Mabel Smith, the first of his two wives, on Dec. 31, 1907 in the parsonage of the Presbyterian Church at Phillipsburg, Kansas. She was a lifelong resident of Long Island, Kansas, and they apparently met while George was working for his father in the area around Long Island. Long Island was, after all, the scene of his father's greatest fossil trove, the Sternberg (Marsh) Quarry. Once married, the couple immediately moved to Scott County in order to settle on a farm that George owned there.

George and Mabel C. Smith Sternberg were to lose their first child at a very early age. Their firstborn daughter, Margery Anna, was born on Nov. 25, 1909 in Lawrence and was to die almost two years later to the day (Nov. 21, 1911). The obituary in the local newspaper, *The Lawrence Journal World*, does not denote the reason for their daughter's death, and in fact mistakenly refers to her as "Marjorie Steinberg." According to Katherine Rogers, however, Margery died "almost immediately after ingesting rat poison she found while exploring the storage cupboards in the family kitchen."[3] Margery was only the second Sternberg buried in Lawrence, and she lies next to Maude Sternberg at the beautiful Oak Hill Cemetery.

George's distinguished career as an independent collector was to begin in the trenches. His dad was not the mollycoddling type and therefore expected that his sons put in a good solid day's work. This early training was to serve him in good stead, for he was to develop substantial and valuable skills in collecting technique at a young age, and he was to also learn preparation skills as a result of his dad's independent collecting business since they could not send "unfinished" fossils to most of their buyers. This was a distinct advantage to him since most of the other collectors of the time only learned the initial skills that involved removal of fossils from their rock graves. These other collectors would ship off the fossils to one of the Leidys, Copes or Marshs to finish and describe. These great men themselves would then farm out the preparation work to other men who did nothing but this facet of the business. You will recall that Charles was to stay with Jacob Geismar, who was Cope's preparator, and who therefore performed this specific function for Cope. One can certainly see the value that a man like George would have provided since he had developed talents in different phases of the business. This, in essence, made him a very unusual kind of collector, and was a prime reason why he was so sought after by a number of museum curators.

Additionally, George and his father were forced to learn how to mount fossils for exhibition as a genuine matter of self-preservation. The techniques for fashioning such mounts (particularly the open mounts) were very specialized skills, and were at that time only really to be found in the bosoms of the world's largest museums. In developing their art they had to assume a great deal of guesswork in this adventure and had to suffer a number of disappointments before their mounts had the quality of design that was needed in order to enhance a fossil's look. This was particularly important since it was much easier to sell a finished product than it was a fossil still in partial matrix. Although this did not really have an impact with the largest museums (who could readily view a fossil for its mounting capabilities), it had a decided affect on the smaller institutions less versed in the art.

The innovation of technique we have already seen in George and his father's successful attempt in first being able to slab mount the fossil invertebrate giant *Inoceramus* perhaps best explains the high levels that their talents were to attain. George was to become the real family expert in the arena of fossil mounting and his father looked often to him for help in this task. When George became museum curator at Teacher's College in Hays, Kansas, this ability was to prove to be of major benefit. The task of simply laying out a museum case can be difficult, but when one attains the broad understanding that George's training had built, the task is made infinitely easier. One can find good examples of his talent in this area at the Sternberg Memorial Museum in Hays, Kansas. This museum, on the campus of Ft. Hays State University, (formerly known as the Ft. Hays Museum), has a wealth of material personally collected by George, and there is even a full-sized mount of an *Inoceramus* shell that is magnificent.

George's real expertise, however, was to remain within the vertebrate world. By his own admission he claimed to "know practically nothing about fossil shells,"[4] and spent his time on the larger and more impressive vertebrate mountings.

George's natural ability was to make him a world-renowned and much sought-after collector. Although his main collecting grounds were to be in the United States (and most particularly the plains states) he was also to collect in Canada extensively, as well as traveling to Patagonia. As we shall see, he was particularly well-suited to travel to out-of-the-way locations to ferret out fossils, and because of these abilities he became a favorite of C.W. Gilmore at the

National Museum (Smithsonian), and also of Childs Frick who frequently made use of his talents. This word of mouth extolling of his capabilities was to allow George to keep busy virtually all of the time if he desired to be working.

NOTES

1. George F. Sternberg, "Thrills in Fossil Hunting," *Aerend*, Kansas State Teachers College, Hays, KS, Vol #1, #3, 1930, Pg. 145

2. Charles H. Sternberg, *The Life of a Fossil Hunter*, Henry Holt & Company, NY, 1909, Pg. 55

3. Katherine Rogers, *The Sternberg Fossil Hunters: A Dinosaur Dynasty*, Mountain Press Pub Co., Missoula, 1991, Pg. 123

4. SDNHM Ltr, Dec 11, 1930 (GFS to Kate Stephens @ San Diego), Materials Courtesy of San Diego Nat. Hist. Mus. Archives

CHAPTER TWENTY-TWO
THE YOUNGER SONS: EARLY YEARS

It is always difficult, even under the best of circumstances, for the children who must follow in the footsteps of an older sibling. This seems particularly hard for men. Living up to what older brothers have accomplished (even when the accomplishments are mythical) can seem an insurmountable obstacle to the younger brothers.

There is also a stressful relationship that occurs between the oldest son and those following, that relates somehow to their relationships with their father. While some might scoff and downplay such generalities it does certainly seem that there really is something in this area with which many must reckon. The younger, but not youngest sons, seem to suffer the most and perhaps most often have to bear the heartache in lack of self-esteem. The youngest generally receives a great deal of attention from both parents and doesn't seem to generate the same difficulties in self-esteem.

If this is so, we have to immediately speculate as to how this interaction must have worked within the Sternberg family. We recognize right off that George, as eldest son, held a special place in his father's heart, and that even though Charles was to love his other sons greatly, there continued to be a special and obvious bond between he and George that was not shared in the same way with his other sons. Knowing this allows us free reign in speculating as to how it is the other sons, Charles and Levi, were able to maintain such a good relationship with their father, as well as with their brother George who certainly might have been the cause of bad feelings due to his prominence in the family.

Son Charlie was to delineate this area somewhat for us. He was to say that the brothers "got along all right," and elsewhere that they "got along fairly well."[1] His luke-warm response perhaps highlights the fact that there was some obvious friction among the various brothers. It does appear, however, that the main problem revolved around George as the oldest. George clearly felt that he was "the boss" and when things didn't work in that manner there were some disagreements. Still, however, they remained a close-knit family, even considering the mileage that separated them.

One does not have to read far to recognize in spite of the possibilities for ill-will or favoritism that each of the children held a special place in Charles' heart, and each was especially gifted in different ways. It is perhaps this clear and evident love by the father for each of the children that overcame all of the natural difficulties.

One might additionally describe this special bond by saying that George was the down-to-earth one, who most closely resembled his father in terms of lifestyle and occupation, and therefore he held a special place because of their similarities in station.

On the other hand, Charles Mortram could be described as being the more aloof one, but the one with burning ambition to do things right. His perseverance and fossil hunting skills would have greatly pleased his father. While Levi's place would have been taken not by force, but by quiet resolution. He had much of his mother in him, although he also had the quiet determination and resourcefulness of his father as well. So it is clear that each of the "boys" was to mean something different and special to their father, and he carefully cultivated their differences also.

Charles Mortram was much more polished than his brother George and was to educate himself better. Charlie was the only one of the brothers to attend high school, and he was in his twenties before he was able to do that. The delay in beginning high school related to his being the principal man around the farm who took care of his mother while his father and George were away. George apparently later did attend business college, but never had any high school training.

Charlie was to acknowledge that he did have some regrets about not being able to pursue higher learning, but he was pragmatic about life's turnings and accepted what came. They just never had the money for him to be able to attend college, and so he took the next best avenue: self-education. He stocked up on books and read all he could get his hands on.

When queried as to whether or not he felt a university education would have helped him he was able to note that he would probably have ended up with a job teaching if he had, and that kind of job, or others like it, could not have compared with the job he was to love for a lifetime.

Charlie was most like his dad in being declared the "eagle-eyed" one, the one who could spot a fossil where even the most gifted would pass right by. Charles' technical writings exceed that of all the other family members, and his credentials will undoubtedly serve to keep him as the most prestigious scientist. His bibliography of professional writings runs over 65 articles in the world's most famous paleontological publications.

Levi, the youngest son, had the unenviable position of being last, and like most situations of this sort, had to suffer the notion from others that he was the baby of the family. This naturally made it more difficult for him to be taken seriously initially, although it perhaps contributed to actually making him tougher than he might normally have been. Levi is also the least well known of the family members, and the one who seems to blend into the background most of the time. Levi, however, could always be counted on to pitch in and help, and his strength seems to have been in service. He was the one who invariably hauled fossils to the railhead, or stayed with his father when the others went off on their own. He proved the stalwart in the long run.

Maude, the lone girl in the family, was the real love of her father's life, and her early death was a source of constant sorrow to him thereafter. Charles' descriptive passages about her are filled with love, and they have a depth to them that is not found with any of the other children. This is somewhat odd because he seems so reticent to discuss in print his wife Anna at all by name, and only infrequently is she even referred to tangentially. Maude, on the other hand, shines. She seemed to be the source of her father's strength as well as the source of his greatest sorrow. His love for her was so great that he constantly found himself dreaming about her, and his dreams form a real blend of life with her and life in past ages when his fossils were living and breathing animals. To say his dreams were a curious mix of fact and fiction would certainly seem to be self-evident.

With that brief recapitulation of the children, it is perhaps easiest to take them singly in order of their age.

Charles Mortram Sternberg, the third-born son and second oldest of the four children who lived, was named after his father. His middle name of Mortram apparently came about in part through his father's mistake. According to Charlie it was his mother's desire to name him Matram after their family doctor. His father, however, in filling out the ministerial card for use at the baptism, wrote the name so poorly that he was mistakenly baptized as Charles Mortram Sternberg, and the name was to stick with him.[2]

Charlie was born on Sept. 18, 1885 in Lawrence, Kansas, and like his brother George he was to be initiated into the realm of fossil hunting at a young age. Because of his need to stay with his mother on the farm, however, he was not to actually begin collecting with his dad until the age of 20. Charlie was to be one of "we boys" as described by George and we are told that at one time Charlie was the official camp cook. Charlie actually was to take his turn as camp cook, just as George had before him. The cooking chore was the bottom rung on the ladder. From there one graduated to doing the dishes, and from that to doing neither. The boys shared these jobs until they eventually left their father's employ.

George was to indicate in his reminiscences that each of them was to leave home at the appropriate age to join their father's expeditions. That age varied by circumstances as we have seen. Charlie was to begin his camp duties in 1906 on this, the first of his fossil hunting expeditions for his father.

Charles Mortram Sternberg had expressed some dissatisfaction over his father's handling of financial matters in Canada, and was probably equally galled over the fact that his older brother George in 1910 was making a salary of $100 a month working for their father while Charles was only making $30.[3]

Charlie (as his father was to call him) was perhaps the most talented of the sons in terms of a traditional style paleontology. Although the other sons were excellent fossil hunters, perhaps even approaching their famous father in this respect, there appears no question that Charlie had a talent that was above and beyond them all in terms of his ability to search out the most elusive of fossils. In his Canadian autobiography, Charles H. gives his second son special due when he points out Charlie's talent for ferreting out fossils in the Belly River series: "Charlie, as I said, was the lucky one, as he found the most complete skull of this strange creature we have ever attained..."[4]

Maude was the third child, and the love of her father. It is evident from his writings that her early death left a vast hole in her father's life, and that even as the years waned he continued to mourn her death as if it were yesterday. Maude was born Oct. 30, 1890 and was to die at the young age of 20, falling victim to scarlet fever on Jan. 15, 1911. She is buried at the beautiful and spacious Oak Hill Cemetery in Lawrence.

The depth of her father's mourning is perhaps most clearly seen in his poetry. In one of his poems, simply entitled, "Maude," he was to label her as the "joy of my life" and his "comfort and pride." Such emotion as she summoned up in her father is worthy of some note. The third and fourth stanzas of this poem in particular bear repeating:

> Her face gave me pleasure, her presence was
> rest;
> She was satisfied with me, and never in jest
> Complained that I was not all that she
> wished; --
>
> I will see you no more in my earthly career,
> So dear to my heart,--God knows you were
> dear!
> I have laid you away in the depth of the tomb,
> Your loved body to rest in its shadow and
> gloom.[5]

One highly suspects, considering the bold manner in which he discusses the loss of his daughter, that Charles never fully recovered from her death. This does seem somewhat ironic when one considers the openly displayed emotion shown for his daughter, in comparison to the presumed intentional lack of even having mentioned his wife by name in either biography. However, it may be due to the fact that Maude was apparently such an open and loving child who exuded this warmth to all who knew her. In this light it is perhaps easy to see how a father would be captivated by her very liveliness, and perhaps more reticent about a wife whom he loved, but who was less approachable.

In his autobiography *Red Deer* Charles was to use an odd literary device (considering the type of book) when he depicts for the reader a dream that he had; a dream that stars both the ancient world and its habitues, as well as his lost daughter. Charles sets the scene by indicating that he often suffered from "strange day dreams," and that "my only daughter died some years ago; though in imagination she is often with me."[6] Then in a curious blending of the real with the fictional he weaves a life-like dream for the reader.

In the first sequence, the first of three, he describes the scene as being a seashore, with limestone terraces that he recognizes as the "old Cretaceous Ocean." In his dream he encounters the beauty of the life that lived there during that particular geologic age (plesiosaurs, mosasaurs, etc.), but feels an immediate longing for his daughter who is missing. He says, "How glorious, but where is

Maud."[7] He follows with an insight that allows us to understand just what it was that made this lost daughter so special, for he states "with her appreciative ear, to listen for what my mind conceived, and my lips uttered, she never contradicted me when I uttered an opinion."[8] She obviously was someone with whom her father felt very comfortable, and one to whom he could share his innermost thoughts, no matter how fantastic. Certainly such a loss would be tough for anyone, but most decidedly so for such a private person as himself.

He goes on to describe this wonderful child of his in glowing terms. Such words and phrases as "eager face," "flashing eyes," and the like fill this work, and he obviously took pleasure in her having called him "papa." When she died there was clearly a void in his life that stayed with him until his own death.

The last of his children was Levi. Levi was born in March 1895 in Lawrence, and he took his name from his paternal grandfather. Levi as the youngest son was the slowest to develop in the fossil trade, and it seems that his father showed him special favor by reason of his being the youngest. When George and Charlie went off to pursue their own fossil careers Levi quietly remained with his father. It appears that it was only the advancement in age of his father, and Charles' sudden moves around the country that finally allowed Levi the opportunity to make a life for himself. This sense of responsibility for aging parents seems to be most closely felt by the youngest children, and Levi was no exception.

Levi liked Canada in a way that his father never did, and certainly in a much more positive manner than his brother George did. Levi was to quietly become a force in Canadian paleontological circles during his lengthy tenure with the Royal Ontario Museum. As Chief Preparator of the museum he was in a good position to use the talents so carefully cultivated by his father, and it helped to provide him with a prestige that his lack of formal education in science couldn't bring. Levi's responsibilities went much beyond what we would normally conceive of as a Chief Preparator's role for he was constantly in the field, most frequently as expedition leader. In the 1920's it was impossible to keep him out of the field, so engrossed in dinosaur hunting was he.

Levi was to marry Ann Lindblad, sister to Gustav Lindblad, who was to be Levi's able assistant for many years as well as a very capable collector. Levi and Ann were to settle permanently in Canada, in the city of Toronto, Ontario. They were to bear no children. Levi found a real home there at the University of Toronto Museum. He worked for W.A. Parks who was the paleontologist in charge, and he and Parks seem to have functioned well together. In later years Levi was to be stricken with the debilitating disease of shingles, and it did limit his activities to some extent.

Levi seems to have published little, almost nothing in fact, and our glimpses of him are always somewhat illusive. He did not have the force in his personality that his father and brothers had, but perhaps it was that shadowy nature of the man that makes him doubly interesting. One gets a small glimpse of the man and his talents only extraneously by reading between the lines in the limited articles he wrote, by looking at personal letters, or by taking as fact the opinions others express about him. A review of all these factors seems to place him in good light as he seems to have been well thought of by almost everyone.

Levi's sense of humor was very evident to those who knew him and is always mentioned as one of his most memorable traits. In his obituary, written by his colleagues, they noted both his professionalism and his fun nature:

> Levi was a meticulous collector with an amazing ability to judge the condition and orientation of a specimen on the basis of a few minutes' work with a Marsh pick. In the field he was a hard worker and a congenial companion, always joking and teasing. He always enjoyed excellent relationships with the local people wherever he was working.[9]

His peers were also aware of his innovation of technique as well; a Sternberg trait it seems:

> In his later years he perfected several latex casting techniques both for replication of fossils and for making flexible casts of modern fishes and reptiles for display.

Ray Martin was also to comment on Levi's playfulness, labeling him as a "practical joker, but his jokes were [not] to hurt one's body or ego. He too was very kindly and helped many people throughout his life."[10]

Levi was a corresponding member of the Brodie Club of Toronto, and although generally keeping himself in the background, he managed a very active schedule both on and off the fossil fields. Levi finally retired from the Royal Ontario Museum in 1962. His position upon retirement was that of associate curator.

There is never too much that can be said about the close-knit nature of the Sternberg family, and about the way they really liked each other. They showed their love for one another in countless ways, and visited together (even from very great distances) whenever possible. They also frequently traveled together, and additionally kept track of one another through other men in their profession. C.W. Gilmore and some of the other greats in the profession often passed on information from one brother to another when exchanging friendly correspondence with one of the others. With such cement they could only prosper.

This sharing of a profession made this family extra special. It formed the double-bond that welded the family together in a manner that simply would not allow it to be pulled apart. Because this is such an infrequent occurrence in our world today, it is a matter of some interest. Each of Charles' sons genuinely loved the profession, and each was a fitting figure to be pointed at as being of some prestige. There was no forcing of an occupation upon these sons; they clearly relished their respective lifestyles and were proud of their family's accomplishments. What a tribute that is to them as a family, and to Charles Sternberg as the patriarch and main motivator.

NOTES

1. Interview with Charles Mortram Sternberg by Free-lance Broadcaster Laurie LaMaguer, Courtesy of Tyrrell Museum, Dept. of Paleo.

2. Ibid

3. Senckenberg Mus Ltr, Sept. 14, 1910 (to Dr. Fritz Drevermann)

4. Charles H. Sternberg, *Hunting Dinosaurs in the Bad Lands of the Red Deer River Alberta, Canada,* Privately pub, Lawrence, KS, 1917, Pg. 122

5. Charles H. Sternberg, *A Story of the Past or the Romance of Science,* Sherman, French & Co., Boston, 1911, Pg. 57

6. CHS, *Red,* Pg. 156

7. Ibid, Pg. 156

8. Ibid, Pg. 157

9. File Obituary at the Royal Ontario Museum by Loris Russell & Gord Edmund

10. Personal Communication

CHAPTER TWENTY-THREE
CHARLES AS AUTHOR AND LECTURER

One of the most rewarding results of reading any of the various writings of Charles Sternberg is that one can savor the full pleasure of his eloquent descriptions of places and things over and over again since the passages have the lilt of poetry and make for easy memorization. He had a knack for making what might seem the mundane and the humdrum on the surface seem vibrant and worthy of our attention. This zest was most likely just a translation of the innate zest that he experienced in just living from day to day, converted into a literary form. The man's writings simply could not be separated from what he himself was down deep, regardless of what he was writing about.

By some standards Sternberg could be considered a prolific writer, particularly so when considering the time frame from which he wrote. He did, afterall, write much of his shorter articles and poetry by the light of many campfires and not from the ease of a cozy study. Many of the early fossil hunters and paleontologists were non-writers except for the technical journals and the job-related necessities expected of them. Charles, on the other hand, by nature of his independency, did not have to write and therefore did so simply out of a love for the craft. Writing to him was a means of expressing the joys that he found in living from day to day. While he certainly would have been paid for some of his articles, one expects that it was not really the compensation which ultimately led him to write. He simply found fulfillment in it.

In the course of his lifetime Charles was to produce three full-length books: *Life of a Fossil Hunter, Hunting Dinosaurs in the Bad Lands of the Red Deer River, Alberta,* and *A Story of the Past or the Romance of Science.* Each of these books was different from the other two, was lengthy and had a broad scope, and he mainly marketed them himself. The last two books were published through his own auspices, and he was to carry the full burden for seeing that money was made on their production. It seems, therefore, that we can safely assume that he was a man willing to gamble on his own literary prowess, and that he expected his books would be well-received, which has generally proved the case over the years.

The fact that his *Red Deer* was just republished again in 1985 by the NeWest Press in Canada, and *Life* by the University of Indiana Press in 1990 certainly seems to bear out this lasting quality.

His books, however, were not published without a great deal of effort and exasperation. Charles himself was to labor over his books. He took pride in their writing and wanted them to be perfect. To that end he always took every precaution to see that "experts" would review his work to make it as right as possible. He showed the same diligence here as he did when he toured the eastern museums to consult experts as to how best to mount dinosaur material.

Charles was to rely quite heavily upon Henry Fairfield Osborn when it came time to seek expert technical assistance. Osborn was to allow some of his staff to use their time to read and edit Charles' book-length works. William King Gregory, on staff at the American Museum, took a great deal of time to look over the manuscript of *Life* at Charles' behest, and he was apparently to interface with Holt and Company personnel who were to eventually publish that work. Gregory was also apparently instrumental in obtaining the services of a Miss Margaret Wagenhalls of New York who was to serve as editor. Holt and Company was not the first publisher to consider publishing Charles' work, and it may well have been through the urgings of Gregory and Osborn that the book was published.

Life is the more vibrant of the two autobiographical works, although both of them have great merit. *Life* has a robust quality, a zest, that makes it exceptionally readable. Because it was his first book Charles seems to have taken more care with it in terms of layout and readability. While the fossils are perhaps not as spectacular as those found within the pages of *Red Deer*, the glory of fossil hunting and the sheer adventure of it all comes across much more. This often seems to be the case with first ventures.

Red Deer, on the other hand, is more innovative. His efforts at capturing the awe of dinosauria through his picturings of how they must have been in real life was both innovative and in keeping with his personal charm as a story-teller. His nephew, Ray Martin, recalled that as children he and his brothers could listen to his stories for hours on end.

This picturing, or breathing of life into bones, was perhaps something borrowed from Edward Drinker Cope. If not borrowed, then certainly honed. One has only to recall Charles' being enthralled by Cope's conversation as they rode side by side on their Cow Island experience:

> On the way he [Cope] fell into one of his frequent absent-minded moods, picturing the land as it must have been at the time of the dinosaurs, when the shale of these black-sided canyons was mud on an ocean floor. So fascinated were we both by his descriptions that the time flew by unheeded...[1]

Charles was to ratify this "picturing" quality in a form letter excerpt dating from 1911:

> In this poem [*Story of the Past*] I have tried to give my own ideal picture of animals and plants, of the different ages, from whose sediments I have collected during the last forty years. Taking my readers into the distant past and allowing them to see through my eyes, not dead fossils alone, but land and water and air teaming (sic) with a strange and varied life. I have tried to put meat on the dry bones and show as in life some of the wonders of creation I have discovered.[2]

In addition to these full-length books, he was to publish a number of other articles and items for other sources. Some were of a limited technical nature and others more of a popularized account of his expeditions and their finds. Some of the more technical (but still readable) articles were produced as reports about the activities of expeditions in the *Transactions of the Kansas Academy of Science*. There were a number of years in the early 1900's where he would specifically relate the finding of various fossils that he and his sons had made in the midwestern chalk, as well as providing anecdotal incidents that had occurred during the previous year's season.

He had early in his career written an article on prehistoric man which was, by Sternberg's claim, published in *Science* under Cope's name. There appears to be no confirmation of this fact in any of Cope's writings, nor does there appear to be any animosity in the way that Sternberg describes it. Under those set of circumstances it can only be assumed that Sternberg had somehow lent his article to Cope to be published in that manner. This was published under the title *Pliocene Man* in the *American Naturalist* in 1878, and it is the earliest of his writings that I have been able to locate.

Additionally, Sternberg was to write popular kinds of articles, such as the article in *Popular Science Monthly* from December 1929 entitled *Fossil Monsters I Have Hunted*. This article was intended for an amateur audience, and was much different from his semi-technical form of writing in that its audience would obviously have been those who would be titillated by articles on "monsters." Even in this kind of writing, though, one has to appreciate Sternberg's verve in making his science fun for the average man. He was always interested in converts to the field, and it is especially evident in his writing.

Appendix B in the rear of this book lists more completely the various articles that Charles Sternberg and his three sons published during their lifetimes, and will make more obvious the breadth of subject matter that each was to cover. Each, with the exception of son Levi, was a reasonably prolific writer in their own individual way. Each made a contribution that will be remembered in the annals of their science as well.

Now that we have at least noted the major efforts that Sternberg made in his writings we can look at what really counts, and that is the quality of his writing. Most people that come to Sternberg's writings for the first time cannot but be impressed with the way in which he is able to take a well-worn picture and bring it to life. Good examples abound in his works, but some of my favorite ones relate to the ways that he is able to take what some might consider dreary old landscape and make it jump into being:

> How often in imagination I have rolled back the years and pictured central Kansas...Go back with me, dear reader, and see the treeless plains of to-day (sic) covered with forests. Here rises the stately column of a redwood; there a magnolia opens its fragrant blossoms; and yonder stands a fig tree. There is no human hand to gather its luscious fruit, but we can imagine that the Creator walked among the trees in the cool of the evening, inhaling the incense wafted to Him as a thank-offering for their being. All His works magnify Him. The cinnamon sends forth its perfume beside the sassafras; linden and birch, sweet gum and persimmon, wild cherry and poplar mingle with each other. The five-lobed sarsaparilla vine encircles the tree-trunks, and in the shade grows a pretty fern. Many other beautiful plant forms grace the landscape, but the glorious picture is only for him who gathers the remains of these forests, and by the power of his imagination puts life into them;...[3]

He also managed to breathe the same life into the bones that he used to hunt, as his imagination again soared off into revery:

> We are back again where the two mosasaurs did battle royal for our enjoyment. Watch that ripple! It is caused by a shoal of mackerel scurrying in toward shallow water, in a mighty column five feet deep. They are flying for their lives, for they have seen behind them their most terrible enemy, a monster fish with a muzzle like a bulldog's, and huge fangs three inches long projecting from its mouth. Two rows of horrid teeth, one above and one below, complete its armature. The great jaws, fourteen inches long and four deep, move on a fulcrum, and when they have dropped to seize a multitude of these little fish, they close with a vise-like power. The crushed and mangled remains pass

down a cavernous throat to appease a voracious appetite.[4]

To me this is a most fiercesome description, and one that cannot be excelled for the picture it brings to one's mind. After looking at the bones of one of these huge denizens of the early deep one cannot but be impressed by their sheer immensity, and their lack of a modern counterpart. Sternberg stages the action for us well, while at the same time providing an apt picture of the horridness of their capabilities. This serves to give us as vivid a picture as that which is served up in any of the modern-day horror films, and certainly one with a great deal more realism.

This early visualization method of writing was trend-setting. Sternberg seems to be the first science writer of whom I am aware that used this technique. It is used with frequency today, and can be found for instance in the deCamps' *Day of the Dinosaur* and Horner's *Digging Dinosaurs*. It appears that Sternberg may have borrowed this, in a sense, from the tongue of Cope who used to talk to Charles in a like manner as they rode or worked.

In addition to the pictorial vignettes that Sternberg is able to provide, we cannot overlook his ability to take emotion-wrought circumstances and translate them into a realistic experience for the reader. When he describes being suspended in the air between two facing cliffs with only his spread-eagled limbs and geology pick between him and certain destruction we can almost envision the ground below us. His hair-raising account of traveling with Cope through the badlands at night gives one a real sense of the fear that was evoked in him personally, and his depiction of the river barge incident whereby one of his helpers is almost decapitated by a wire strung across the water is exceptional. This is true adventure, and it is this kind of writing that enthuses a whole new generation of adventurers and that brings new readers and writers to the fold. Sternberg was an expert at this kind of writing, and there are few who would prove his better. His grasp of the genre is much like that of the more widely known Roy Chapman Andrews, but his writing excels Andrews' works in technique.

Sternberg's "ordeal of the cliffs" is too choice a piece of writing not to be given coverage here, as it instills real feeling for the situation in the reader:

> When I was halfway over I began to slip, and confidently raising my pick, struck the rock with all my might. God grant that I may never again feel such horror as I felt then, when the pick, upon which I had depended for safety, rebounded as if it had been polished steel, as useless in my hands as a bit of straw. I struck frantically again and yet again, but all the time I was sliding down with ever-increasing rapidity toward the edge of the abyss, safety on either side and certain and awful death below.[5]

I found myself pulling at the edges of my armchair to keep from falling down with him into that great and yawning abyss, and found myself relieved to learn he escaped injury. That is good writing!

The power of persuasion is substantial and Sternberg clearly has a major impact on his reader even when he's wrong. In the few instances where he takes up an issue that is clearly contrary to what we know (or assume to be) true today, you still cannot fail but to be impressed with his logic, and by the depth and fervor in which he has approached the issue. This was not a wishy-washy man, and he took firm umbrage when he felt something was incorrectly being fostered upon the unsuspecting public. This is the case, for instance, with Charles' assumptions regarding the Hadrosaurs. Since Charles was to discover with his sons that webbed feet were a part of the integument of their famous mummy finds, the natural assumption was logical in suggesting an aquatic lifestyle. Although the point is by no means settled it does seem that there is now contrary evidence (mainly from stomach food contents) that seems to refute this interpretation.

One also cannot neglect to make mention of the poetical quality of Charles' writings since it is an area where he also excelled. Some of the more evident blendings of prose and poetry are found in the language of his book-length writings:

> On we journeyed, through what seemed an interminable expanse

> of sage-brush, greasewood, and sand. The bunches of sage-brush topped conical mounds of sand, whose sides were scoured and polished by the winds that howled in and out through the labyrinth of hills, laden with drifting sand. If one could have gained an elevation above the level of these sandhills, and looked out over the landscape, one would have gazed over a scene of even greater desolation than that afforded by the parched short-grass plains of western Kansas,--a dreary, monotonous waste of olive green, stretching away north, east, and south, as far as the eye could reach, and shut in on the west by the great ranges of the Sierras, whose flanks, dark below the timber line with heavy forests, were deeply scarred above with glistening white glaciers.[6]

Without wanting to place too much emphasis on the fact one would have to say that his poetic efforts (published mainly in the volume *The Story of the Past*) fail miserably in relation to the lilt of his prose-poetry, perhaps because its very structure forces it to be something it is not. It is as good as many other offerings that have been lauded as "good poetry" erroneously, but at the same time it does not have the luxuriousness that the prose has. It is at least as good as the poetry of Rod McKuen and the like which has general appeal, even though they could not stand the test of any real criticism.

While it may not be great poetry, it nevertheless remains interesting, and tells us much about Sternberg the man and what he believed to be true. Many of the things that allow me to be able to guess at his true views on evolution, for instance, are taken from this volume, since he discusses it in a way that he cannot discuss things in his other writings. These stanzas make areas of his life open to us that would simply have been guesses had we not had them available to us.

Sternberg was to try his hand at most kinds of poetry, and we even find a manner of "love poetry" in his works. It is not the more traditional love poetry as that of Donne, Herrick or Marvell, but more in the vein of Swinburne. He didn't publish love poetry designed for his wife, or poetry even dedicated to his wife, but wrote his poem to his beloved daughter lost early in life.

An investigation into Charles Sternberg's writing ability would not be complete without some kind of examination of his letter-writing skills. His letters, particularly his business letters, are written in haste (or seemingly so) and therefore lack the polish that his books show so clearly. This leads one to speculate that writing was a task for him that was not as easy as it might appear at first reading. While it is true that the same vehemence and concern can be seen in the way that things are stated, and while one recognizes that business letters are not ready vehicles for translating philosophical or emotional thoughts, one still finds more stiffness and lack of fluidity in these letters.

In a letter to William Diller Matthew of the American Museum Charles was to state just how important he felt letter writing was to his very existence:

> I suppose you are also very busy, but I live and breathe in the atmosphere of my correspondence and I long for it as a hungry man for food. I have no one to talk with on the subject so near my heart, and though I may prove a nuisance by my importunity, always remember my whole heart is with the advancement of paleontology and bear with me.[7]

Over his many fossil collecting years, Charles was to find need for various styles of business letterhead. There are at least 10 major examples of fundamentally different letterhead, and many others with more minor variations. He was fond of having a picture-plate of some fossil skeleton or plant leaf as the focal point of his letterheads, and his son George was to follow in that same mode.

It appears that Charles would keep some of his old letterhead around and would reinstitute it at later dates in order to freshen up things a bit. Some of the major "pictured" letterheads were:

1902 - 1904	*Dolichorhynchops osborni* Williston
1904	*Diospyros rotundifolia* #26 Lesquereux
1903 - 1907	*Betulites westii*, oblonga #21, Lx
1904	*Betulites westii*, var. latiffolius #10
1905 - 1906	*Hedra ovalis*, Lesq. #39
1906 - 1909	*Betulites westii*, var. crassus, Lesq.

Then there also was the commercial letterhead for his secondary business of

"Sternberg's Silicious Soaps" (see Chapter 24). Each of these various letterheads was clearly intended to catch the eye of museum curators and potential fossil purchasers, and was an attempt to legitimize his credentials since he was not a degreed paleontologist. To this end they certainly were effective.

One suspects that letters he would have written to Anna in their courtship days would have been filled with both forethought and also the same kind of "lilt" we find missing in these business letters, and yet they are helpful in that they give us a clear sense of the importance he placed on the enforced "business" aspect of his livelihood.

The Smithsonian National Museum was kind enough to provide copies of about 200 or so letters between the museum curators (mainly Gilmore and Gidley) and Charles, George and Charles M. Sternberg. These letters provide a wealth of biographical information and allow us real insight into the nature of how this family transacted business. They give a clear indication as to the confidence that the museum had in the family, the closeness of the relationships that developed in writing and the genuine fondness that each had for each. George's communications with C.W. Gilmore certainly go well beyond a simple business-related letter and make it evident to the reader that he felt a great fondness for Gilmore, and Gilmore for him. This certainly accounts in part for the fact that the Smithsonian continued to use George as expedition chief year after year.

The letters also make mention of the real hand-to-mouth existence that sometimes befell the Sternbergs. The fossil business did not always show great profit, and sometimes the finest of fossils could not be sold because of the economics of the time or the nature of the breadth of certain collections. In these times of hardship the Sternbergs still managed to maintain a sense of decorum that was enviable.

You might well ask how a sense of hardship pertains to his abilities as an author, and the answer lies in the way that it shows us how he maintains his style even when under adversity. It also gives us a clear sense of his pride. One cannot read these lines without sensing the difficulty he has in writing about these hard times, and how his heart was hurt to the quick to have to mention them. This valor under duress is a trait any author would aspire to.

In a 1929 letter to C.W. Gilmore he was to write:

> THE GREAT DISAPPOINTMENT of my last years expedition was the failure of Dr. M to secure my great skull As I had run out of money and was in debt, there at Thoreau some $500 But for the assistance of Mr. B.I. Staples I could not have finished my work here, or brought in to the station the result of my discoveries. For months I have been helpless, and Levi has been obliged to go to work as a common laborer to help keep off the wolf.[8]

Besides the grammatical errors (which suggest speed and perhaps agitation), the most evident thing about the passage is the openness that Sternberg expresses. This is decidedly different than the reticence we normally find in his revealing such matters. Sternberg showed the ability to translate his emotions into words with ease.

Sternberg was justifiably proud of his writings, and since he was forced (due to the economics of the situation) to pursue the marketing of his own books, he was to send out flyers proclaiming the merits of his books to museums, natural history institutions and agencies with a scientific bent. In one such flyer he was to sample some of the press comments that he found the most provocative. His choice of the numerous comments available is interesting, and tells us something about what he considered important. The choices are not really surprising, but simply bear out what we know from other sources:

From the *Chicago Herald*, March 20, 1909

> Any body will instantly feel the spell of interest in Mr. Sternberg's autobiography '*The Life Of A Fossil Hunter.*' Mr. Sternberg writes simply, unpretentiously, entertainingly, and there runs all through his book a curious union of scientific devotion and religious reverence that is as unusual as it is charming.

From the *San Francisco Argonaut*, June 5, 1909

> There are few hunters of live game who can tell so good a story, who has seen so much adventure, or experienced so many escapes. Such a record would in any case be interesting, but it becomes

fascinating from the exuberance of its style and hearty enthusiasm that animates every page.

From *The Interior*, June 17, 1909

But he not only stuck to his self-imposed task but raised a whole family of boys, every one of whom took to fossil hunting as a duckling does to water. Best of all, to the Christian reader, it will seem the author kept his faith in God and the Bible unimpaired, and his pages are full of descriptions of praise to the Maker of heaven and earth.

From *The Lawrence Gazette*, March 8, 1909

A remarkable book. The author has a way of telling things that is charming because of its simplicity. He uses scientific terms only when necessary, and a child could read and understand this book.

Lest the last press release be considered negative (in the light of its perhaps suggesting that the book was too simple and therefore unscientific) be sure to understand that Sternberg was capable of a much more weighty form of writing, but chose not to because he well understood his audience.

These brief press reviews highlight the things that we would note as indicative of him as a man: religious, a family man, a father who passed on his trade, a lucid and vivacious writer of adventure stories, a self-made man, and most importantly a man happy with his life and the purpose he saw to that life.

So as to be fair in assessing Sternberg's book in the light of book reviewers it is probably best to look at some other reviewers not picked as samples by Sternberg himself:

From *The Nation*, May 13, 1909

It may be said that, although the author does not pose as an original and scientific investigator, one cannot read his book without getting a good general view of current paleontology, and of many of its chief promotors, both of the present and the past.

This reviewer is both right and wrong about Sternberg's merits as an author. He is certainly right in saying that Sternberg did not "pose" as a scientific investigator, but his manner of saying so implies that Sternberg was not in actuality scientific. Here I would part company, as it is obvious to me that Sternberg, while unschooled in the higher learning aspects of science, still had a much firmer grasp of what was really going on in the science than did some of the "promoters" who are given the credit.

From the *Journal of Geology*, May 1909

> A few errors of a minor character which subtract little from the general readability of the book should be mentioned.

Obviously this reviewer was taken by the book, but felt there were areas where Sternberg was not scientifically correct. This is rather interesting since William King Gregory had personally read over the work himself! However, no one would claim his writings to be errorless, and the same can be said of the "promoters" of the science. By its very definition a true scientist must be wrong sometimes. He can only rely on what is suggested to him to be true at any given point in time. Whether that truth is the same truth as held by others either at the same time or at a later date is basically irrelevant since knowledge changes and new information becomes available from which to form one's opinions.

Sternberg made very little money for all of the effort he put into his various writings. Many of the articles he wrote paid either paltry sums or nothing. He was to note, for instance, that the *American Inventor* had only offered him the sum of $3.00 per page for a series of illustrated articles that he apparently never completed.[9]

J. Mark Cattell, editor for *Science* and *The Popular Science Monthly*, wrote Charles in 1903 to indicate that they could not pay for his illustrated articles, but that as an offset they could provide him with free advertising space in the *Popular Science Monthly*.[10]

However, one must note that things were considerably cheaper in those days. Charles was, for instance, to market his *Story of the Past* through the publishers Sherman, French and Company of Boston at a price of only $1.10 a copy. His edition of *Life of a Fossil Hunter* actually sold for $1.75 per book.

Although Charles was not a jealous man by nature he did at the same time jealously guard his name and always sought assurance that it would remain affixed to the fossils he collected, and the books he would write. At the birth of his third son Charles, he most likely never gave thought, nor even suspected, that name problems might result in later life. Who at that date would even suspect that the son would follow in his father's professional footsteps? Yet clearly the first same name did cause some problems. In a letter of Jan. 10, 1915 Charles was to note to Charles Gilmore of the National Museum:

> As I wrote you my name has been taken off the list of those who receive paleontological papers from the National Museum. C.M. Sternberg now receives the papers I used to get...I do not begrudge him copies of these papers and think he certainly deserves them but I do not think when all my life I have received these papers I should be deprived of them by an evident mistake and confusion of names...if only one copy comes to one family I leave it with you to decide who is the one most entitled to it.[11]

Evidently Charles also found himself upset over the name confusion between he and his son as it related to his own books. This appears evident if one looks at some of the copies that Charles peddled himself. I have personally seen at least three title pages where Charles had apparently penciled in his full middle name in the frontispiece of copies of *Life of a Fossil Hunter*. He also curiously placed two small hash-like marks below the "S" in Sternberg. I have, however, also seen copies where these changes have not been made. Perhaps these other copies were sold directly through the publishers or Charles only made these distinctions after a certain date.

Closely allied to Charles' writing was his lecturing. It seems that he enjoyed this form of communication as well. Being a gifted story-teller this must have been a natural form for him, and surprisingly enough, even his more introverted sons were to frequently make use of this method of espousing the profession.

Charles often spoke (perhaps taught is a better word?) at small group gatherings such as church-sponsored youth groups, boy scout packs and the like.

He often supplemented his talks with slide lantern scenes, more commonly known as stereopticon slides. It is not clear just how early Charles made use of these slides, but their use was common during the period from about 1900 to 1940. These slides were the precursor of today's movie projectors. It is also not clear as to whether or not Charles actually had his own illuminating equipment or whether he simply relied upon the various organizations to provide the equipment. It is sure, however, that he had his own set of slides.[12]

One suspects that the quality of Charles' slides would have been excellent considering his reputation for exactness. In December of 1914 at the 6th Annual Meeting of the Paleontological Society in Philadelphia Charles read his paper on *The Evidence Proving that the Belly River Beds of Alberta are Equivalent with the Judith River Beds of Montana*. He spiced up the reading of this fine paper with both lantern slides and photographs of the Bear Paw Shales.

In retrospect one has to say that Charles Sternberg was talented at both writing and lecturing, and that this talent was an offshoot of his story-telling abilities. Because he had personally lived out many of these adventures he was always received well and audiences of all types, professional or no, appreciated what he had to say. This simplicity of style and entertaining story-telling always attracts a willing audience. Clearly he was both an excellent speaker and an author.

NOTES

1. Charles H. Sternberg, *The Life of a Fossil Hunter,* Henry Holt & Co, NY, 1909, Pgs. 88-89

2. Form Ltr, USNM, Sept. 18, 1922 (Sent to C.W. Gilmore)

3. CHS, *Life,* Pgs. 16-17

4. Ibid, Pg. 59

5. Ibid, Pgs. 72-73

6. Ibid, Pg. 155

7. AMNH Ltr, Mar 27, 1903 (To William Diller Matthew from Lawrence, KS), From the Archives of the Dept. of Vert. Paleo.

8. USNM Ltr, 1929 (To C.W. Gilmore)

9. AMNH Ltr, Apr 23, 1903 (To Henry Fairfield Osborn from Lawrence, KS), From the Archives of the Dept. of Vert. Paleo.

10. Ltr, Mar. 25, 1903 (From J. Mark Cattell [ed of *Science*] to CHS), Materials Courtesy of the San Diego Nat. Hist. Mus. Archives

11. USNM Ltr, Jan 10, 1915 (To C.W. Gilmore)

12. Paul F. Long, "The Stereopticon; Magic & Light," *Kanhistique*, Vol #16, #9, Ellsworth, KS, Jan 1991, Pgs. 2-4

CHAPTER TWENTY-FOUR
SIDELIGHTS

It should probably not be too surprising to us to find that Charles Sternberg and his sons were all multifaceted individuals, and actually had a life outside the fossil arena. Their resourcefulness and their ability to take discarded material and put it to practical use, honed in their early years at farming, has already been described. This native ability slipped over into their personal lives as well and perhaps made it inevitable that there would be side occupations that would take up a fair amount of their time.

Among the more evident diversions in Charles' life were his farming episodes, his fruit tree and juice "business," his preoccupation with clays and silicious rock uses and his hand soap manufacturing business. Each of these activities was to take up some of his time for extended periods, often helped to fill up those time frames where fossil collecting had to be curtailed, and each brought either the potential for funds or actual dollars into the household.

Charles was always cognizant of the value of a dollar, and has been criticized by a number of people for what they considered his preoccupation with money. This is patently unfair. Although it must be granted that money was frequently mentioned in most of Sternberg's letters, one must also temper any criticism by pointing out that he was never "flush," (and if we are to believe his son Charlie, he was generally underpaid for his services). Since this was the case, he was literally forced to constantly harp on the subject. He can, perhaps, be judged as foolhardy in the sense that he should have asked for fairer wages rather than allowing others to set these parameters for him. Being beneficent in the name of science is commendable, but not always the wisest course of action.

A hint of Charles Sternberg's personal dissatisfaction with his financial position can be found in an extract of a letter to Henry Fairfield Osborn of the American Museum. Anticipating a couple of important sales of some large vertebrate fossils Charles was to write:

> If I can make that sale it will enable me to make my credit good in my own town, where every one has trusted me on my honor alone as I have nothing of this worlds goods that could be sold to pay bills. Only a little home for wife and children.
>
> As you know I am tired of the struggle of trying to sell fossils, as the effort is greater than the expense of collection and preparation, and I hope that I can spend my remaining days in the service of a paleontologist.[1]

As early as 1884 we find plentiful evidence of Sternberg's diversions. Charles' penchant for fancy letterhead had already resulted in a bold printing that heralded his manufacturing business of "Sternberg's Silicious Soaps." This was to be the earliest of his letterhead designs (see chapter 23) and yet it was the most provocative. The letterhead itself had a much more "modern" look to it than those that were to follow, and has the look of being professionally designed. The bold repetition of the "S's" is really quite striking.

In a letter to Othniel C. Marsh Charles was to highlight this secondary business, and made it obvious that he had every intention of pursuing that work as well:

> I send you by this mail a sample of fine silica that I discovered in this formation (Loup Fork). I am very anxious to have it analyzed & thoroughly [tested] & its value for commercial purposes noted. I manufacture in Lawrence a silver soap for polishing silverware & bright surfaces & also a laundry and toilet soap [comprised?] of this sand.

Will you kindly request a competent chemist to look into the matter. Prof. Bailey gives the following analysis...

I think it will be found valuable to be used in the manufacturing of common soap instead of Rosin because of its detersive qualities. I am slowly building up a trade in my soaps (sic) give satisfaction to all who have used it. I would like to know whether it is possible to convert it into stone without injuring the grit, as it would make a fine oil stone. Or for many things like the (place?) of emery wheels.[2]

In later correspondence to Marsh that same year Charles intimated that the business was not yet really off the ground, but soon would be: "This winter will be likely the only time I could go to Texas as my manufacturing business would prevent it. I will go into business this winter unless I go into the field."[3]

It is clear that Charles' mind was not held to the simple restrictions of handsoap, but looked at a multiplicity of other uses for the rocks and clays he was to encounter. He looked to oil stones, emery wheels, silverware polish, fuel components, and other uses yet for these materials.

In 1910 Charles was to write to Professor Edward Orton Jr. of Ohio University requesting a testing of a particular clay sample.[4] It seems likely that somehow this testing was perhaps connected with this same manufacturing business. Charles always seemed to be aware of the possible different ways in which the land could be utilized.

Also in 1910 Charles was to query about and eventually receive a set of fifty different specimens of industrial clays from various parts of Germany. These were supplied him by Dr. Fritz Drevermann of the Senckenberg Museum. The set also included a series of "metamorphic stones from the Taunus, sediments and eruptiva" properly labeled.[5] Presumably Charles' request for these items was commercially related as well.

Up until the later part of the 1800's Charles clearly considered his manufacturing business and related work as supplementary income (and perhaps fun too), but always gave precedence to his fossil business. Charles Mortram

(Charlie) Sternberg was to indicate that his father, however, grew very disenchanted with the state of things around the turn of the century and actually seriously considered a complete change of occupations. Charlie indicated that his father had discovered a location that promised to provide much valuable volcanic ash that could be used in the "old dutch cleanser," and a deal was apparently struck in order to provide the ash at a given price. This deal, however, fell through when another individual patented another process that somehow caused the value of ash to fall drastically.[6] Charles therefore went back to his beloved fossil work, and perhaps with renewed vigor. Considering how much more Charles was to discover in fossil material after this date it is intriguing to speculate as to how great a loss this dalliance could have cost the science of paleontology.

Before the turn of the century Charles had shown evident interest in what he labeled as "artificial fuel." A search of patent records shows that patent #316,580 was issued to Charles Sternberg on Apr. 28, 1885 for a "Composition for Fuel." This fuel was intended to take "materials heretofore nearly useless for this purpose" and to suggest their use as materials for use in wood or coal stoves to produce heat.[7]

His patented fuel was a 10 percent combination of rosin and asphaltum mixed with a 90 percent combination of slacked or powdered coal, sawdust, wood shavings or chips, ground corncobs, chopped weeds and dry manure. These ingredients were then pulverized, heated to a "sticky consistency" and then made into bricks or blocks through a mold process. The finished blocks would then be dipped into a solution of hot rosin which apparently set the product and made the bricks transportable.

I have found no added reference in any of Sternberg's voluminous correspondence to this process so one can only assume that it gained no favor as a fuel source. The most interesting fact about the process, it seems to me, is the fact that it sharply defines Sternberg for us. He was a man used to looking at natural things in relation to their potential for use in other ways as long as such could be done cheaply and easily. So not only did he look at the land in terms of its fossil potential but in terms of its mineral potential as well. This too, considering Cope's mining debacles, may have been suggested by Cope's forays into these other areas.

Since we're on the subject of patents here it should be noted that Charles was hardly the only Sternberg family member to hold a patent. As previously mentioned, his older brother George was to patent an anemometer and also a heat regulator, his cousin James Hervey Sternbergh was to patent an iron-nut machine, and another cousin A. Irving Sternberg of Grouverneur, New York, was to hold patent #581,315 for another type of heating apparatus.[8] There was also a Levi Sternberg from St. Louis (who may or may not have been related) who held patent #421,326 for an innovative type of suspenders. In any event there is clearly no question that the whole of this family was oriented to thinking about better ways of doing things.

There are also many other diversions hinted at in the letters to and from Charles Sternberg, but research has so far turned up little background as to their significance. A letter to Charles from a D.E. Lantz of the U.S. Department of Agriculture, for instance, hints at Sternberg having had some governmental connection with coyote and gopher bounties. Was Charles perhaps a county official at one point with duties that involved tabulation of bounties?[9] If so I have been unable to verify such a fact after extensive research.

Sternberg also received a letter from a D.H. Otis who was affiliated with the animal nutrition section of the college of agriculture for the University of Wisconsin who was inquiring about a canning device. Otis wanted the canner as an example of a labor-saving device to be used in a course in what he termed as "domestic science."[10] The fact that he was referred to Sternberg by others seems to bolster the idea that Sternberg was thought to be very much up on those farm and product machineries that saved time and money.

The involvement Sternberg had with canners might well be a result of his early farming experiences, as well as his experience with fruit trees. In January

1893 Charles was to purchase 20 acres of land near Lawrence from an Alfred A. and Susan A. Greene. This land purchase was to prove to be a good investment in a number of different ways. Not only did the land provide a place of residence for his burgeoning family but it provided both an area to garden and another source of family income with the fruit trees that Greene had apparently planted. The fruit itself both helped feed the family, and provided added income for the family when they sold the fruit and its byproducts locally.

Alfred A. Greene, from whom Charles was to purchase this property, was quite famous in his own right, having shown a true "green thumb" when it came to fruit trees. An early and prolific property manager, Greene was to own seven different farms in Jackson County alone, and on each of these farms he had planted orchards. A biographical sketch of Greene was to claim "it is conceded that he has done perhaps more than any other man in this region in encouraging fruit growing, as he was the first to forward this industry among the people of the new state."[11] From this, although Charles doesn't tell us so specifically, we can assume that the Lawrence property he purchased from Greene was liberally stocked with fruit trees. This certainly would have made the property more valuable to a growing family, and would have contributed extensively to their family income possibilities.

Son Charlie was to substantiate the fact of the family gaining some of its income from apple trees on their property for he was to note in an interview that they would take the "wind-fall" apples to a cider mill for processing into cider. He indicated that a bushel of apples produced two gallon lots of cider. Charlie would then sell the cider to the locals at around 25 cents a gallon.[12]

Charlie also acknowledged selling both apples and grapes to the local Indians attending a neighborhood school (Katherine Rogers, another biographer has called this the Haskell Institute)[13], the cost being a nickel for an unspecified amount of fruit. Apparently the school was segregated since he mentions preferring going to the girl's side where he was always paid. When he sold to the boys they often took advantage of him by taking packages of apples and not paying for them.

Charlie also was to acknowledge further that the family (i.e. other than Charles H. Sternberg) "made a lot of the money" through its operation of selling apples and apple cider, and their other side jobs (such as his side job of working at the local tanning factory). This was a necessity apparently for Charles often could not sell his collected specimens for an amount exceeding what it actually cost him to collect. These sidelights therefore often sustained the family and allowed Charles to keep collecting even though he might have occasionally worked at a loss.

While the sons were certainly less diverse than their father in terms of actually performing other kinds of work (once they had actually settled into their permanent occupations) they still maintained that native ability to innovate, and to make do with things at hand. George's homebuilding ingenuity as previously described is a good example of this, and there are a multitude of available examples of how each of the brothers used these types of skills to overcome problems on the fossil fields. Problems with cars, animals, and terrain were all easily overcome because they had a good understanding of these things and were not afraid to improvise. These were indeed very self-sufficient men who in a very real sense lived off the land.

NOTES

1. AMNH Ltr, Feb 22, 1904 (To Henry Fairfield Osborn), From the Archives of the Dept. of Vert. Paleo.

2. Yale Univ Ltr, Jul 9, 1884 (To O.C. Marsh), Othniel Charles Marsh Papers, Manuscripts & Archives

3. Yale Univ Ltr, Sept 28, 1884 (To O.C. Marsh), Othniel Charles Marsh Papers, Manuscripts & Archives

4. SDMNH Ltr, Oct. 27, 1910 (To CHS @ Lawrence from Prof. Edward Orton, Jr., Ohio State Univ, Columbia, OH), Materials Courtesy of San Diego Nat. Hist. Mus. Archives

5. Senckenberg Mus Ltr, Oct 10, 1910

6. Interview with Charles Mortram Sternberg by Free-lance Broadcaster Laurie LaMaguer, Courtesy of Tyrrell Museum, Dept. of Paleo.

7. US Patent Ofc, Sunnyvale, CA

8. A. Irving Sternberg was also a student at Hartwick College

9. U.S. Dept of Agri Ltr, Mar 24, 1909 (To CHS from D.E. Lantz), Materials Courtesy of San Diego Nat. Hist. Mus. Archives

10. Univ of Wisconsin Ltr, College of Agri, Jan 2, 1909 (To CHS from D.H. Otis, Materials Courtesy of San Diego Nat. Hist. Mus. Archives

11. *Portrait and Biographical Album of Jackson, Jefferson and Pottawatomie Counties, Kansas*, Chapman Bros, Chicago, 1890, Pg. 537

12. Interview with Charles Mortram Sternberg

13. Katherine Rogers, *The Sternberg Fossil Hunters: A Dinosaur Dynasty*, Mountain Press Pub Co., Missoula, 1991, Pg. 79

CHAPTER TWENTY-FIVE
GEORGE MAKES IT ON HIS OWN

George's first real job completely unrelated to his family was that of a field collector for the American Museum of Natural History. During this period of time (the year 1912) he worked under the direct authority of Barnum Brown, one of the most prolific fossil hunters who ever lived. He worked for Brown in the Red Deer of Alberta, along with Brown's assistant Peter Kaisen. It appears that it was Henry Fairfield Osborn, however, who was to solicit George's services, for Charles was to comment in a letter to Dr. A. Smith Woodward that "Dr. Osborn saw him and wanted him."[1]

There is no question that George and Brown each respected the other for their substantial abilities, and yet there is also no question that they were as like as night and day. George was exceptionally unassuming, and "midwestern" in his lifestyle, if you will. George, although not naive by any means, was hardly the kind of dandy that Barnum Brown was, even though Brown was a born and bred midwesterner. Brown was the other extreme. Brown was well known for his penchant for wearing "fancy clothes" on the fossil fields. A bow tie was frequently part of his uniform, and he would more often than not wear an african safari pith hat of the type that was so common in early movie making, when the white hunter was the leading man's role. Numerous pictures of Brown attest to his unusual mode of dress for field work, but there was never any question as to anyone forgetting they had met Brown, for his dress and exuberance were noteworthy. It is hard to imagine two personalities so different.

Not only was Brown noteworthy himself, but he used to often bring along his noteworthy wife Lilian Brown. The author of several books of a popular nature on the subjects of fossil hunting and travel as seen through the eyes of a companion and confidant to her famous husband, she was a frequent camp visitor who would then assist in the cooking. Although George Sternberg would sometimes bring his wife and child into camp also when he was working for his father, it was under much different circumstances. George was not the dandy that Brown was, nor did he have the traveled jet-set aura about him that Brown always had. Pictures of Brown are always stately, and those of George are always homely. They were certainly opposites, except when it came to their abilities in tracking the elusive dinosaur.

One could at the same time point to some similarities between Brown and George's father Charles since Brown was a Kansan at heart, and was brought up working hard on the family farm. Brown's fledgling interests in fossils were stirred when he found to his delight that they could be ferreted out from the overburden turned up by the plow. It was in this way that he found his first two fossils, a "petrified cornucopia and a piece of fossil honeycomb."[2] Because of his mounting interest he was to study under Charles' friend Samuel W. Williston, and was to work under the tutelage of both J.L. Wortman and James Bell Hatcher, who had both briefly been students of Charles Sternberg.

Roland T. Bird in his biography, *Bones for Barnum Brown*, was to call his boss, Barnum Brown, one of the "three Kings of Dinosauria." The other two in his estimation were Cope and Marsh. He labeled Brown as the "biggest" of the three, although we must certainly admit he had to have been prejudiced in the matter. If he were speaking strictly of personal collecting ability we would have to wholeheartedly agree with him, since Brown was always at heart a field man. Neither Cope nor Marsh, however, no matter how great their overall claims for greatness, could match any of the four Sternbergs for field experience and technique, and it is interesting that Roland T. Bird mentions these three particular men as kings since the Sternbergs were intimately involved with all three men at one time or another. Perhaps then we should call the Sternbergs knights?

Bird himself is of some interest because he is very similar to George Sternberg in many respects. They both had the same country boy air about them, both learned their trades on their own and both had a penchant for their field vehicles (Bird his motorbike/home and George his trucks). They both also held a great respect for the talents of Barnum Brown. Both Bird and George were field men at heart, and the other functions of paleontology were more tolerated than sought after.

The reasons for George leaving the employ of the American Museum are not really clear, but they most likely relate to George's inbred desire for freedom. Working for any institution that restricted his personal freedom would have been difficult for him. Organizations (such as museums) that rely on public and private funding always find that there are a number of financial watchdogs around to ensure that their monies are being spent according to their wishes, and the organization therefore always remains somewhat accountable. This tendency towards control and encumberment would have gone against George's grain, and it is something we also saw in his father as when he parted company with the Canadian Geological Survey in his dispute with Dr. Lambe.

In the diaries kept by Peter Kaisen, Brown's assistant, there is reference to a building division between the camps of Charles Sternberg and Barnum Brown, which Kaisen attributed to the Sternbergs not keeping their word when it came to the superficial boundaries that were established by the two camps. It was apparently a ticklish affair, and heat was fanned by the fact that there was the feeling that George, as a relative, was leaking information to his father's camp. Renie Gross, in his book *Dinosaur Country; Unearthing the Badlands' Prehistoric Past*, provides a very readable and fair appraisal of the whole theatrical episode for those interested in further detail.

George, perhaps as well, was also wanting to be associated with his family again, since upon leaving the American Museum he was to immediately join up with his father, who was then working the Dead Lodge Canyon area in the Red Deer of Canada. This occurred in 1912. George was to spend five years with the Canadian Geological Survey (for whom his father also worked) at the position of chief collector and preparator, before leaving in 1917 to return to independent collecting. His father had already left the survey by then (1914), along with Levi, in a dispute of some sort with Dr. Lambe, and George was to lead his first party in the North the following year.

In an early letter to C.W. Gilmore George was to express dissatisfaction with things in Canada claiming he was "not satisfied with conditions here," and he at the same time inquired about an opening at the Field Museum in Chicago. The text of the letter implies he was encountering difficulty too with Dr. Lambe, and that it was this which prompted the job search. George was particularly incensed over Lambe's refusal to allow George to send photographic copies of a new Trachodont skull to Gilmore. Lambe's reasoning was petty in this instance it seems, for George says in the same letter "he [Lambe] in so many words said he would not o.k. the recquisition for me as the letter [requesting copies] was sent to me and not to him..."[3]

George's brother Charlie, however, was to assess the reason for George's departure quite differently. Charlie was to claim that Lambe had given him [Charlie] complete financial control of both expedition parties in 1917. Since George was to head the second party he was particularly upset at being passed over for his younger brother. Charlie said that Dr. Lambe "had to arbitrate" and he further added that there "was some hard feeling; a little bit, couldn't help it."[4]

The present curator at the Sternberg Memorial Museum, Dr. Richard Zakrewski (pronounced Zak-chev-ski), points with some amusement to the fact that George was very specific on his official permit to leave Canada in filling in the slot labeled as to the reason for leaving Canada with the words "for good." One suspects an underlying vehemence there that goes beyond the words themselves.

George was to indeed join the Field Museum in 1922. He was to remain at this famous institution (founded by the same man, Marshall Fields, who founded the specialty department store of the same name) until late 1924. George's certification to collect for the Field Museum (now on display in the Sternberg Memorial Museum in Hays, Kansas) was dated Sept. 27, 1922 and was signed by then director D.C. Davies. It notes George's Chicago address as 5515 Blackstone Avenue. George's passport that would take him to Patagonia is also there, it being issued on Oct. 13, 1922.

George's family life in the early days was also interesting. His first wife Mabel was the one who used to join him in fossil camps along with their second child, Charles W. Sternberg. They were eventually to have a third child whom they were to name Ethel.

Certainly the life of a fossil collector was a rough business as well for the wives. In more recent years paleontologists have the benefit of being able to bring wives with them on some trips, which does help to defray some of the tension that can develop with marital separation. George and his father, however, only infrequently could afford such a luxury. The divided loyalties caused when a wife is along on a true field trip might perhaps prove a plus as far as the marriage is concerned, but more likely negative as far as the science is concerned. Since the Sternbergs had to pay out of pocket all expenses it would have been normally quite difficult for them to have brought their wives regularly.

In 1919 George was to work for a short period again in Canada and its Cretaceous fossil fields. This time, however, it was for the University of Alberta. In late November 1922, however, George was to have his greatest opportunity: the chance to go to Patagonia! Patagonia was the spawning ground of some of the greatest fossil hunting talent ever to be. Carlos and Florentino Ameghino were the reigning monarchs there of course, but John Bell Hatcher had plied his craft there (1896-1899) as well as Barnum Brown (1898), Dr. Friedrich von Huene (1924), and Frederic B. Loomis (1911). George was to travel there under the auspices of the Field Museum, and under the immediate direction of Elmer Samuel Riggs. Although this man, and his successor, Bryan Patterson, are not quite so famous in scientific social circles, they were men of great integrity and talent. In his book, *Discovers of the Lost World*, George Gaylord Simpson touts the abilities of both men, and attributes them with being what he describes as sort of Mutt and Jeff types; Riggs as "pedestrian" and "predictable," and Patterson as "more interesting" and "more congenial."

George was working for Elmer Riggs and the Field Museum in Canada when the call came for them to head for Patagonia. With almost no notice the men headed out on this expedition. When given a chance to work in this fossil hunter's paradise nothing could have held them back. They took with them the chief preparator at the Field Museum, John B. Abbott, who was also working with them in Canada.

Patagonia (land of "Big Feet") is split from north to south, with the western half being owned by Chile and the eastern half by Argentina. It was really the Argentine areas that George and party were to work.

The extremely primitive terrain of the Patagonian pampas is matched by the primitive nature of its inhabitants. The rough and ready existence of life in these parts is very similar to that of the early settlement days of the western United States. It took a hardy individual to survive the rigors of life there.

During their lengthy stay in Patagonia the party met with considerable success. Since the barren and denuded terrain of Patagonia is so open to scouting, fossils are in a sense easier to find. One cannot, however, say this without mentioning the flip side of the matter; and this is the physically demanding nature of the terrain and the weather which is so rough and unpredictable. For those interested in reading about the rigors of fossil hunting in Patagonia they could consult either J.B. Hatcher's *Bone Hunters in Patagonia* or F.B. Loomis' *Hunting Extinct Animals in the Patagonian Pampas*. Both of these narratives give one an excellent feel for the countryside that George and the other party members had to work in.

Although there were many fossils uncovered on their trip they have been given less recognition here in the United States because the fossils belong to that fuzzy group of animals often labeled as "South American." Since these fossils are not really analogous to our fossils in this country and the north they don't seem to have the same recognition quality.

Frederic B. Loomis of Amherst College, who headed a party to Patagonia a decade before the Field Museum party, commented on the special nature of past Patagonian fauna, and the peculiar fact that the terrestrial fossils were of more immediate importance than the marine fossils:

> The animals represented by our collections of bones represent a very specialized fauna comparable to none found on any other continent. It is from the relationships of the land animals that conclusions as to whether they have migrated from one continent to another are formed, and thus as to what connections a continent had.[5]

George was to be the one to uncover the greater part of the upper dentition of an Adianthine Litoptern named *Proadiantus ameghino* (1897). George made this find of the hoofed mammal and ungulate family in the Deseado formation of Cabeza Blanca, Chubut, Patagonia. It was labeled as Field Museum #P14698 and was subsequently described by one of the party principals, Bryan Patterson, for the Field Museum.

George was to remain in Patagonia until late fall of 1923 when he returned to the states along with his fossil hunting partner, John B. Abbott. When they left, they thought, as did Riggs who remained, that they would return after the winter had passed to resume the work they had been forced to quit due to the dictates of the insensitive Patagonian climate. Alas, however, again budgets got in the way and the return trip was never to be for the transportation costs just proved too prohibitive for the museum at that time.

Riggs tried to stop George and Abbott from leaving when he learned of this snafu, but they were already on their way and could not be recalled. Undoubtedly if they had been given the option to remain they would have chosen to do so, even though they had already been a year and a half apart from their families. The expedition's success, like that of the parties who had gone before them, was eminently successful. One cannot ignore the irony in the situation, since George and the rest of the party followed directly in the footsteps of John Bell Hatcher, the same Hatcher who was briefly schooled by George's father, and who attempted to take the elder man to task for his "carelessness."

Then in 1925 George was to return to Kansas, and he was to spend most of his time from then forward in Kansas (and its surrounding states) until he was finally to quit the field. On his initial return to Kansas George made his residence in western Kansas in the town of Oakley. It was from here that he began to promote his independent fossil collecting business.

Just why George decided to settle in Oakley is not quite clear, but is most probably related to Oakley's proximity to great fossil fields. Oakley, which is situated some 90 miles due west of Hays is a jumping off point to some major fossil-bearing beds of the Cretaceous period. Most of these are in a southerly direction from Oakley.

George himself was particularly enamored of the "pyramids" that lie about 25 miles southeast of Oakley. These sedimentary deposits, carved by the natural elements into figures resembling castles, faces and other rarities rise 60 feet or more from the Kansas prairie. To the southwest of Oakley lies a great river bluff called "Jerusalem" or the "Chalk Bluffs" that line the Smoky Hill River. This formation was also a great favorite of his.

In 1927 George was to become affiliated with the Fort Hays State University Museum as Curator, and it is with this institution that he was to complete his long career. In agreement with the university, George, who was on a meager salary, was allowed ample time to collect as an independent. This loose arrangement, somewhat similar to that of his father at the San Diego Museum, was ideal for both George and the university.

This change in employment made it necessary for George to move east to Hays. After his death the museum was to change its name to the Sternberg

Memorial Museum in a fitting tribute to the man who was to design it almost single-handedly, and who was to fill its cases with top-notch fossils of every description. The Sternberg Geology Club, an offshoot of George's expertise, was to make the proposal to the Board of Regents after his death, and the action was quickly taken. The museum continues to be a major draw to anyone interested in paleontology, and is perhaps the best museum of this type in the midwest. The college has certainly benefited, and that perhaps accounts in part for why the museum was named in his honor after his death.

If one visits the Sternberg Memorial Museum nowadays one cannot help but be impressed by a number of factors. The campus is clean, well kept and has an Ivy League look about it in large part because things are so green. Some of the brick work, however, is distinctly midwestern since it is literally filled with the shells of area fossils. The museum is spare in its look, but stocked with a wealth of first class material that would compliment any museum in the world. The main display room, although small contains the wonderful skeleton-within-a-skeleton that helped to make George famous as a collector. This magnificent find, from beds near Castle rock in Lane County, was a giant fish by the name of *Portheus molossus* (now known as *Xiphactinus audax)*, and if the fossil were not great enough on its own it had the added benefit of housing within its 12-foot frame the then freshly eaten skeleton of a six-foot fish called *Gillicus arcuatus*.

This fish within a fish has inspired many, and in one case it has even inspired the writing of a children's book on fossils. This book by Aliki Brandenburg, *Fossils Tell of Long Ago,* written from a child's vantage point, lends credence to our contention that this has to be one of the greatest fossil finds ever discovered. If this is so (and there are many who would agree with me), one would also have to say that the Sternbergs as a family were at least responsible for finding two of the top 10 fossil discoveries ever made. Perhaps this is doing the Sternbergs a disservice in limiting them to two, but then again there have been many fossils discovered since the profession began.

There is real irony in this whole situation since George himself certainly underestimated the value of the *Portheus* find. This is easily seen when one considers that he was only asking a total of $500 for this fossil in 1926, and had offered it to C.W. Gilmore of the National Museum at that sum, while asking substantially more for equally fine but less noteworthy fossils.

Although it is not the same animal, George does note the taking up of another *Portheus molossus* in late July of 1926; it is what he claims is the third of three that season. His description of the notoriety it caused is interesting for he notes in a letter of Aug. 5, 1926:

> We found it necessary to camp on the ground night and day as there was a continuous stream of cars bringing people from all directions to see it, I therefore had to stick close to the job of taking it up. There were over 50 people there the last Sunday though it was entirely cased in plaster slabs...[6]

The summer expedition of 1928 for the Smithsonian was again to include George as the main field assistant, with Gilmore in charge and a man from Hays, Edwin Cooke, as cook and helper. The summer season's work was to be spent in the northern region of Montana, just below the border. The area of badlands in the northwest corner of the state was on a Blackfeet Indian Reservation, and it was not an unknown site since Gilmore himself had collected there before. This was back in 1913 when he had extracted a skeleton of a *Brachyceratops* from the area while working for the United States Geological Survey.

The *Brachyceratops* was a smallish ceratopsian about six feet in length ("short horned face") that was a close relative of *Monoclonius*. This was to be the type skeleton for this particular animal, and hence, proved of major importance. The general area in which it was found is Upper Cretaceous, and has become known as the Two Medicine Formation after a river which cuts through the region near Cut Bank. This is the same area that Charles H. Sternberg was to complain to Gilmore about after being denied entrance to the reservation by the bureaucrat Mather, and hence, there is some irony in all this. The area was particularly intriguing since it was right across the border from the Milk River Formation in Alberta, and all available data suggested there would be valuable material to find, and that the two formations were related.

Gilmore was to note in his recap of the season that their camp was established on the south side of the Milk River where they were to work from May 16 through June 2. He further notes that their work provided "only fair results." Of the things found, the most important was the partial skeleton of a *Hypacrosaurus* (high crested duck-bill) which included part of the skull. This was found on the north side of the Two Medicine River by George and was to become museum catalogue #11893 at the National Museum. The skull of this important find was described by Gilmore in a paper he was to write in 1937 on behalf of the museum. Additionally they were to find a section of *Styracosaurus* frill (first occurrence outside of Alberta), and another skull of an armored dinosaur.

Once they had worked out the south side of the river they moved to the north side of the river, and worked there from June 2 until the end of the season in July. Gilmore was more pleased with the finds on this side of the river. The finds here included a complete skull and a substantial but partial skeleton of a little known armored dinosaur called *Panoplosaurus*. *Panoplosaurus*, or "fully armored lizard," was a Nodosaurid ankylosaur from the late Cretaceous approximately 23 feet in length.

They also discovered a partial skull and skeleton of a *Monoclonius*, and a veritable treasure trove of hadrosaurian material, particularly valuable because it was that of a small duck-bill and small skeletons are a real rarity in the science. By the time of their quitting work at this site they had found nine different lower jaws at this same spot, so it was quite evident that there were a number of different individuals involved.

The following season's work (summer of 1929) found George again working for the Smithsonian, except that on this trip they were to travel to New Mexico for work. The results of this trip, as well as his father Charles' trips there, are in chapter nineteen. I placed the emphasis together since it was one area, outside of Kansas, where the family worked independently, and not together as a team as was their earlier practice. For that reason it is of additional interest.

The summer season of 1930 was another Smithsonian summer, with George again as field assistant for Gilmore's expedition into the Bridger Basin of Wyoming. George's crew this year included George B. Pearce as both collector and cook. Their season began in the latter part of May, where Gilmore met them in Green River, Wyoming and from whence they traveled first to Fort Bridger for supplies and then on to their first campsite at a locale known as Smith's Fork, which Gilmore notes was in the vicinity of the small town of Mountain View.

The first campsite that year proved to be in poor hunting grounds so they moved to a site at the head of Little Dry Creek where they were to meet with considerable success. Gilmore indicates that the creek area was paralleled by an escarpment, and that this provided perfect rock within which they could search. Gilmore also recounts an amusing story that is reminiscent of the Sternberg stories of chapter fifteen:

> One day in crossing a small water course the car became stalled in the soft mud of the creek bottom. Looking about for stones with which to block up the wheels, we noticed a rocky layer protruding from the bank and Pearce was instructed to get the pick and pry out some of it for ballast. A stroke or two with the pick brought an exclamation of surprise, for on the under side of the first slab detached was the complete skull of a crocodile in excellent preservation. A most happy surprise and a valued addition to our then small accumulation of fossils.[7]

On June 11 they moved camp to Leavitt Creek where they found good material without having to be "surprised." Between this site (where they spent a full month) and the next site on Henry's Fork (where they began work on July 16) they found considerable material. This material really made the expedition worthwhile since the initial pickings had been poor. Gilmore was pleased with both the quantity and quality, and went so far as to label a few of the specimens as of "an outstanding character." The magnitude of the finds is best left to the imagination when one realizes that they filled 24 large cases with material, and that their total weight came to 7,430 pounds.

Gilmore notes that their season finds included an articulated skeleton of the rhinoceros-like animal *Hyrachyus,* another nearly complete small skeleton of what they thought in the field was *Orohippus* and two substantially complete crocodile skeletons. They were to find later that the "horse" skeleton presumed to be *Orohippus* was actually a small mammal named *Helaletes* from the tapir family. The mount made from this skeleton proved to be the first of this little animal.

They also found a wealth of complete turtle material that often included the rare skull finds as well as other often non-preserved bones. Gilmore notes that there were 38 different specimens found, and that they represented a number of different genera. This area was so prolific in turtle remains that Gilmore was to say "they were packed together so closely that it was impossible to remove one specimen from the mass without damaging a number of others."[8] This area was apparently as wealthy in turtle material as was Charles' material from the New Mexico sites.

George was to remarry later that year in November around the same time that his oldest son Charles was to enroll at Kansas University. George's second wife was Anna Ziegler, the daughter of a prominent Hays family, the T.G. Reeds. They were married in the home of her brother, T.M. Reed, Jr. on West 17th Street in a small private ceremony attended only by the principal parties. The service was performed by a local minister, Dr. C.F. Wiest. Immediately after the spare service the two went on a honeymoon trip that included a visit to Carlsbad Caverns in New Mexico, and a visit to his son Charles at school in Lawrence, and his daughter Ethel in Lincoln, Nebraska.

George and Anna had a good and solid relationship, something missing in his first marriage. He and Anna were to eventually share 35 years of married life before her death in 1965.

C.W. Gilmore in a letter to George around the time was to reference the fact that George had talked about the happy event taking place sometime that year, while they were on the expedition in the summer of 1930. Gilmore was to write further in a letter dated Nov. 29, 1930, "Congratulations on your marriage, I am sure you will be happy and it will be a much more satisfactory way to live than formerly."[9] Such a statement implies that Gilmore as a friend knew much of what was happening to George at the time, and that the circumstances of the first marriage had perhaps been not too favorable. This first marriage of George's was to end in divorce sometime after 1922, purportedly the final straw having been the prolonged Patagonian stay.

Later in December of that same year George was to write to Gilmore inquiring further about the following season's work, and he was to state that his son Charles was home for Christmas vacation and assisting him, and that Charles wanted to be considered for part of the crew on the next expedition.[10] In order to make ends meet Charles had been cooking at a restaurant while attending school. George graciously gave Gilmore plenty of room to say no, and Gilmore responded honestly in return stating that he felt Charles "has no especial interest in the work" and that he therefore had to say no. He reiterated that he found the matter "not [an] altogether agreeable task to perform."[11] So, at last we have found a Sternberg who was not to find his life's work to be that of a fossil hunter! It is also a measure of the relationship between George and C.W. Gilmore that each could know the other so well that they could handle such a potentially volatile matter so sensibly.

As a footnote to this, however, Charles W. Sternberg was to stay relatively close in professions, as he was to become a successful petroleum geologist, working out of North Carolina. Of this his father could certainly be proud. Charles also was to publish articles on a few different fossils in national periodicals before deciding to move into the petroleum field.

In late May of 1931 George was to be part of an expedition with the National Museum into the Madison River country of Montana. George acknowledged, "I know nothing about that locality so will be glad to see it."[12] They were also that same season to travel to Lance Creek where they were to pick up fragments of a *troodon*, and from there they moved on to Lusk. George's wife, Anna, was to take "her son and with the Pontiac coach make the trip at the same time we do"[13] and was to stay near Belgrade before making a visit to relatives in Idaho.

It was also in 1931 that George was part of a Smithsonian expedition to the Big Horn Basin in Wyoming. Gilmore was again to be in charge. In a paper entitled, *A Taeniodont Skull from the Lower Eocene of Wyoming*, from April 1936, C. Lewis Gazin was to note that George was to find an extremely important fossil skull of the little known taeniodont, *Ectoganus gliriformis* Cope, that had previously been described only by fragments which had been found by J.L. Wortman back in 1881. The "rat-like" specimen was located in the Gray Bull beds of the Wasatch, about eight miles north and west of Worland, in Washakie County, Wyoming. The specimen (USNM #12714), although partially damaged was almost complete, and hence it made it possible to be more exacting about the nature of this little animal. They proved definitively that the teeth were the "deciduous and little worn permanent teeth of what had been called *Calamodon*."[14]

During this period George's wife Anna was to prove an invaluable assistant and she did the majority of his correspondence. Letters, and other writings before their marriage were generally handwritten or were typed rather poorly. After their marriage the quality of his correspondence was vastly improved, and she was to assist further by working diligently on his fossil lists. There is every evidence of her devotion to him, and he to her.

During the summer of 1932 George was to again work for the National Museum, and part of the season was spent in the Judith River area on the Missouri. George had requested a shorter season due to his desire to build a house that summer, and therefore would only commit to about two and one half months of work. Gilmore himself admitted his time was limited also as he was "head over heels in the Mongolian dinosaur work now..."[15] and so it fit in well with his plans as well. The former assistant curator at the National Museum, James Gidley, was to be replaced that season (he had died in Washington on Sept. 25, 1931) and George's new field partner was to be Dr. C. Lewis Gazin. Gazin started with the museum in March 1932, and was to quickly become a fast friend and associate of George.

In March 1932 George was to actually begin building the home that he had set his sights on earlier. As with most things, problems immediately arose when the weather refused to cooperate. After digging out the foundation they were beset upon by a terrific blizzard that quickly filled the hole they had dug out with snow. And this calamity was not the only source of difficulty, for additionally he encountered cash flow problems when he couldn't cash his "giltedge securities" as anticipated. For that reason he begged to be able to start the season later, and this wish was granted.

George's house, by his description, was a five room, one bath English-style abode of wood frame, with a four room and one bath apartment in the basement. His initial estimates were that the house was to cost about $5,000, but it undoubtedly came to more than that before completed.[16]

To show George's versatility one could cite many different instances, but it is a measure of his resourcefulness that we find he traded fossils to receive in return an old building that was located on the Hays College grounds, plus $125 in cash.[17] This deal was apparently struck with the College President, William A. Lewis. George then gave half the cash to the carpenter who was helping him build on his property, and between them they tore down the old building to use for wood for his new home. Some of the extra wood he was able to sell for $25 and he still had much wood left over. This fine deal allowed him to build a garage at little cost so that he could finally house his cars that had always had to stand out in the elements. This incident makes it obvious that George's resourcefulness was not simply restricted to the fossil business.

The 1932 season was to begin on June 15th or so from Harrison, Nebraska. The season was to end in September with George coming back from Sydney in a light rain and with only a few detours to detain him.

Field work in 1933 differed in that Gilmore finally stepped aside from active summer work, and Dr. C. Lewis Gazin took over the reins. Gilmore was to describe Gazin to George as a "fine young man but with limited field training."[18] George was certainly the man to teach him. Like his father before him, he had a real gift for teaching the trade. Before the season arrived, however, the season was canceled by Dr. Alexander Wetmore, assistant secretary of the institution. The reason apparently was connected with the money troubles brought on by the country's economic woes.

The year 1933, however, was still to prove a good year for George. Kansas State College at Ft. Hays was to honor George with an honorary degree for his fine work in vertebrate paleontology. George had finally reached the nadir of his profession, a distinguishing mark similar to that which his father had been able to attain in 1911 when Charles had received an honorary degree from Midland College in Atchison, Kansas. George was particularly pleased with the tribute to his life's work, and had posted conspicuously the program for the festivities in his memento scrap book.

1933 also was to provide George with a two-month stint in the field for Childs Frick. Considering the tight money situation in the country this was indeed providential. It also again says something about the regard in which George was held by the industry. Gilmore was to comment upon the difficulties that led to the canceling of the 1933 field season in a letter to George dated June 20, 1933: "Congress has at last adjourned and we now can breathe a little easier. For all reports the Museum may be able to get by with short payless furloughs to its personnel -- and perhaps none."[19]

George was to have only a moderately successful season that summer, and found some of the material disappointing. He collected in the vicinity of Oberlin, Kansas down near Phillips and Norton Counties. Gilmore had a healthier respect for the season's output and indicated to George by letter that he considered the collection of their Miocene horses to be of interest; enough so that he would try to trade with Frick for the remains from their excess of *Plesippus* material.[20] This material was taken at the latter part of the season, and helped to make the season a more valuable one.

In the summer of 1934 George also worked for the National Museum, again for Gazin. Childs Frick had attempted to get George for another season, but George's longterm commitment to the National Museum took precedence. George was to advise Gilmore how pleasant he found it to work with Gazin, calling him a "prince of a fellow."[21] The 1934 season was to bring a plethora of fossil horse skulls from a site in southern Idaho. In July George had already tallied the total to be around 50 in number. The west quarry at Hagerman he described was exceedingly difficult, as he noted that around 40 to 45 feet of loose sand had to be removed in order to establish a floor from which to work. After removal of this, however, he indicated that the bone was exceptionally easy to extract. They apparently spent their off hours playing rummy to pass the time.

In November of that same year George was to travel northward to Canada with his son Charles at the invitation of the University of Alberta. Their job was to prepare mounts of the dinosaurs George had left there some 14 years prior. His estimate was that it would take four to five months to complete the job. They were working under the direction of Dr. Allan. George denoted noticeable surprise that his son Charles finally showed a spark of interest, "I am somewhat surprised at the interest Charles is showing in the work, guess letting him hustle for himself for a while didn't hurt him. I hope he continues to show interest."[22]

In a letter a few months later where Gilmore opted to use George B. Pearce again instead of George's son Charles, George was to comment in return, "It's entirely satisfactory to me that you use George instead of Chas. However, I would not have been surprised if he had preferred to stay at home this year."[23]

Obviously the hoped for burst of activity and interest from his son Charles was short-lived, and things were back to status quo. Although Charles had some talent for fossil work he just didn't have the longterm interest that makes for a good paleontologist.

George wouldn't give up, however, for he mentioned to Gilmore in a letter in May 1935 that he had two "prospects" for work for Charles, and that he felt it would be valuable for him to spend some time with another institution. This apparently culminated in Charles working for Childs Frick in the summer of

1935, under the direction of C. Falkenbach. The party was to spend a substantial amount of time in Clarenden, Texas.

In October of 1935 George's wife was to finally seek medical attention for a breast lump and was forced to an operation in Rochester, Minnesota where her right breast was removed. Not only was the operation and its affect on his wife of turmoil, but the cost of the operation forced George to adjust his field plans. It additionally caused him to have to cancel seeing his two brothers, Levi and Charlie, who were to be visiting Hays along with their wives. In late November Anna was "getting on fine," but George noted that she had to have additional x-ray treatments, treatments which upset her stomach. She was also to have further treatment in Denver during December.

On Dec. 6, 1935 George's daughter Ethel, who had married Lloyd Beeman, was to give birth to a baby boy and George seemed delighted at the new addition to the family. George had remained close to his daughter over the years, having traveled with her as well, and as with his father before him in the loss of Maude, he missed being around her.

In April 1935 George was to visit his father in Pasadena, California. George was to visit Berkeley and Los Angeles enroute. Probably as a result of his father's example he always used these jaunts to visit prominent paleontologists and their museums, and geologic sites that were in the areas visited. These visits undoubtedly helped to keep the family up on the latest of the profession.

The years after 1935 were important ones for George too. He continued to spend many of the summer seasons working for the National Museum. As late as 1947 he was still plugging away, after the enforced interlude for the war years. In 1941 he again worked for Gazin, this time in the Knight Beds of the Green River in western Wyoming. Their crew that year included F.L. Pearce.

The war years were lean years for field work for all of the countries' museums, and it was next to impossible for anyone not employed directly by a museum to work for them during this period. Since a great deal of George's income was based on his ability to continue servicing the major museums this came as an unexpected shock, and George definitely lost much in opportunities to collect. After the war, however, things seemed to quickly get back to a form of normalcy and he was to again work for the Smithsonian. The 1947 season was, however, his first after the war, and his last in their employ. The 1947 season was again to New Mexico, except this time in the vicinity of Lamy.[24]

Something should certainly be said of George's ability to recognize and develop men with real potential in paleontology. He would help to mold their natural ability into a real talent. The men that he was to take under his wing would become very proficient collectors, also sought after by the major museums. Since George was to be an independent collector for most of his life, and headed up expeditions, it was essential that he have valuable men who could be called upon to serve in the various camp capacities. This is a talent his father had also been able to master.

One of George's mainstays, who could generally be relied upon to provide good collecting abilities was Myrl V. Walker. Sternberg had met Walker...In a letter dated February 1928 George had tried to get Walker on with a summer expedition for the National Museum. George was to say:

> Walker is an enthusiastic collector with a keen eye. He found several good specimens for me, and all through his high school and college work has collected minerals and fossils. No doubt he would make good though he does not care for the cooking and camp work.[25]

George was to further say that Walker planned to major in paleontology, and that he was somehow connected with Dr. Wooster who was willing to relinquish Walker. Gilmore, however, was not able to use Walker that season.

Walker was to be a friend for many years, both to George and the rest of the family. He had married a Miss Wilda Opdycke. He was to temporarily leave the profession when he joined the National Park Service and assumed duties at the Grand Canyon National Park as park naturalist. George was to make at least one visit to see him there. It is only fitting that upon George's retirement that Walker was to assume the curatorship of the museum, a position he held until his own death. He was replaced by one of his students too, Richard Zakrewski. It was Walker who was to write George's obituary for the Society of Vertebrate Paleontology.

In October 1928 George indicated he had taken on an assistant by the name of George Pearce. Pearce was a graduate of Hays and came highly recommended as a "careful worker, honest and willing."[26] He had apparently initially intended upon being a teacher, but had changed his mind. This was a second man, since George had already maintained a part-time assistant by the name of Edwin Cooke, who had been on the previous year's expedition. Cooke was dropped in early 1929 because he could not give George the necessary hours, and from that point onward George Pearce became his real assistant.

When George's health started to fail it was naturally tragic for all who knew him as such an active and hardy man. George spent his last few years in Hillcrest Manor, a convalescent home in Hays. His failing health had been of concern to all his family and friends. In a letter to Myrl Walker from Ontario in June 1969 George's brother was to write, "Thank you very much for your letter of the tenth inst. Sorry to hear that George is failing mentaly (sic). I suppose something has to go sooner or later."[27]

George Sternberg was to die on Oct. 23, 1969, and is buried at the local cemetery in Hays. The heart of this man, however, continues to live in the halls of the wonderful museum that he helped build and which is filled with the fabulous fossils he managed to extract from their earthen hiding places all around the world and brought for all to see.

In an obituary on the death of George's younger brother Levi, Dr. Loris Russell was to call George "a legend." Considering his accomplishments it would be hard to argue against that assessment.

NOTES

1. BMNH Ltr, June 23, 1912 (To A. Smith Woodward from Lawrence, KS), Courtesy Natural History Archivist, London

2. Geo Soc of Amer Bull, Feb 1964, Vol #75, #2, Pg. P-20

3. USNM Ltr, May 21, 1917 (To C.W. Gilmore from Ottawa, Ont)

4. Interview with Charles Mortram Sternberg by Free-lance Broadcaster Laurie LaMaguer, Courtesy of Tyrrell Museum, Dept. of Paleo.

5. Frederic B. Loomis, *Hunting Extinct Animals in the Patagonian Pampas*, Dodd, Mead & Co., NY 1913, Pg. 133

6. USNM Ltr, Aug 5, 1926 (To C.W. Gilmore from Oakley, KS)

7. Charles W. Gilmore, "Fossil-Hunting in the Bridger Basin of Wyoming," *Explorations and Field-Work of the Smithsonian Institution in 1930*, Wash, 1931, Pgs. 13, 15

8. Ibid, Pg. 18

9. USNM Ltr, Nov 20, 1930 (To G.F. Sternberg at Hays, KS from C.W. Gilmore)

10. USNM Ltr, Dec 29, 1930 (To C.W. Gilmore from Hays, KS)

11. USNM Ltr, Feb 9, 1931 (To G.F. Sternberg at Hays, KS from C.W. Gilmore)

12. USNM Ltr, Mar 9, 1931 (To C.W. Gilmore from Hays, KS)

13. USNM Ltr, May 21, 1931 (To C.W. Gilmore)

14. C. Lewis Gazin, "A Taeniodont Skull From the Lower Eocene of Wyoming," *Abstract Proceedings Amer Philosophical Soc*, Vol IXXVI, #5, 1936, Pg. 597

15. USNM Ltr, Jan 5, 1932 (To G.F. Sternberg at Hays, KS from C.W. Gilmore)

16. USNM Ltr, Mar 23, 1932 (To C.W. Gilmore from Hays, KS)

17. USNM Ltr, Dec 10, 1932 (To C.W. Gilmore from Hays, KS)

18. USNM Ltr, Feb 23, 1933 (To G.F. Sternberg from C.W. Gilmore)

19. USNM Ltr, Jun 20, 1933 (To G.F. Sternberg at Hays, KS from C.W. Gilmore)

20. USNM Ltr, Nov 15, 1933 (To G.F. Sternberg at Hays, KS from C.W. Gilmore)

21. USNM Ltr, Jul 18, 1934 (To C.W. Gilmore from Hagerman, ID)

22. USNM Ltr, Dec 16, 1934 (To C.W. Gilmore from Edmonton, Alb)

23. USNM Ltr, Feb 20, 1935 (To C.W. Gilmore from Edmonton, Alb)

24. Annual Report Smithsonian Institution, Pg. 22

25. USNM Ltr, Feb 27, 1928 (To C.W. Gilmore from Hays, KS)

26. USNM Ltr, Oct 10, 1928 (To C.W. Gilmore from Hays, KS)

27. Univ of Kansas Ltr, Jun 24, 1969 (Myrl Walker to GFS), Courtesy of John Chorn

CHAPTER TWENTY-SIX
THE BOYS IN THE NORTH

The affiliations of both Charles Mortram and Levi Sternberg in the north began at different times, and took different paths. Since the two brothers were very different in their chemical makeup it was natural enough that they would be different in their ultimate vocations, even though they would both stay in the field of paleontology.

Charles Mortram, or Charlie, the elder of the two, began collecting for his father in 1906. In that first season he indicated he was still uncommitted to the science (in fact disliked the thought of it) and went on the expedition only because it was the family business and he needed to do his part. During that first season, however, the Sternberg in him would begin to blossom and the love of fossils began to take hold.

Part of the reason for his developing an interest in fossils took place that first season with him discovering his first fossil. He was to uncover in the Kansas chalk an almost complete fossil skeleton of a bird, one that lacked only the skull. This fossil turned out to be one of some prestige, and it eventually was sent to the American Museum for preparation and mounting. According to Charlie this bird had been previously described as a "small ostrich," but his find proved it to actually be more "loon-like," and contrary to popular previous opinion it turned out to be "not able to stand up at all."[1] With Charlie it was not the romance of the bones that drew him to the science, but rather the detective work that gave him an opportunity to work on the great puzzle.

During the summer of 1907, while on expedition, he was to meet his future wife, Myrtle Martin, while at the Martin family farm. This farm was located about three miles from Castle Rock in the Niobrara chalk, near Utica, Kansas and the Sternberg encampment was nearby. Myrtle was the youngest of the two Martin girls, and she was the same age roughly as Charlie. Myrtle also had a younger brother, who was to later work for a short period with Charlie in the fossil fields.

Charlie had been sent by his father to resupply their water stores, and the Martins apparently graciously allowed Charlie to fill half a barrel from their cistern, even though they were on a limited supply themselves. Myrtle was at the breakfast table with her sister, and Charlie was immediately taken with her. Just a short four years later (Sept. 18, 1911) he and Myrtle were married at the Anglican church in Lawrence. She then moved with him to the Wyoming homestead he had owned for over a year already.

Charlie's homestead was a piece of property 150 choice acres in size, that skirted a bluff near the mouth of Old Woman Creek. He had built himself a log cabin on the property, which he built by using drift logs from the creek itself. He did minimal work on the property, allowing just enough work to comply with prevailing homesteading laws, and he plowed a very limited portion of the acreage. He was to maintain some horses on the property, which were probably used to some degree in the family business, but maintained no other animals. He eventually sold the property (after two or three years) and was elated to find that he could sell it for $1000.

During this period, like most of the Sternberg women, Myrtle was to prove her resilience. She apparently showed no qualms about moving to this isolated area, and adapted quickly. Wanting to contribute to their newly started family she found a job teaching school. Although it only lasted for about three or four months, she did travel daily to a place called Seaman's Ranch, where she taught a limited number of students (variously said to be between two and six). She rode back and forth on a horse named Bob, and earned $60 a month for her teaching abilities.

Myrtle presumably did not teach further because they were to move after a year, and the balance of the year they were in Wyoming was during the growing season when children could not be spared from their chores at home and so school had to wait. Myrtle further showed her adaptability and cleverness by fashioning a serviceable family couch out of orange crates and a plank. Such innovations alleviated much of the financial strain so common with newly married couples newly out on their own.

In November 1912 Charlie and Myrtle moved their household to Ottawa, Canada. Myrtle, who was seven months pregnant, managed the train trip in fine order, and they moved in with Charles and Anna at their house at 101 Gilmour Street. They were to live there with Charlie's parents until early 1914, when they purchased their own home on O'Connor Street between Third and Fourth Avenues.

Before their move to O'Connor Street, however, they were to be blessed with the birth of their eldest son, Raymond McKee Sternberg. Raymond was born on Christmas Day 1912, not long after the family Christmas Day dinner was completed. Raymond, the only remaining paleontologist left in the Sternberg family is really only a paleontologist in name since he found himself having to change professions.

Ray Martin (AKA Ray Sternberg) was to note that there was no real pressure placed on him by any family members in order to see he pursued a career in paleontology. He was to get his degree in the science around the start of World War II. Because of the war there were no jobs available in the field so he became an Air Force meteorologist. At a later date he was to change affiliations and worked for the Royal Canadian Navy, Department of Personnel Selection. After the war he joined the navy and was given the rank of Lieutenant Commander, permanent force. Ray was to note that he never regretted his decision to join the navy since the pay was better and the job stable.[2]

Ray apparently received his middle name of McKee from the man who was to serve as best man at Charlie and Myrtle's wedding. Raymond, as with Charlie's other children, was to change his name later to Martin (adopting his mother's maiden name). He is presently in ill health and suffering from a bone cancer.

Ray was to describe his relationship with his father in much the same way that his father was to describe that with his father. Ray said:

> My father was a quiet man with a good sense of humour. He was an ardent Mason and went through all the chairs in his Ottawa Lodge. He loved my mother very dearly and would do anything for her. He was a good father but didn't really know us very well -- he didn't skate or ski or play football etc. so we missed that kind of contact. He seemed shy at home, but mixed easily in groups. We were all very fond of him![3]

In 1920 the sternbergs were to move to another home on First Avenue, to another again in 1924 in Ottawa South, and then finally back again to Ottawa in 1938.

Both Myrtle and Charlie were to be very proud of their three sons. Second son Stan was to gain some prominence as a microbiologist, after completing his education at Guelph. Glenn, also a professional man, gained his stature as a bacteriologist. Charlie was particularly pleased that each of his three sons had completed advance university training and had gotten their degrees, which was an opportunity that had not been open to him personally.

Charlie began his real tenure with the Geological Survey of Canada in 1917 when he made his first independent collecting trip into the Red Deer around the area of Little Sandhill Creek. That season there were four separate expeditions into the Red Deer itself of which two were led by Sternbergs. Charles Hazelius led his successful expedition into the area around Little Sandhill Creek as well, along with his son Levi, but they were no longer with the survey and were collecting independently. The aforementioned expedition was led by Charlie into the same general vicinity.

The year 1917 was to prove to be the last for Charles H. in the Red Deer, even though he was again very successful. That year he and Levi collected a fine skeleton of a carnivore of the Tyrannosaurus family, which was named at that point as *Gorgosaurus sternbergi* but has since become identified as being the same animal as the previously named *Albertosaurus libratus*. Although smaller than the *Tyrannosaurus rex* this fierce animal lacked nothing in terms of killing efficiency and sheer power. It was sold to the American Museum where it is on prominent display as AMNH #5664. It is labeled as coming from Quarry 54 in C.M. Sternberg's mapping in 1950 of Canadian dinosaur quarries.

Charles Mortram Sternberg's success that year was equally of value, for his party made some spectacular finds as well. There were finds of a partial skeleton of the dinosaur *Panoplosaurus mirus*, an armored nodosaurid with a rather massive head of thick bone, a hooded duck-bill skull of *Lambeosaurus lambei* and a *Centrosaurus longirostris*, or horned dinosaur skull. While these did not represent the entire finds of the season they were the principal successes. It is both a measure of Charles' strength as a teacher, and Charlie's as a student, that Charlie was to prove to be such a capable expedition leader right from the start. Charlie's "book-learning" was certainly of help as well, as it gave him insight into areas he could not known of otherwise.

In what seemed to be the tradition in the Sternberg family, wife Myrtle was to proceed her husband in death. She died in 1977. Charlie was to live for another four years, dying in Ottawa on Sept. 8, 1981, just short of his ninety-sixth birthday.

NOTES

1. Interview with Charles Mortram Sternberg by Free-lance Broadcaster Laurie LaMaguer, Courtesy of Tyrrell Museum, Dept. of Paleo.

2. Personal Communication

3. Ibid

CHAPTER TWENTY-SEVEN
THE STERNBERG LEGACY

The question of legacy, at least as it relates to the Sternberg family, is a multi-faceted one because one has to qualify one's answer by asking "what area or category are we talking about?" We have certainly seen that the family made a substantial contribution to future generations in the areas of fossil collecting, medicine, government, exploration and innovation in fossil technique among other things, but there are other areas worthy of note also, although they are perhaps less known.

While there will perhaps be those that will continue to scoff at Charles Sternberg's unschooled interpretations of the evidence left at fossil sites, these should diminish. In light of recent years, and new discoveries, they bear up as well under time as do those of more professional credentials. Sternberg's interpretations, based upon what he found around his fossils, led him to conjecture about what kind of death or environment he was investigating, and it is exceedingly clear that he thought long and hard about what he was seeing. It is due to this fact that he subsequently made solid contributions by setting a precedence of sorts that suggested to other individuals that they needed to think more deeply about the available evidences also.

This forethought, which voiced itself in the way he examined things carefully before working a fossil site, displayed itself in other areas as well. Similar forethought was very evident, for instance, in his taking a rather lengthy trip to various museums in the United States after working in Alberta so he could consult with a great number of different experts so that the Sternbergs did not approach mounting the valuable *Albertosaurus libratus (Gorgosaurus)* find from Alberta in a way that did not maximize its presentability. This is not a small thing, particularly for a man who has been described (most unfairly) as a plodder of little intelligence. It is perhaps this care and forethought that has led to him being described as a plodder, when the trait is certainly commendable. Ah, but such is the nature of criticism!

In a more personal vein there must be mentioned the importance of the family's emphasis of many years on their religious beliefs. The tradition, as we noted earlier, was established very early in the family, and it flourished. Levi's father John was a minister, as was his wife Margaret's father, George B. Miller and Charles' wife Anna's father, Charles Reynolds. We noted too that Charles' twin brother, Edward Endress Sternberg, was also an itinerant preacher to complete the family religious. It is certainly safe to say that the family had a plethora of pastors and preachers! This being the case, one can recognize how important it was to the family that this tradition be passed on to each successive generation. In that, it appears one would say they would perhaps rate their success at best as moderate.

Charles' children appear to have turned away somewhat from their religious upbringing, and their children also. George, married twice, although at least nominally practicing his faith, was certainly not the vocal and outspoken Christian that his father was. He certainly lacked his father's religious verve. He did, however, continue to support his church and his wife was active in church circles.

Ray Martin, son of Charles M. Sternberg, was to refer to himself and his brothers as "backslid Christians, and not regular churchgoers."[1] Yet Martin was to further say that "my grandfather, father and uncles were all Christians in the best sense of the word."

Sadly, there are few Sternbergs in name left to carry on the traditions of this great family. The Sternberg name has been set aside by many of the family members. Charles Mortram Sternberg was to recount a number of different instances where family members were discriminated against simply because of the Jewishness of their name. Rather than putting up with the bigotry, they apparently chose to change their name to their mother's family name of Martin. All of the three sons of Charlie Sternberg were to do so. Considering the pride that Charles Hazelius Sternberg had in his name, it is sad to see it discarded.

Ray Martin, when asked about his father's reaction to the name change, stated, "I know my father was hurt when my brothers and I changed our name to Martin, but he understood the problems. There was a lot of anti-semitism in Canada at that time and people believed we were Jewish in spite of anything we could say. My brothers and I are nothing but happy with the change."[2]

A very important contribution made by Charles was the fact that he was apparently the first fossil hunter or collector to publish an autobiography in the United States. His *The Life of a Fossil Hunter* was published in 1909, and was in a limited sense at least a new art form for America. To show that it was of obvious benefit one need only look at those who would follow. William Berryman Scott, George Gaylord Simpson, Edwin H. Colbert, Roy Chapman Andrews, Mrs. Barnum (Lilian) Brown, and Roland T. Bird would all publish autobiographical works in the years that followed that borrowed from what Sternberg started.

As was mentioned briefly in the chapter on his abilities as an author Sternberg seems to have begun a form of writing that was new to America. This involved the unique and original style of attempting to flesh out or animate the dinosaurs and other paleolithic denizens in order to give the reader a much better sense of what they were really like. He began this genre in an early article, *Ancient Monsters of Kansas*[3], and perfected it in his more lengthy autobiography *Red Deer*. This type of writing has been frequently copied since. Most recently John (Jack) Horner has used the device in his fine book, *Digging Dinosaurs*,[4] to good effect.

Another area of legacy can be found in the paleontological speculations of Charles Mortram Sternberg. An expert on dinosaurs he had a tremendous influence on Canadian paleontology in general, as well as on the establishment of Dinosaur Provincial Park and its fine museums. There has just recently been confirmation of son Charlie's fine sense of being able to "read" paleontological signs. In a brief article of 1955 entitled *A Juvenile Hadrosaur from the Oldman Formation of Alberta*,[5] Charlie was to suggest that the pattern of juvenile dinosaur finds (as limited as they were at that point) intimated that the nesting of dinosaurs occurred only in the uplands, and that only the mature animals would venture to the more dangerous lowlands. Jack Horner gives Charlie credit for recognizing this fact:

> The hypothesis is simple enough. I think Charles [Mortram] Sternberg was right. He said to look in the uplands, the upper coastal plains. His notion was that the lower coastal plains were too wet and acidic, so that dinosaur eggs would rot or be eaten away, I think he's probably right about the reason. But I know he's right about the result.[6]

Since this revelation comes from Horner, who has found more juvenile dinosaur and egg material than any other paleontologist in the world this is a compliment indeed!

The real lasting legacy with the Sternbergs, however, remains the fossils. One cannot enter any major museum in the United States or in Europe without a direct confrontation with a behemoth of some sort that has been collected by the Sternberg clan. We've already remarked about the wonderful dinosaur mummy that is prominently housed in the American Museum, and on George's magnificent fossil fish-within-a-fish (*Portheus molossus*) that is the focal point of interest at the Sternberg Memorial Museum in Hays, Kansas, but there are certainly many others.

We haven't mentioned, for instance, the large skull of the duck-billed dinosaur *Prosaurolophus maximus* Brown [#12712] from the Belly River Series near Steveville, Alberta that Levi collected in 1931 and is prominently housed at the Smithsonian, nor the numerous invertebrate fossils in the Smithsonian collection that his father contributed. Part of the legacy the Sternberg's left can be seen in the newly revamped dinosaur exhibit at the Philadelphia Academy of Science where on prominent display are two famous Sternberg dinosaur skeletons: a sixteen foot frilled *Chasmosaurus belli* collected by Charles H. in 1914 and a thirty-two foot *Corythosaurus casuarius* collected in 1927 by Levi. These two magnificent fossils add much to a prominent corner in this oldest of American Museums.

In 1927 while working in New Mexico Charles was to find the skull of another Ceratopsian. This was to prove to be the first horned dinosaur found west of the Rocky Mountains. The frilled beast was named in his honor by Henry Fairfield Osborn who was to describe it, *Pentaceratops sternbergii*.

Charles Sternberg prided himself on the many museums that housed his vertebrate fossils. The letter-plate he used (there were several versions) proudly noted that his fossils were to be found in these great museums:

> U.S. National Museum (Smithsonian)
> British Museum of Natural History
> Paleontological Museum of Munich
> American Museum of Natural History
> National Museum of France (Paris)
> National Museum of Germany (Berlin)
> Senckenberg Natural History Museum
> (Frankfurt, Germany), University of
> Tubingen, Roemer Museum (Hildersheim),
> California, MIT, Yale, Harvard, Michigan,
> Chicago, Toronto, Minnesota, Ohio, Iowa,
> Cornell, Princeton, Poukeepsie, Kansas, etc.

This is a small list of those museums in the world holding material collected by Charles Sternberg, and it doesn't begin to recognize the vast coverage by volume in Canadian museums and stateside for items found by George Sternberg and his Canadian brothers. Important to note too in consideration of numbers is that these are priceless fossils we are most often talking about. These are fossils that are both rare and valuable because they form links with related material, because they set precedents for certain areas of research, because they provided material that overturned traditional thinking on certain fossil life, or because they were simply new to science. They were and are, after all, Sternberg fossils!

NOTES

1. Personal communication

2. Personal communication

3. Charles H. Sternberg, *Popular Science News*, Vol XXXII, Dec 1898, Pg. 268

4. John R. (Jack) Horner & James Gorman, *Digging Dinosaurs*, Workman Publishing, NY, 1988, Pgs. 114-115

5. Charles M. Sternberg, *Canadian National Museum Bulletin*, #136, Pgs. 120-122

6. John R. (Jack) Horner & James Gorman, Pg. 194

APPENDIX A
DATE SEQUENCING RELATING TO THE STERNBERGS

Legend: LSS = Levi Sternberg (Sr)   LS = Levi Sternberg (Jr)
  CHS = Charles Hazelius Sternberg
  CMS = Charles Mortram Sternberg
  GMS = George Miller Sternberg
  GFS = George Fryer Sternberg
  MCZ = Museum of Comparative Zoology
  CAS = California Academy of Sciences
  PMM = Paleontological Munich Museum
  ROM = Royal Ontario Museum
  USNM = United States National Museum (Smithsonian)
  AMNH = American Museum of Natural History
  BMNH = British Museum of Natural History

Major	CHS	Other
1710	Major Palatine landing in New York (Jun & Jul)	
1804-06	Lewis & Clark Expedition	
1806		Leo Lesquereux born (11-18)
1807		Louis Agassiz born (05-28)
1810		Asa Gray born (11-18)
1814		Levi Sternberg (Sr) born (02-16)
1815		Hartwick Seminary established (12-15)
1817		Benjamin F Mudge born (08-11)
1823		Joseph Leidy born
1828-32		LSS student @ Hartwick College
1831		Othniel C Marsh born
1835		Alexander Agassiz born (Dec)
1837		LSS married Margaret (09-07)
1838		GMS born (06-08)
1840		Theodore Sternberg born (09-15)
1842		Clarence King born (01-06)
1843		J Fred Sternberg born (03-12)
1845-49	James Polk President	
1845		Rosina Sternberg born (03-08)
1847		Lesquereux to America
1848		Emily Sternberg born (02-29)
1848		Ft Kearny established
1848-50		LSS @ St Marks, Middleburgh, NY
1849-50	Zachary Taylor President	
1849		Military took over Ft Laramie
1849		1st Expedition into badlands (J Evans)
1850-65		LSS Principal @ Hartwick Seminary
1850	CHS born (06-15) in Middleburgh, NY	
1850		Edward Endress Sternberg born (06-15)
1850		FV Hayden 1st explored Judith badlands

1850-53 Millard Fillmore President
1853-57 Franklin Pierce President
1853 William A Sternberg born (03-14)
1854 Kansas-Nebraska Act enacted (05-30)
1854 James H Sternberg asst teacher Hartwick
1855-57 GMS asst teacher @ Hartwick
1855 Albert Sternberg (11-09)
1857-61 James Buchanan President
1857 Williston to Kansas
1858 Gold discovered in Colorado
1858 1st dinosaur in America discovered @ Haddonfield
1858 William B Scott born (02-12)
1858 Frank Sternberg born (03-31)
1858 Marais de Cygnes Massacre (05-19)
1859 *Origin of Species* by Darwin published
1859 Chas Reynolds founded Trinity Church (Aug)
1860 GMS gradulated College of Physicians
1860 Robert Sternberg born (01-19)
1860 CHS injured leg in barn fall
1860 1st train into Kansas
1861 GMS Asst Surgeon National Army
1861 Benjamin Mudge 1st to Kansas
1861-65 Abraham Lincoln President
1861 Civil War began
1861 Kansas Territory became a state (01-30)
1863 Quantrill's "bushwackers" raid Lawrence (08-21)
1863 John F Sternberg enlisted (11-16)
1863 KS State Agri College established (03-03)
1863 Chas Reynolds church @ Ft Scott
1864 Ft Ellsworth [Harker] established
1865 LSS family to Albion, Marshall County, Iowa
1865 Civil War ended
1865 Quantrill killed in Kentucky (06-06)
1865 Thomas Condon 1st to John Day (Fall)
1865 GMS married Louisa Russell (10-19)
1865 Mudge to KS State Agri College
1865 Theo Sternberg admitted to bar in NY
1865-69 Andrew Johnson President
1866 Chas Reynolds Chaplain @ Ft Riley
1866 GMS to Ft Harker
1867 City of Lawrence established (May)
1867 1st Texas longhorns to Kansas
1867 Chisholm Trail to Kansas in full use
1867 Kansas-Pacific Railroad reached Ellsworth
1867 Gen Winfield Hancock's abortive Indian campaign
1867 Louisa R Sternberg died cholera (07-15)
1867 CHS & brothers moved to Ft Harker ranch
1867 CHS "slung-shot" during robbery
1867 GMS @ FtRiley (08-67 to 10-70)
1868 Battle of Arickaree (09-22)
1868 Emily Sternberg wed Frank Humlong (12-03)
1869-77 Ulysses S Grant President
1869 John Wesley Powell's Colorado River trip (05-24)

1869	GMS wed Martha L Pattison (09-01)
1869	George B Miller died (04-05)
1870	BF Mudge 1st expedition in Kansas
1870	OC Marsh 1st expedition in Kansas
1870	CHS fossil leaves to Smithsonian
1871	William Diller Matthew born
1871	OC Marsh trip to John Day
1871	Edward Drinker Cope 1st Kansas trip
1872	Ft Harker "officially" abandoned (Apr)
1872	Lesquereux @ Ft Harker (Nov-Jan '73)
1872	CHS hospitalized 3 months @ Ft Riley
1873	Barnum Brown born (02-12)
1873	Louis Agassiz died (12-14)
1874	GM Dawson discovered 1st dinosaur in Canada
1874	Mudge fired KS State Agri College (Feb)
1874	Mudge 1st with OC Marsh
1874	ED Cope joined Wheeler Survey
1875	CHS studied @ KS State Agricultural College
1876	Huxley's American Tour
1876	CHS ltr to Cope & 1st exped to chalk (Judith)
1876	Custer Massacre @ Little Big Horn (06-25)
1876	SW Williston 3 year pact w/ OC Marsh
1876	CHS wintered w/ Cope @ Haddonfield
1877-81	Rutherford B Hayes President
1877	Morrison Beds discovered by Lakes
1877	CHS in Scott County, Kansas
1877	Thomas Condon 1st to Silver Lake (Jun)
1877	CHS Loup Fork of Nebraska (July)
1877	CHS Silver Lake in Oregon (Aug)
1877	CHS wintered @ Pine Creek, Washington
1878	CHS Ft Walla Walla (Apr)
1878	*Iguandondon* cache discovered @ Bernissart, Belgium
1878	Marsh elected VP Ntl Academy of Sciences
1879	Cope to Fossil Lake in John Day (Fall)
1879	Yellow fever commission to Cuba
1879-80	CHS @ John Day, Bridge Creek (S Oregon)
1880	WB Scott PHD @ Princeton
1880	CHS wed Anna M Reynolds (07-07)
1880	CHS last year w/ Cope
1881	James Garfield President
1881	Sitting Bull surrendered
1881	CHS worked for MCZ (Mar-Apr) Dakota of KS
1881	CHS to Cambridge Mass
1881	Charles Reynolds Sternberg born (05-25)
1882	Cope in southeastern Oregon desert
1882	Charles Darwin died
1882	Charles Reynolds Sternberg died (09-12)
1882	CHS in Texas Permian for MCZ @ Harvard
1881-85	Chester A Arthur President
1882	CHS discovered mastodon on Sappa Creek
1882	CHS discovered the "Sternberg" Quarry @ Long Island
1882	CHS worked Loup Fork beds of Nebraska (thru 1884)
1883	GFS born @ Lawrence (08-26)

451

1884	CHS worked for OC Marsh (Mar to Oct)	
1884	Cope pub *Vertebrata of the Tertiary Formations of the West* ("Cope's Bible")	
1884	Joseph Tyrrell discovered 1st dinosaur bones Red Deer	
1885	CHS patent #316580-"Composition for Fuel" (02-17)	
1885	CMS born (09-18)	
1885	Chas Reynolds died Junction City (12-28)	
1885-89	Grover Cleveland President	
1887	JB Hatcher wed Anna M Peterson	
1887	Ferdinand V Hayden died	
1887	CHS independent in Dakota group - plants	
1888	Asa Gray died (01-30)	
1888	Margaret Sternberg died Ellsworth (12-07)	
1889	Leo Lesquereux died (10-25)	
1889	CHS sold pottery pieces from mounds of Arkansas	
1889-93	Benjamin Harrison President	
1889	Cope Prof of Geology @ Univ of Pennsylvania	
1890	Battle @ Wounded Knee	
1890	Major Cope/Marsh blowup in New York Herald	
1890	Maude Sternberg born (10-30)	
1890	SW Williston on staff @ Univ of KS	
1891	Barbour 1st worked Agate Springs Ranch	
1891	Joseph Leidy died	
1891	CHS in Gove County for PMM (*Lamna*)	
1892	Clarence Denby Sternberg died (09-24)	
1893	GMS Surgeon General United States Army	
1893	JB Hatcher left Marsh for Princeton	
1897-1901	William McKinley President	
1897	CHS in Texas Permian (Spring)	
1897	CHS spent 3 months in Dakota Group	
1897	Wm A Sternberg City Treasurer, Tacoma, WA	
1897	Edward Drinker Cope died (04-12)	
1897	OC Marsh received Cuvier Prize	
1897	Gustav Eric Lindblad born in Sweden	
1898	CHS Gove County KS (Plum Creek)-*Protostega gigas*	
1898	Theodore Sternberg to Philippines (10-03)	
1899	Othniel C Marsh died (03-18)	
1899	AMNH purchased Cope Collection	
1900	CHS in Elkader, KS	
1900	CHS Niobrara, Logan County @ Butte Creek-*Portheus*	
1900	CHS collecting in Kansas for German Museum	
1901-09	Theodore Roosevelt President	
1901	CHS in Texas Permian for Zittel	
1901	CHS tried to sell fossils to Louisiana Purchase Expo	
1901	Joseph LeConte died (07-06)	
1901	Clarence R King died (12-24)	
1902	GMS retired (06-08)	
1902	CHS Niobrara Cretaceous, Logan County (Oct)	
1903	CHS Hay Creek, Logan County, Scott City (Jun-Aug)	
1904	Theodore Sternberg retired active svc (09-15)	
1905	CHS Long Island (Aug), Loup Fork, NW Kansas	
1906	CHS Quinter, KS Niobrara (July-Sept)	
1906	CMS 1st collected w/ CHS	

1907	CHS Dakota plants near Ellsworth	
1907	CHS Deadwood, South Dakota (Jan 10)	
1907		Thomas Condon died (02-11)
1908	CHS Converse County, Wyoming (Jul-Sept)	
1908	CHS Greasewood Creek-*Trachodon* locality	
1908		Dinosaur quarry discovered in Utah
1908		1st Tengaduru Expedition to East Africa
1909-13	William H Taft President	
1909	CHS Lawrence, KS (Jan)	
1909	CHS published Life of a Fossil Hunter	
1909	CHS Laramie Beds, Converse County Wyo-Niobrara	
1910	CHS in Warren, Wyoming (07-11 to 09-16)	
1910	CHS elected to Society of Vertebrate Paleontology @ Boston, Mass meeting	
1910		Barnum Brown 1st to the Red Deer, Alberta
1911		Maude Sternberg died @ Lawrence (01-15)
1911		CMS wed Myrtle Martin (09-18)
1911		Margery Anna Sternberg died (11-21)
1911	CHS published A Story of the Past	
1911	CHS in Texas Permian near Seymour	
1911	Wm D Matthew Curator VP @ AMNH	
1912	CHS Lawrence laboratory destroyed by tornado	
1912	CHS & GFS eastern trip to museums	
1912	CHS to Red Deer of Alberta (July to Oct 10)	
1912	CHS to Victoria Mem Mus Ottawa to mount dinosaur	
1912		Raymond McKee Sternberg born (12-26)
1913-21	Woodrow Wilson President	
1913	CHS in Red Deer (06-12 to 10-03)	
1914	CHS in Red Deer (06-02 to 09-26)	
1914-18	World War I (07-28-14 thru 11-11-18)	
1914		Bertha M Sternberg died (02-01)
1914	CHS short trip to Judith River, MT	
1915	CHS in Red Deer (06-01 to 09-23)	
1915		GMS died (11-03)
1915		Barnum Brown finished in the Red Deer
1916	CHS @ Hartwick Seminary for visit (Apr)	
1916	CHS quit Canadian Geological Survey (05-31)	
1917	America entered war against Germany (Apr)	
1917	CHS in Red Deer w/ LS for BMNH	
1917	CHS pub Hunting Dinosaurs on the Red Dee, Alberta	
1917	CHS in the Texas Permian	
1917		CMS 1st independent exped in Red Deer
1918	CHS Texas briefly, then Logan County, KS	
1918	CHS visited GFS @ Hays, KS (Apr)	
1918		GFS quit Canadian Geological Survey
1919		LS on staff of ROM
1919		CMS @ Little Sandhill Creek
1919	CHS Quinter, KS, Hackberry Creek (Apr)-*Uintacrinus*	
1919	CHS Gove County near Monument Rocks	
1919	CHS Smoky Hill, North Utica, KS	
1919	CHS Gidley Horse Quarry, Bristow County, TX	
1920	CHS moved to California	
1920		LS @ Little Sandhill Creek

1921-23 Warren G Harding President
1921 CMS Morgan Creek badlands, Saskatchewan & then @ Little Sandhill Creek
1921 LS @ Little Sandhill Creek
1922 GFS to Patagonia w/ Elmer Riggs (Jul)
1922 LS in Edmonton @ Morrin Ferry
1922 1st American Expedition to Mongolia under RC Andrews
1922 GFS joined Field Museum (09-30)
1921-23 CHS explored San Juan Basin, New Mexico
1923-29 Calvin Coolidge President
1923 LS w/ CHS in San Juan Basin
1923 CMS in Edmonton @ Drumheller
1924 CMS in Edmonton @ Munson Ferry North
1924 LS in Pleistocene of Saskatchewan
1924 GFS resigns Field Museum (09-30)
1924 CHS Imperial County desert, California
1925 CHS @ Ensenada, Baja California
1925 CHS @ McKittrick in Central California (Aug to Dec)
1925 GFS moved back to Kansas
1925 GFS @ Logan City -*Portheus* find
1925 CMS ten day visit to USNM
1925 CMS Edmonton @ Tolman Ferry (W of Rowley)
1926 CHS & GFS 3 week trip to Yellowstone & beyond
1926 CMS Red Deer Valley
1926 LS Red Deer, Steveville in Oldman
1926 CHS hit by motorcycle in San Diego-Hospitalized
1927 GFS joined Ft Hays State Univ (03-27)
1927 Theodore Sternberg died
1927 LS Oligocene, Cypress hills, Steveville
1928 CMS SW Saskatchewan, then Steveville Ferry
1928 LS in Wyoming & Nebraska
1929-33 Herbert Hoover President
1928 J Fred Sternberg died (02-16)
1928 CHS visited GFS @ Ft Hays (01-26), then went to Canada
1928 GFS N Montana for USNM
1928 GFS independent @ Crawford, NE & White River
1929 William A Sternberg died (06-27)
1929 GFS visit to San Diego (Fall)
1929 GFS to San Juan, NM for USNM
1929 CMS in southern Saskatchewan
1929 LS in Wyoming, Nebraska & S Dakota
1930 William D Matthew died (09-24)
1930 GFS wed Anna Ziegler (10-11)
1930 CMS @ Peace River, NE British Columbia
1930 LS @ Little Sandhill Creek
1931 Handel T Martin died (01-16)
1931 GFS Douglas, Wyo (Nov), Holyrood & Wakeeney
1931 LS in Edmonton @ Munson Ferry
1931-35 CMS no trips due to "Great Depression"
1932 GFS Harrison, Nebraska (Jun)
1933 GFS w/ Frick @ Republican River Beds, then Norton, Phillips City & Oberlin, KS
1933 LS Steveville, Munson & Morrin Ferrys

1933-45	Franklin D Roosvelt President
1934-35	GFS mounted dinosaurs @ Univ of Alberta
1934	GFS Hagerman, ID (Jul), then w/ Frick @ Clark County (Oct)
1934	LS south of Little Sandhill Creek, Oldman
1935	CMS mapping Steveville, Deadlodge
1935	CWS w/ Frick @ Clarendon TX all winter
1935	GFS visited CHS (Apr)
1935	CMS & LS visited GFS (Nov)
1935	LS east of Little Sandhill Creek
1937	GFS w/ Frick Clark City, Florrisant
1927	CMS Se Alberta, Mayberries & Comrey
1938-45	War break for Dinosaur expeditions in Canada
1938	CHS's wife Anna died in San Diego (12-24)
1939	CHS visited GFS @ Ft Hays (02-02)
1943	CHS died in Toronto (07-20)
1945-51	World War II (09-01-39 thru 10-19-51)
1946	CMS Upper Edmonton, W of Big Valley
1947	CMS NE of Elnora in Red Deer
1947	WB Scott died (03-29)
1948	LS @ Cypress Hills
1949	CMS made a Fellow of Royal Society of Canada
1950	CMS retires (10-16)
1950	LS Upper Milk River beds, Deadhorse Coulee
1951	Myrl Walker to the Grand Canyon
1953-61	Dwight D Eisenhower President
1954	LS last dinosaur exped to Little Sandhill
1956	CMS consultant to Dinosaur Provincial Park
1960	CMS LL.D @ University of Alberta
1961-	John F Kennedy President
1962	LS retires from ROM
1962	Gustav Lindblad died in Toronto
1963	1st Polish Dinosaur Expedition to Mongolia
1963	Barnum Brown died (02-05)
1965	Anna Reed Ziegler Sternberg died
1969	GFS died (10-23)
1969	Sternberg Memorial Museum established
1974	CMS D.Sc Degree from Carleton University
1976	Myrtle Martin Sternberg died (Jan 1)
1976	LS died (10-21)
1981	CMS died (09-08)

APPENDIX B
STERNBERG FAMILY TREE

Johan Jacob Sternberg   Catherine ?
b.   1665   b. 12-13-????
d. 10-09-1742   d. 02-18-1748

Eva
Anna Catharin
Nickolas

Adam
b. 02-17-1711

John Jacob
b. 12-13-1712

Elisabeth
b. 10-09-1726

Abram

                               Spouses

         Lambert Sternberg   Catharine Teller
         b. 10-07-1713   b. 02-09-1712
         d. 01-25-1784   d. 08-11-1777
                 (m) 10-07-1730

14 children plus: Nicholas Sternberg   Catharine Rickert
                     b. 02-04-1736   b.
                     c. 12-12-1809   d. 10-06-1807
                          (m) 10-04-1757

8 children plus: Johanis Sternberg   Anna Shafer
                     b. 02-01-1768   b.
                     d.   d.
                              (m)

10 children plus: Levi Sternberg   Margaret Levering Miller
                     b. 02-16-1814   b. 01-02-1818
                     d. 02-13-1896   d. 12-09-1888
                           (m) 09-07-1837

See Next Page

Levi Sternberg Margaret Levering Miller
b. 02-16-1814 b. 01-02-1818
d. 02-13-1896 d. 12-09-1888

Spouses/Children

George Miller Sternberg Louisa Russell (1st Wife)
b. 06-08-1838 (Middleburgh, NY) b.   (Cooperstown, NY)
d. 02-13-1896 (Washington DC) d. 07-15-1867 (Ft. Harker, KS)

Martha Pattison (2nd Wife)

Theodore Sternberg Bertha (Bessie) Margaret Schmidt
b. 09-15-1840 (Pennsylvania) b. 09-??-1855 (St. Louis, MO)
d. 01-05-1927 (Kanopolis, KS) d. 02-01-1914 (Ellsworth County, KS)
 Charlotte (Lottie) Margaret
 John Levi

John Frederick Sternberg Never Married
b. 03-12-1843 (New York)
d. 02-16-1928 (Kanopolis, KS)

Rosina (Rose) Sternberg (Phelps) Ira Warner Phelps
b. 03-08-1845 b.
d. d.

Emily Sternberg (Humlong) Frank Humlong
b. 02-29-1848 b.
d. d.

Charles Hazelius Sternberg Anna Musgrove Reynolds
b. 06-15-1850 (Middleburgh, NY) b. 03-05-1853 (Brooklyn, NY)
d. 07-20-1943 (Toronto, Canada) d. 12-23-1938 (Sa Diego CA)
 Charles Reynolds
 George Fryer
 Charles Mortram
 Maude (Maud)
 Levi

Edward Endress Sternberg Eliza A. (Scates?)
b. 06-05-1850 (Middleburg, NY) Herbert
d. Frank

William Augustus Sternberg Never Married
b. 03-14-1853 (Ellsworth, KS)
d. 06-27-1929 (Tacoma, WA)

Albert Sternberg Never Married
b. 11-09-1855 (Ellsworth, KS)
d.

Francis (Frank) Sternberg Nellie

b. 03-31-1858 (Ellsworth, KS) b. 08-??-1865 (England)
d.   (San Francisco) d.

Robert Sternberg Never Married
b. 01-19-1860 (Ellsworth, KS)
d.

IMMEDIATE FAMILY

Charles Hazelius Sternberg   Anna Musgrove Reynolds
b. 06-15-1850 (Middleburgh, NY)   b. 03-05-1853 (Brooklyn, NY)
d. 07-20-1943 (Toronto, Canada)   d. 12-23-1938 (San Diego, CA)
. (m) 07-07-1880   Ft. Riley, KS .

Charles Reynolds Sternberg   Never Married
b. 05-25-1881
d. 09-12-1882 (Cambridge, MA)

George Fryer Sternberg   Mabel Smith
b. 08-26-1883 (Lawrence, KS)   b. 12-07-1882 (Long Island, KS)
d. 10-23-1969 (Hays, KS)   d. 04-30-1965
   Emily
   Charles W.

Charles Mortram Sternberg   Myrtle A. Martin
b. 09-18-1885 (Lawrence, KS)   b. 01-19-1865 (Utica, KS)
d. 09-08-1981 (Toronto, Canada)   d. 01-01-1976 (Torono, Canada)
   Raymond McKee
   Glenn
   Stanley

Maude (Maud) Sternberg   Never Married
b. 10-30-1890 (Lawrence, KS)
d. 01-15-1911 (Lawrence, KS)

Levi Sternberg   Ana Lindblad
b. 03-10-1894 (Lawrence, KS)   b.   1893
d. 10-21-1976 (Canada)   d. 01-19-1976 (Canada)

BIBLIOGRAPHY

Charles Hazelius Sternberg

1878 "Pliocene Man," *American Naturalist*, Vol XII, #2 (Feb) Pg. 125-126 [referenced in *Life of A Fossil Hunter*, Pg. 159 and published under Edward Drinker Cope's name]

1881 "Miocene Beds of the John Day, Oregon," *Kansas City Review of Science & Industry*, Vol IV: 540-542

1881 "The Pliocene Beds of Southern Oregon," *Kansas City Review of Science & Industry*, Vol IV: 600-601

1881 "The Quaternary of Washington Territory," *Kansas City Review of Science & Industry*, Vol IV: 601-602

1881 "The Dakota Group," *Kansas City Review of Science & Industry*, Vol IV: 675-677

1881 "The Judith River Group," *Kansas City Review of Science & Industry*, Vol IV: 730-733,

1881 "The Niobrara Group," *Kansas City Review of Science & Industry*, Vol V:1-4

1881 "The Fossil Flora of the Cretaceous Dakota Group of Kansas," *Kansas City Review of Science & Industry*, Vol V, 243-244

1881 "Museum of Comparative Zoology, at Harvard College," *Kansas City Review of Science & Industry*, Vol V, #8, 492-494

1881 "Pliocene Formation of Southern Oregon," *Kansas City Review of Science & Industry*, Vol V, #8, 491

1881 "Miocene Fauna of Oregon," *Kansas City Review of Science & Industry*, Vol V: 416-417, 491

1882 "Floating Stones," *Kansas City Review of Science & Industry*, Vol V, #9,

528

1882 "The Loup Fork Group of Kansas," *Kansas City Review of Science & Industry*, Vol VI, #4, 205-208

1883 "Explorations in the Judith River Group," *Kansas City Review of Science & Industry*, Vol VII, #6, 325-330

1883 "The Trissac Beds of Texas," *Kansas City Review of Science & Industry*, Vol VII, 455-457

1884 "The Fossil Fields of Southern Oregon," *Kansas City Review of Science & Industry*, Vol VII, 596-599

1884 "Explorations in Northeastern Oregon," *Kansas City Review of Science & Industry*, Vol II, 674-678

1885 "The Flora of the Dakota Group," *Kansas City Review of Science & Industry*, Vol VIII, 9-12

1885 "Directions For Collecting Vertebrate Fossils," *Kansas City Review of Science & Industry*, Vol VIII, 219-221

1885 "Practical Studies in Geology," *Kansas City Review of Science & Industry*, Vol VIII, 481-485

1889 "The Young Fossil-Hunters: A True Story of Western Exploration and Adventure," *The Swiss Cross*, N.D.C. Hodges, NY, [a series] Vol #V

1898 "Cretaceous Leaf Nodules," *Popular Science News*, XXXII, Feb, 26

1898 "Mine of Mammoths," *Popular Science News*, XXXII, July, 169

1898 "Pliocene Man," *Popular Science News*, XXXII, Apr, 182

1898 "Ancient Monsters of Kansas," *Popular Science News*, XXXII, Dec, 268

1899 "The First Great Roof," *Popular Science News*, XXXIII, June, 126-127

1899 "Kansas Rhinoceros Deposit," *Popular Science News*, XXXIII, 105

1899 "A Kansas Mosasaur," *Popular Science News*, XXXIII, 259-260

1900 "Fossil Collector's Experiences," *Popular Science News*, XXXIV, 34

1900 "The Sharks of Kansas," *Popular Science News*, XXXIV, 38

1901 "A Cretaceous Monster Fish," *Popular Science News*, XXXV, 29-30

1901 "A Kansas Pleisosaur," *Popular Science News*, XXXV, 104

1901 "Field Work in the Cretaceous," *Popular Science News*, XXXV, 127-128

1902 Letter to Editor of Science, "The 'Prickly Pear,'" *Science*, (n.s.), XV, #383, 714-715

1902 Letter to Editor of Science, *Science*, (n.s.), XVI, 989-990

1902 "The Permian of Texas," *Popular Science News*, May, XXXVI, #5, 106-107

1902 "A Pliesosaur of the Cretaceous Ocean," *Popular Science News*, Nov, XXXVI, #11, 248

1903 "Elephas Columbi and Other Mammals in the Swamps of Whitman County, Wash.," *Science*, (n.s.), #17: 511-512

1903 "The Life of a Fossil Hunter," *The American Inventor*, #10, 311-313 [Not to be confused with his book of same title]

1903 "Notes on the Judith River Group," *Science*, (n.s.), #17: 870-872

1903 "Experiences with Early Man in America," *Kansas Academy of Science Transactions*, #18: 89-93

1903 "The Permian Life of Texas," *Kansas Academy of Science Transactions*, #18: 94-98

1904 "Dr. Karl A. vonn Zittel [Obit]," *American Geologist*, XXXIII, 263-265

1905 "Protostega gigas and Other Cretaceous Reptiles and Fishes From the Kansas Chalk," *Kansas Academy of Science Transactions*, #19: 123-128

1906 "The Loup Fork Miocene of Western Kansas," *Kansas Academy of Science Transactions*, #20 (#1): 71-74

1907 "Portheus Molossus Cope and Other Fishes From the Kansas Chalk," (Abst), *Science*, (n.s.), #25: 295

1907 "The Great Inferior Tusked Mastodon of the Loup Fork Miocene," *Science*, (n.s.), #25: 971-972

1907 "Some Animals Discovered in the Fossil Beds of Kansas," *Kansas Academy of Science Transactions*, #20 (#2): 122-124

1908 "My Expedition to the Kansas Chalk For 1907," *Kansas Academy of Science Transactions*, #21: 111-114

1909 *The Life of a Fossil Hunter*, New York, Henry Holt & Co, 286 pp

1909 "Expedition to the Laramie Beds of Converse County, Wyo.," *Kansas Academy of Science Transactions*, #22: 113-116

1909 "An Armored Dinosaur From the Kansas Chalk," *Kansas Academy of Science Transactions*, #22: 257-261

1909 "A New Trachodon From the Laramie Beds of Converse Co., Wyo.," (abst), *Science*, (n.s.), #29: 753-754

1909 "In the Laramie Beds of Wyoming," *Guide To Nature*, July, Vol II, #4: 123-130

1911 "In the Niobrara and Laramie Cretaceous," *Kansas Academy of Science Transactions*, #23-24: 70-74

1911 "Still in the Laramie Country, Converse County, Wyo.," *Kansas Academy of Science Transactions*, #23-24: 219-223

1911 *A Story of the Past or the Romance of Science*, Sherman, French and Co., Boston

1913 "Expeditions to the Miocene of Wyoming and the Chalk Beds of Kansas," *Kansas Academy of Science Transactions*, #25: 45-49

1914 "Notes on the Fossil Vertebrates Collected on the Cope Expedition to the Judith River and Cow Island Beds,Mont.,in 1876," *Science*, (n.s.), #40: 134-135

1915 "Evidence Proving That the Belly Rvier Beds of Alberta are Equivalent with the Judith Riverdsof Montana," *Science*, (n.s.) #42: 131-133 (abst- *Geological Soc of America*, #26:1:49)

1917 *Hunting Dinosaurs in the Badlands of the Red Deer River, Alberta,*

Canada, Lawrence, KS, World Company Press, 232 pp

1918 "Five Year's Explorations in the Fossil Beds of Alberta," *Kansas Academy of Science Transactions,* #28: 205-211

1918 "Sternberg's Expedition to the Red Deer River, Alberta, 1917," *Kansas Academy of Science Transactions,* #29: 88-91

1921 *The Acacia Tree,* (A poem) published in unknown Philadelphia periodical with a partial date of -- 31, 1921 [article found in George F. Sternberg's scrapbook at the Sternberg Memorial Museum]

1922 "Explorations of the Permian of Texas and the Chalk of Kansas, 1918," *Kansas Academy of Science Transactions,*#30: 119-120

1922 "Field Work in Kansas and Texas," *Kansas Academy of Science Transactions,* #30: 339-341

1928 "Extinct Animals of California," *Scientific American,* #139: 225-227

1929 "Fossil Monsters I Have Hunted," *Popular Science Monthly,* Dec, #56-57: 139-140

1930 "Denies Fossil Remains Found," *San Diego Union,* Mar 31, 1930, [Article "written" by Clinton G. Abbott, but he excerpts letter by CHS that he requested CHS write. Letter is in response to a fallacious article written by a Capt. W.I. Penny entitled "Fossils of Rare Monsters Found on Ocean Front" in the *San Diego Union* of Mar 23, 1930]

1930 [Russian Hunters For Fossils], a chapter in *Life of a Fossil Hunter,* Moscow, Russian translation by T.L. Khitkov, 239-289

1931 *The Life of Fossil Hunter,* San Diego, CA, Jensen Printing Co., 286 pp

1932 *Hunting Dinosaurs in the Badlands of the Red Deer River, Alberta, Canada,* San Diego, CA, privately published by the author, 261 pp

1936 [Introduction & a chapter in *The Life of a Fossil Hunter,* Moscow, 2nd Edition, Russian translation by T.L. Khitrov], 7-13

1976 *The Life of A Fossil Hunter* (2 chapters), in *Fossil's Magazine,* Vol #1, Issue #1, May, 82-102

1985 *Hunting Dinosaurs in the Bad Lands of the Red Deer River, Alberta,*

Canada, Edmonton, Alberta, Canada, NeWest Press, 221 pp

1990 *The Life of a Fossil Hunter*, Indiana University Press, (paper), Bloomington & Indianapolis, 281 pp

---- *Poem in Honor of Jennie McKee's Wedding Day* [referenced in *Hunting Dinosaurs in the Bad Lands of the Red Deer River, Alberta Canada*, Pg 176] n.d. [no other data known]

....

George Fryer Sternberg

1921 "Fossil Hunters Find Prehistoric Reptiles in Red River Bad Lands," *The Calgary Daily Herald*, Oct, #8: 1921

1921 "Search for Prehistoric Monsters in Red Deer Bad Lands Is Success," *Edmonton Journal*, (Saturday), Oct 15,1921, Illus

1922 "Collecting and Preparing Dinosaur Specimens for the University of Alberta," *Univ of Alberta Press Bull.*, Edmonton, Alberta, #13, Vol VII, Jan, 134

1926 "Where Birds as Big as Aeroplanes Once Flew," *Union Pacific Magazine*, Nov

1929 (with Louis Hussakof) *A New Teleostean Fish from the Niobrara of Kansas*, **New York City, American Museum of Natural History

1930 "Thrills in Fossil Hunting," *Aerend*, Kansas State Teacher's College, Hays, KS, Vol #1, #3: 139-153

1935 (with Erwin Hinckley Barbour), *"Gnathabelodon thorpei*, gen. et sp. nov., a new mud-grubbing mastodon," *Nebraska State Mus Bull*, May, Vol #1, #42: 395-404

1942 (with George Robertson), "Pliocene Waterhole in Western Kansas," *Science*, (abst), #95: 97

1942 (with George Robertson), "Fossil Mammal Tracks in Graham County, Kansas," *Kansas Academy of Science Transactions*, #45: 258-261

1952 News Bulletin, *Society of Vertebrate Paleontology*, Feb

1957 (with Myrl Vincent Walker), "Report on a Pleisosaur Skeleton from Western Kansas," *Kansas Academy of Science Transactions*, Vol #60, #1: 86-87

1958 (with Myrl Vincent Walker), "Observation of Articulated Limb Bones of a Recently Discovered *Pterandon* in the Niobrara Cretaceous of Kansas," *Kansas Academy of Science Transactions*, Vol #61, #1: 81-85, illus

1960 "Fossil Collecting Still Fascinates After 62 Years in Paleontology," *Ft. Hays State College Leader*, May 17, 1960, illus

1962 (with Halsey W. Miller & Robert Reyer), "A New Locality For Acanthoceras? amphibolum Morrow in the Graneros Shale of Kansas," *Journal of Paleontology*, Vol #36, #4: 836-837

....

Levi Sternberg

1937 "Graveyard of 50 Million Years; Red Deer Valley, Alberta," *Illustrated Canadian*, July

1955 "When Dinosaurs Ruled the Land [Alberta]," *Canadian Nature*, Vol #17, (Jan-Feb), #1: 12-20, Illus

1956 "The Use of "Gelva" in the Vertebrate Laboratory," *New Bull Soc of Vertebrate Paleontology*, Vol #48: 33

....

Charles Mortram Sternberg

1917 "Notes on the Feeding Habits of Two Salamanders in Captivity," *The Ottawa Naturalist*, Vol XXX, #10, Jan, 129-130

1918 "Notes on the Burrowing Habits of Frogs," *The Ottawa Naturalist*, Vol XXXII, #3: 56-57

1921 "A Supplementary Study of Panoplosaurus mirus [Belly River Beds, Alberta]," *Royal Soc of Canada, Proc & Trans*, Ser #3, Vol #15: 93-104

1921 "A Popular Description of Dinosaurs," *Canadian Field-Naturalist*, Vol

#35, #4: 61-66

1922 "Not a Plesiosaur," *Canadian Field-Naturalist*, Vol #36, May, 97

1924 "Notes on the Lance Formation of Southern Saskatchewan," *Canadian Field-Naturalist*, Vol #38, #4, Apr, 66-70

1925 "Integument of *Chasmosaurus belli*, *Canadian Field-Naturalist*, Vol #39, May, 109-110

1925 "The Bison and its Relations," *Canadian Field-Naturalist*, Vol XXXIX, May, 111

1926 "A New Species of Thespesius from the Lance Formation of Saskatchewan," *Canada Geo Survey Bull*, #44: 73-84

1926 "Dinosaur Tracks from the Edmonton Formation of Alberta," *Canada Geo Survey Bull*, #44:85-87

1926 "Notes on the Edmonton Formation of Alberta," *Canadian Field-Naturalist*, Vol #40: 102-104

1927 "Homologies of Certain Bones of the Ceratopsian Skull," *Royal Soc Canada Trans*, Ser #3, Vol #21, Sec 4, May, 135-143

1927 "Horned Dinosaur Group in the National Museum of Canada," *Canadian Field-Naturalist*, Vol #41, Apr, #4: 67-73

1928 "A New Armoured Dinosaur [Edmontonia longiceps] from the Edmonton Formation of Alberta," *Royal Soc Canada Trans*, Ser #3, Vol #22, Sec 4, 93-107

1929 "A Toothless Armoured Dinosaur [*Anodontosaurus lambei*] from the Upper Cretaceous of Alberta," *Canada Ntl Mus Bull*, #54, 28-33

1929 "Dinosaurs in the National Museum of Canada," *The Civil Service Review*, #1 (4), 288-292

1929 "A New Species of Horned Dinosaur [*Anchiceratops*] from the Upper Cretaceous of Alberta," *Canada Ntl Mus Bull*, #54, 34-37

1930 "New Records of Mastodons and Mammoths in Canada," *Canadian Field-Naturalist*, Vol #44, #3, Mar, 59-65

1930 "Miocene Gravels in Southern Saskatchewan," *Royal Soc Canada Trans*, Ser #3, Vol #24, Sec 4, May, 29-30

1930 "Garter Snakes as Fishermen," *Canadian Field-Naturalist*, Vol XLIV, Sept, 149

1930 "Dinosaur Tracks from Peace River," *Ntl Mus Canada*, Ottawa, 59-85

1931 "Two Ichthyosaur Localities in British Columbia," [Abst], *Geo Soc Amer Bull*, Vol #42, #1, 363, Mar 31 and *Pan-Amer Geo*, Vol #55, #2, Mar, 157

1931 "Dinosaur Tracks from Peace River, British Columbia," *Ntl Mus Canada Bull*, [Abst], #68, 59-85, *Geo Soc Amer Bull*, #42, #1, 362-363, Mar 31 and *Pan-Amer Geo*, Vol #55, #2, Mar, 157-158

1932 "Two New Theropod Dinosaurs from the Belly River Formation of Alberta," *Canadian Field-Naturalist*, Vol #46, #5, May, 99-105

1932 "The Skull of *Leidyosuchus canadensis* Lambe," *Amer Midland Naturalist*, Vol #13, #4, July, 157-169

1932 "A New Fossil Crocodile from Saskatchewan," *Canadian Field-Naturalist*, Vol #46, #6, Sept, 128-133

1932 "Dinosaur Footprint Bird Bath," *Canadian Field-Naturalist*, Vol #46, #9, Dec, 203-204

1933 "Relationships and Habitat of Troodon and the Nodosaurs," *Annals & Mag Ntl History*, 10th ser, Vol #11, Feb, 231-235

1933 "Prehistoric Footprints in Peace River," *Canadian Geo Journal*, Vol #6, #2, Feb, 92-102

1933 "A New Ornithomimus with Complete Abdominal Cuirass," *Canadian Field-Naturalist*, Vol #47, #5, May, 79-83

1933 "Carboniferous Tracks from Nova Scotia," *Geo Soc Amer Bull*, Vol #44, #5, Oct, 951-964

1934 "Notes on Certain Recently Described Dinosaurs," *Canadian Field-Naturalist*, Vol #48, #1, Jan, 7-8

1935 "Hooded Hadrosaurs of the Belly River Series of the Upper Cretaceous,"

Canada Ntl Mus Bull, #77: 1-37 [Abst], *Geo Soc Amer Procs* 1934, 403

1936 "The Systematic Position of Trachodon," *Journal of Paleo*, Vol #1, #7, Oct, 652-655 and *Geo Soc Amer Procs* [Abst], Jun, 403

1936 "Preliminary Map- Steveville Sheet," *Canada Dept of Mines, Geo Survey Paper*, 36-38

1937 "Classification of Thescelosaurus," *Geo Soc Amer Procs*, 1936, 375

1938 "Monoclonius from Southeastern Alberta Compared with Centrosaurus," *Journal of Paleo*, Vol #12, #3, 284-286

1938 "Were There Trunkbearing Dinosaurs?, *Geo Soc Amer Bull*, Vol#29, #12, Pt 2, Dec, 1923

1939 "Were There Proboscis-Bearing Dinosaurs?," *Annals & Mag Ntl Hist*, #11, #3, 556-560

1940 "Ceratopsidae from Alberta," *Journal of Paleo*, Vol #14, #5, 468-480

1940 "*Thescelosaurus edmontonensis*, n. sp., and Classification of the Hypsilophodontidae," *Journal of Paleo*, Vol #14, 481-494

1942 "A New Restoration of a Hooded Duck-Billed Dinosaur, *Journal of Paleo*, Vol #16, 133-134 and [Abst] *Geo Soc Amer*, #52, 1991

1943 [with R.S. Lull], "Obituary of Charles H. Sternberg," *Amer Journal of Science*, Vol #241: 647-648

1945 "Pachycephalosauridae Proposed for Dome-Headed Dinosaurs, *Stegoceras lambei*," n. sp. described, *Journal of Paleo*, #19: 534-538

1945 "A New Species of Selachian from the Lower Carboniferous of Canada," *Journal of Paleo*, #19: 539-540

1945 "Canadian Dinosaurs; Their Place in Nature & Conditions Under Which They Lived," *Canada Geographical Jorunal*, #30, #4, 186-199

1946 "Canadian Dinosaurs," *Can Ntl Mus Bull*, #103, Geo Ser, 54

1947 "The Edmonton Fauna & Description of a New Triceratops from the Upper Edmonton Member; Phylogeny of the Ceratopsidae," *Can Ntl Mus Ann Report*, 1947-1948, Ottawa, 33-46

1947 "The Upper Part of the Edmonton Formation of Red Deer Valley, Alberta," *Canada Geo Survey*, paper 47-1, 11 pp

1947 "New Dinosaur from Southern Alberta, Representing a New Family of the Ceratopsia," [Abst], *Geo Soc Amer Bull*, Vol #58, #12

1948 "Canadian Dinosaurs," *The Instit Journal*, Prof Instit Civil Service, Canada, Dec, 6 pp

1949 "The Edmonton Fauna & Description of a New Triceratops from the Upper Edmonton Member," *Can Ntl Hist Bull*, #113, 33-46

1950 "Notes on the Dinosaur Quarries," [on map 969A Steveville], *Geo Survey of Canada*

1950 "Leptoceratops, the Most Primitive Dinosaur from the Upper Edmonton of Alberta, (abs), *Royal Soc Canada Proc*, 3rd Series, Vol #44, Pg. 229

1950 "Pachyrhinosaurus Canadensis, Representing a New Family of the Ceratopsia, from Southern Alberta, Canada," *Can Ntl Mus Bull*, #118, Pgs. 109-120

1951 "White Whale and Other Pleistocene Fossils from the Ottawa Valley [Ontario]," *Can Ntl Mus Bull*, #123, Pgs. 259-261

1951 "Complete Skeleton of Leptoceratops Gracilis, Brown from the Upper Edmonton Member on Red Deer River, Alberta," *Can Ntl Mus Bull*, #123, Pgs. 225-255

1951 "The Lizard Chamops from the Wapiti Formation of Alberta-- Polyodontosaurus Grandis Not a Lizard," *Can Ntl Mus Bull*, #123, Pgs. 256-258

1952 "Discussion of Classification of Canadian Upper Cretaceous Hadrosauridae," *Proc of Royal Soc of Canada*, Series #3, Vol #46, Pg. 149 (abs)

1953 "A New Hadrosaur from the Oldman Formation of Alberta- Discussion of Nomenclature," *Can Ntl Mus Bull*, #128, Pgs. 275-286

1953 "Classification of American Duck-Billed Dinosaurs, *Geo Soc of America*, Bull #64, Pg. 1478 (abs)

1954 "Classification of American Duck-Billed Dinosaurs," *Jour Paleo*, Vol #28,

#3, May, Pgs. 382-383

1955 "A Juvenile Hadrosaur from the Oldman Formation of Alberta," *Can Ntl Mus Bull*, #136, Pgs. 120-122

1956 "A Harp Seal from the Leda Clay West of Hull, Quebec," *Canadian-Field Naturalist*, #70, #2, Pg. 97

1963 "Additional Records of Mastodons & Mammoths in Canada," *Ntl Mus Can Bull*, #19, Pgs. 1-11

1963 "Early Discoveries of Dinosaurs," *Nat Hist Paper*, Ntl Mus Can, #21, Pgs, 1-4

1964 "Function of the Elongated Narial Tubes in the Hooded Hadrosaurs," *Jour Paleo*, #38, Pgs. 1003-1004

1965 "New Restoration of Hadrosaurian Dinosaur," *Ntl Mus Can*, Nat Hist Papers, #30, Pgs. 1-5, (abs) Huene in *Zbl Geo Pal*, Teil 2, 1966, Pg. 307

1966 "Canadian Dinosaurs," 2nd ed., *Ntl Mus Can Bull*, #103, Geo Series, #54, 38pp

1970 "Comments on Dinosaurian Preservation in the Cretaceous of Alberta and Wyoming," *Ntl Mus Can Pub*, #4, vi & 9pp

Bibliography

Adams, Alexander B., *Sitting Bull*, G.P. Putnam's Sons, NY, 422 pp, 1973

Agassiz, Louis, "Prof. Agassiz on the Origin of Species," *Amer Journ of Sci*, 2nd Series, XXX, Jul 1860

Alcock, F.J., "A Century in the History of the Geological Society of Canada," Dept of Mines & Resources, *Ntl Museum Contribution*, #47-1, Ottawa, 1947

Anderson, F.M., "Upper Cretaceous of the Pacific Coast," Memoir 71, *Geological Soc of Amer*, 1958

Andrews, Henry N., *The Fossil Hunters*, Cornell Univ Press, Ithaca & London, 398 pp, 1980

Andrews, Roy Chapman, *Ends of the Earth*, Garden City Pub Co., Garden City, NY, 1929

------, *Under A Lucky Star*, Blue Ribbon Books, Garden City, NY, 1945

Arnold, Ralph & Robert Anderson, "Geology and Oil Resources of the Coalinga District," *U.S. Geological Survey Bull*, #398, Wash DC, 1910

Augspurger, Marie, *Yellowstone National Park*, The Naegele-Aver Printing Co., Middletown, OH, 1948

Ayres, Anne, *The Life and Work of William Augustus Muhlenberg*, Harper & Bros, NY, 524 pp, 1880

Baker, Charles Laurence, "Depositional History of the Red Beds and Saline Residues of the Texas Permian," in "Contributions to Geology," *Univ of Texas Bulletin*, #2901, Austin, 1929

Bakker, Robert T., *The Dinosaur Heresies*, William Morrow & Co., NY, 1986

Beebe, Chas P., *Kansas Facts*, Chas P. Beebe Pub, Topeka, 1929

Bigelow, E.F., ed., *The Observer*, Portland, CT, Vol VI, 1895

Bird, Roland T., *Bones for Barnum Brown*, Texas Christian Univ Press, Ft. Worth, 1985

Birdsall, Ralph, *The Story of Cooperstown*, Charles Scribner Sons, NY, 1925

Brandenberg, Aliki, *Fossils Tell of Long Ago*, Thomas Y Crowell Co., NY, 1972

Brightwell, L.R., *Buckland's Curiosities of Natural History*, The Batchworth Press, London, 1948

Brown, Barnum, "The Mystery Dinosaur," *Natural History*, Vol XLI, #3, Mar 1938

------, "Hunting Big Game of Other Days," *National Geographic*, Vol XXXV, #5, Wash DC, May 1919, Pgs. 407-429

Brown, Lilian, *I Married A Dinosaur*, Dodd, Mead & Co., NY, 1953

Camp, Charles L. and G. Dallas Hanna, *Methods in Paleontology*, Univ of Calif Press, Berkeley, 1937

Camp, Charles, L., ed., "The Letters of William Diller Matthew, " *Journal of the West*, Vol VIII, #2 (Apr 1969) & Vol VIII, #3 (Jul 1969)

Castel, Albert, *A Frontier State at War: Kansas 1861-1865*, Cornell Univ Press, Ithaca, NY, 1958

Clark, Robert D., *The Odyssey of Thomas Condon*, Oregon Historical Soc Press, Portland, 459 pp, 1989

Clark, Ronald W., *The Huxleys*, McGraw-Hill, London, 1968

Clary, David A., "The Role of the Army Surgeon in the West: Daniel Weisel at Fort Davis, Texas, 1868-1872," in the *Western Historical Quarterly*, Logan, UT, Jan 1972

Cloos, Hans, *Conversation with the Earth*, Routledge & Kegan Paul, Ltd., London, 409 pp, 1954

Cobb, Sanford H., *The Story of the Palatines*, G.P. Putnam's Sons, NY & London, 1897

Colbert, Edwin H., *The Dinosaur Book*, McGraw-Hill, NY, 1951

------, "Battle of the Bones; Cope and Marsh the Paleontological Protagonists," *Geotimes*, Vol #2, 1957

------, *A Fossil Hunter's Notebook*, E.P. Dutton, NY, 1980

------, *Men and Dinosaurs*, E.P. Dutton, NY, 1968

------, *Digging into the Past*, Dembner Books, NY, 446 pp, 1989

------, *William Diller Matthew: Paleontologist*, Columbia Univ Press, NY, 236 pp, 1992

Cook, Harold J., *Tales of the 04 Ranch*, Univ of Nebraska Press, Lincoln, 1968

Cook, James H., *Fifty Years on the Old Frontier*, Yale Univ Press, New Haven, n.d.

Cope, Edward Drinker, *Origin of the Fittest*, D. Appleton & Co., NY, 1887

------, "The Silver Lake of Oregon," 162, #1, Jul 1982, *The Country of Willa Cather*, Wm. Howarth

Cushman, Dan, "Monsters of the Judith: Dinosaur Diggings of the West Provided Competitive Arena for Fossil Discovery," *Montana, Magazine of Western History*, Vol #12 #4, 1961

Dary, David, *True Tales of the Old-Time Plains*, Crown Pub, Inc, NY, 1979

Davidheiser, Bolton, *Evolution & Christian Faith*, Presbyterian & Reformed Pub Co., New Jersey, 371 pp, 1969

deCamp, L. Sprague and Catherine C. deCamp, *The Day of the Dinosaur*, Doubleday & Co., NY, 1968

Desmond, Adrian J., *Archetypes and Ancestors, Paleontology in Victorian London 1850-1875*, Univ of Chicago Press, Chicago, 1982

------, *The Hot-Blooded Dinosaurs*, The Dial Press, James Wade, NY, 1976

Dick, Everett, *The Sod-House Frontier 1854-1890*, Univ of Nebraska Press, Lincoln, 1979

Drago, Harry Sinclair, *Wild, Woolly & Wicked*, Clarkson N. Potter, Inc., NY, 1960

------, *The Steamboaters*, Dodd, Mead & Co., NY, 1967

Duke, Donald, ed., *Water Trails West*, Doubleday & Co., NY, 1978

Ferguson, Walter Keene, *Geology and Politics in Frontier Texas 1845-1909*, Univ of Texas Press, Austin & London, 1969

Geikie, Sir Archibald, *The Founders of Geology*, Dover Pub, NY, 2nd ed, 473 pp, 1962

Gibbons, Boyd, "The Itch to Move West," *National Geographic*, Vol #170, #2, Aug 1986, Pgs. 147-177

Gibson, John M., *Soldier in White*, Duke Univ Press, Durham, NC, 1958

Godfrey, Laurie R., ed., *Scientists Confront Creationism*, WW Norton & Co., NY & London, 315 pp, 1983

Gillespie, Charles C., *Genesis and Geology*, Harper & Bros, NY, 1951

Gilmore, Charles W., "On the Reptilia of the Kirtland Formation of New Mexico, With descriptions of New Species of Fossil Turtles," *Smithsonian*, Wash DC, Vol #83, #2978

------, "Hunting Dinosaurs in Montana," in *Explorations and Field-Work of the Smithsonian Institution in 1928*, Wash DC, 1930

------, "Fossil Hunting in the Bridger Basin of Wyoming," in *Explorations and Field-Work of the Smithsonian Institution in 1930*, Wash DC, 1931

------, "Edmontonia Rugosidens (Gilmore) An Armoured Dinosaur from the Belly River Series of Alberta, *Univ of Toronto Studies*, Geo Series #43, Univ Press, 1940

Goetzmann, William H., *Exploration and Empire*, Vintage Books, NY, 1966

Gould, Charles N., *Covered Wagon Geologist*, Univ of Oklahoma Press, Norman, 1959

Grinnell, George Bird, *Pawnee, Blackfoot and Cheyenne*, "An Old-Time Bone Hunt," Chas Scribner's Sons, NY, 1961

Gross, Renie, *Dinosaur Country; Unearthing the Badlands' Prehistoric Past*, Western Producer Prairie Books, Saskatoon, Saskatchewan, 1985

Hart, Herbert M., *Old Forts of the Southwest*, Superior Pub Co., Seattle, 1964

Hatcher, John Bell, *Bone-Hunters In Patagonia*, Ox Bow Press, Woodbridge, CT, 1985

Heins, Henry Hardy, *Throughout All the Years: Hartwick 1746-1946*, Lutheran Pub House, Blair, NB, 1946

Hendrix, Anne W. and Lester E. Hendrix (ed.), *Sloughter's Instant History of Schoharie County*, Schoharie County Historical Soc, Schoharie, NY, 1988

Hibbard, Claude W., *Studies on Cenozoic Paleontology & Stratigraphy*, Univ of Mich, Ann Arbor, 1975

Hibben, Frank C., *Digging Up America*, Hill & Wang, NY, 224 pp, 1960

Hieb, David L., *Fort Laramie*, National Park Service Handbook, Series #20, Wash DC, 1954

Hoagland, Clayton, "They Gave Life to Bones," *Scientific Monthly*, Feb 1943, Pgs. 114-133

Hook, Robert W. (ed.), *Permo-Carboniferous Vertebrate Paleontology, Lithostratigraphy, and Depositional Environments of North-Central Texas*, Field Trip Guidebook #2, SVP, Austin, 1989, Pgs. 40-46, entitled "An Overview of Vertebrate Collecting in the Permian System of North-Central Texas" by Kenneth W. Craddock & Robert W. Hook

Hopkins, Sarah Winnemucca, *Life Among the Piutes*, Chalfant Press, Bishop, CA, 248 pp, 1969

Horner, John (Jack) R. & James Gorman, *Digging Dinosaurs*, Workman Pub, NY, 198 pp, 1988

Hotton III, Nicholas, *Dinosaurs*, Pyramid Pub, NY, 199 pp, 1963

Howard, Helen Addison, *Saga of Chief Joseph*, The Caxton Printers, Caldwell, ID, 376 pp, 1941

Howard, Robert W., *The Dawnseekers*, Harcourt-Brace-Jovanovich, NY, 1975

Howarth, William, "The Country of Willa Cather," *National Geographic*, Vol #162, #1, Jul 1982, Pg. 71

Hsu, Kenneth J., *The Great Dying*, Ballantine Books, NY, 1986

Hummel, Charles E., *The Galileo Connection*, Intervarsity Press, Downers Grove, IL,

Irvine, William, *Apes, Angels and Victorians*, Time Inc, NY, 1955

Isley, Bliss & W.M. Richards, *The Story of Kansas*, Kansas State Printing Ofc, Topeka, 1953

Jackman, E.R. and R.A. Long, *The Oregon Desert*, The Caxton Printers, Caldwell, ID, 1964

Jelinek, George, *Ellsworth, KS 1867-1947*, Salina, KS, 1947

Jenkins, Olaf P., *Early Days Memoirs*, Ballena Press, Ramona, CA, 1975

Jones Jr., Henry Z., *The Palatine Families of New York; 1710*, Universal City, CA, 1985

Klassen, Michael, "Hell Ain't a Mile Off; The Journals of Happy Jack," *Alberta History*, Spring 1990

Kreider, Harry J., PHD, DD, *History of the United Lutheran Synod of New York and New England*, Muhlenberg Press, Vol I, Philadelphia, 1954

Kurten, Bjorn, *How to Deep-Freeze a Mammoth*, Columbia Univ Press, NY, 116 pp, 1981

Lanham, Url, *The Bone-Hunters*, Columbia Univ Press, NY, 1973

Lewin, Rogers, *Bones of Contention*, Simon & Schuster, NY, 1987

Lewis, G. Edward, "Memorial to Barnum Brown," *Geological Soc of Amer Bulletin*, Feb 1964

Long, Paul F., "The Stereopticon; Magic and Light," *Kanhistique*, Vol #16, #9, Ellsworth, KS, Jan 1991, Pgs. 2-4

Loomis, Frederic B., *Hunting Extinct Animals in the Patagonian Pampas*, Dodd, Mead & Co., NY, 1913

Lull, R.S. and Charles M. Sternberg, "Obituary of Charles H. Sternberg," *Amer Journal of Science*, Vol #241

Madsen, Brigham D., *The Bannock of Idaho*, The Caxton Printers, Caldwell,

Idaho, 362 pp, 1958

Matthew, William Diller, "Early Days of Fossil Hunting in the High Plains," *Natural History*, Sept-Oct 1926, Pgs. 449-454

------, *Amer Museum Journal*, Vol IX, #4, Apr 1909, Pgs, 91-95

McIntosh, John S., "Marsh and the Dinosaurs," *Discovery*, Vol #1, Pgs. 31-37

McPhee, John, *Rising From the Plains*, Farrar, Straus Giroux, NY, 1986

McNeal, T.A., *When Kansas Was Young*, The Macmillan Co., NY, 1922

Millar, Ronald, *The Piltdown Men*, St. Martin's Press, NY, 237 pp, 1962

Miller, Hugh, *Testimony of the Rocks*, Gould & Lincoln, Boston, 502 pp, 1867

Miller, Loye (ed. J. Arnold Shotwell), *Journal of First Trip of University of California to John Day Beds of Eastern Oregon*, Bulletin #19, Mus of Nat History, Univ of Oregon, Eugene, 1972

Miller, Max, *Land Where Time Stands Still*, Dodd, Mead & Co., NY, 1943

Nelkin, Dorothy, *The Creation Controversy*, Beacon Press, Boston, 197 pp, 1982

Norman, David, *The Illustrated Encyclopedia of Dinosaurs*, Crescent Books, NY, 1985

North, Luther, *Man of the Plains: Recollections of Luther North 1856-1882*, Univ of Nebraska Press, Lincoln, 1961

Osborn, Henry Fairfield, *Cope: Master Naturalist*, Princeton Univ Press, Princeton, NJ, 1931

------, "A Mounted Skeleton of Naosaurus, A Pelycosaur From the Permian of Texas," *Bull Amer Mus of Nat Hist*, Vol XXIII, 1907, Pgs. 265-272

------, "A Dinosaur Mummy," *Amer Mus Journal*, Vol IX, #1, Jan 1911, Pgs. 6-11

------, "A New Genus and Species of Ceratopsia From New Mexico, Pentaceratops Sternbergii," *Amer Mus Novitiates*, Vol #93, 1923, Pgs. 1-3

Parker, Ronald B., *The Tenth Muse: The Pursuit of Earth Science*, Chas

Scribner's Sons, NY, 1986

Penick, James, "Prof. Cope vs Prof. Marsh," *American Heritage*, Vol #22, Aug 1971

Paul, Gregory, *Predatory Dinosaurs of the World*, Simon & Schuster, (NY Academy of Sciences)), NY, 410 pp, 1988

Plate, Robert, *The Dinosaur Hunters; Othniel C. Marsh and Edward D. Cope*, David McKay Co., NY, 281 pp, 1964

Preston, Douglas J., *Dinosaurs in the Attic*, St. Martin's Press, NY, 1986

------, "Sternberg and the Dinosaur Mummy," *Natural History*, Vol #91, Jan 1982

------, "Barnum Brown's Bones," *Natural History*, Vol #93, Oct 1904

Pride, W.F., *The History of Ft. Riley*, (Privately Published), 1926

Rehwinkel, Alfred M., *The Flood: In the Light of the Bible, Geology and Archeology*, Concordia Pub House, St. Louis, 350 pp, 1951

Reiger, John F., ed., *The Passing of the Great West, Selected Papers of George Bird Grinnell*, Chas Scribner's Sons, NY, 1972

Roenigk, Adolph, *Pioneer History of Kansas*, (Privately Printed), 1933

Rogers Katherine, *The Sternberg Fossil Hunters: A Dinosaur Dynasty*, Mountain Press Pub Co, Missoula, MT, 259 pp, 1991

------, "The Dinosaur Dynasty," *Environment West*, Summer 1990, Pgs. 15-19

------, "The Incredible Sternbergs; They Left Their Mark on Kansas," *Kanhistique*, Vol #16, #4, Aug 1990, Pgs. 1-5

Rudwick, Martin J.S., *The Meaning of Fossils*, Univ of Chicago Press, Chicago & London, 2nd Ed., 267 pp, 1976

Russell, Dale A., "Tyrannosaurus From the Late Cretaceous of Western Canada," *Ntl Mus of Natural Sciences*, Pubs in Paleo #1, Ottawa, 1970

Sarton, George, "Lesquereux (1806-1889)," *Isis*, #34, 1942, Pgs. 97-108

Schneer, Cecil J., ed., *Two Hundred Years of Geology in America*, Univ Press of New England, Hanover, NH, 1979

Schram, Frederick R., "People & Rocks: Geologists at the Museum," *Environment Southwest*, SDMNH, #507, Autumn 1984

Schuchert, Charles & Clara M. LeVene, *O.C. Marsh: Pioneer in Paleontology*, Yale Univ Press, New Haven, 1940

Scott, William Berryman, *Some Memories of a Palaeontologist*, Princeton Univ Press, Princeton, NJ, 319 pp, 1939

Shor, Elizabeth N., *Fossils and Flies*, Univ of Oklahoma Press, Norman, 1971

Simpson, George Gaylord, *Concession to the Improbable*, Yale Univ Press, New Haven, 1978

------, *Discoverers of the Lost World*, Yale Univ Press, New Haven, 1984

Smith, Michael L., *Pacific Visions; California Scientists and the Environment 1850-1915*, Yale Univ Press, New Haven & London, 198 pp, 1987

Spielhagen, Frederick, *The German Pioneers; A Tale of the Mohawk*, (translated by Rev Levi Sternberg, DD), Donohue, Henneberry & Co., Chicago, 1891

Stegner, Wallace, *Beyond the Hundredth Meridian*, Houghton Mifflin Co., Boston, 1954

Stephens, Lester D., *Joseph LeConte; Gentle Prophet of Evolution*, Louisiana Univ Press, Baton Rouge, 1982

Sternberg, Levi, "Geology and Moses," *Evangelical Quarterly Review*, Aughinbaugh & Wible, Gettysburg, Vol #19, 1868

Sternberg, Martha L. Pattison, *George M. Sternberg*, Amer Medical Assoc, Chicago, 1920

Stock, Chester, "Recent Excavations in California," *Carnegie Institution of Wash News Service Bull*, Vol IV, #32, Nov 20, 1938

Stratton, Joanna L., *Pioneer Women*, Simon & Schuster, NY, 267 pp, 1981

Strobei, Rev. P.A., *Memorial Volume on Semi-Centennial Anniversary of the Hartwick Lutheran Synod, of the State of New York*, Lutheran Pub Soc,

Philadelphia, 1881

Sullivan, -----, "Schoharie Loyalists, 1775-1784," *Schoharie County Historical Review*, Vol XXXXVIII, #1, 1984, Pgs. 15-16

Swartzlow, Carl R. & Robert F. Upton, "Badlands National Monument," *Natural History Handbook*, Series #2, Wash DC, Revised 1962

Swinton, W.E., *The Dinosaurs*, Wiley & Sons, NY, 1970

Tarpy, Cliff, "Home to Kansas," *National Geographic*, Vol #168, #3, Sept 1985, Pgs. 353-383

Thayer, Thomas P., *The Geologic Setting of the John Day Country, Grant County, Oregon*, Ntl Parks & Ntl Forests of the NW Brochure, 1990

Utley, Robert M., *Life in Custer's Cavalry; Diaries & Letters of Albert & Jennie Barnitz, 1867-1868*, Univ of Nebraska Press, Lincoln, 1977

Vestal, Stanley, *Queen of Cow Towns [Dodge City]*, Harper & Bros, NY, 1952

Walker, Myrl V., "Obituary of George Fryer Sternberg," *News Bull Soc Vertebrate Paleontology*, Vol #88, Feb 1970

Webb, William Edward, *Buffalo Land*, E. Hannaford & Co., Cincinnati & Chicago, 1872

Wedel, Waldo R., "An Introduction to Kansas Archeology," *Bull Smithsonian Institution*, #174, Bureau of Ethnology, Wash DC, 1959

Wendt, Herbert, *Before the Deluge*, Paladin/Granada Pub Co., London, 1968, Pgs. 270-304

Wheeler, Walter H., "The Uintatheres and the Cope-Marsh War," *Science*, Vol #131, 1960, Pgs. 1171-1176

Wheelock, Walt & Howard E. Gulick, *Baja California Guide Book*, Arthur H Clark Co., Glendale, 1975

White, Andrew D., *A History of the Warfare of Science with Theology in Christendom*, George Braziller, NY, 1955

Whittemore, Margaret, *Historic Kansas; A Centenary Sketchbook*, Univ of Kansas Press, Lawrence, 1954

Wilford, John Noble, *The Riddle of the Dinosaur*, Alfred A. Knopf, NY, 1986

Wilkins, Thurman, *Clarence King*, Univ of New Mexico Press, Albuquerque, NM, 413 pp, 1988

Willard, Julius T., *History of the Kansas State College of Agriculture & Applied Science*, Kansas State College Press, Manhattan, KS, 1940

Williams, T. Harry, Richard N. Current & Frank Freidel, *A History of the United States (Since 1865)*, Alfred A. Knopf, NY, 1965

Willis, Bailey, "Index to the Stratigraphy of North America," *U.S. Geological Paper*, Prof Paper #71, Wash DC, 1912

Winchell, Alexander, *Sparks From a Geologist's Hammer*, S.C. Griggs & Co., 1893

Wood, L.N., *Walter Reed; Doctor in Uniform*, Julian Messner, Inc., NY, 1943

Worley, Rebecca Baker, "James Hervey Sternbergh; Industrialist, Inventor, Gentleman," *Historical Review of Berks County*, Fall 1984, Pgs. 141-149

Young, Davis A., *Christianity & the Age of the Earth*, Zondervan Pub Corp, Grand Rapids, 164 pp, 1982

n.a., "A Medical Picture of the United States," *American Heritage*, Oct/Nov 1984

n.a., *American Museum Journal*, Vol IX, #3, Mar 1909, Pgs. 68-69

n.a., *Boots & Bibles*, Riley County Genealogical Soc Pamphlet

n.a., "Geologic Maps of California, 1839-1939," *California Geology*, Sacramento, CA, Jan 1989, Pg. 21

n.a., "Disclosures of Ancient Life in the Grand Canyon," *Carnegie Institution of Wash News Service Bull*, Vol II, #9, Aug 17, 1930

n.a., *DAR Bible Records*, Vol #7, 1949

n.a., "Air Towns and Their Inhabitants," *Harper's Magazine*, Vol LI, Nov 1875

n.a., "Life Before Man; The Emergence of Man," *Time-Life Books*, NY, 1972 n.a., *Lineage Books, Daughters of the American Revolution*, Wash DC

n.a., *The Natural History Museum Bull*, #130, Apr 1, 1938

n.a., "The Agassiz Association," *The Observer: The Outdoor World*, Vol VI, #3, 1895, Pg. 43

n.a., *Records of the Families of California Pioneers*, Vol #6, San Francisco, 1938

n.a., "Exploring the American West, 1803-1879," *U.S. Department of the Interior*, Handbook #116, Wash DC, 1982

INDEX

Abbott, Clinton, 308, 310
Abbott, John B., 245, 418, 419
Abernathy, 77, 143
Abbott, John B.,
   Adams, Nathaniel A., 27, 28
Adocus, 335, 338
Aerend, 130, 372
Agate Springs Quarry, 64, 164, 193
Agassiz, Louis, viii, 19, 60-61, 123, 134-135, 165, 185, 229-230, 234 257-258, 344, 357
Agassiz, Alexander 142
Albion, IA, 11, 271, 290
Allan, Dr., 430
Altpeter, Peter, 324
Amauropsis alveatus, 327
Ameghino, Carlos, 245, 417, 419
Ameghino, Florentino, 245, 417
American Geologist, 150
American Inventor, 349, 401
American Museum of Natural History, 60, 76, 89-90, 97, 145, 154, 161, 176-177, 193, 210-216, 221-222, 229, 236, 259, 289, 336, 390, 397, 406, 413, 437, 446-447
American Naturalist, 86, 183, 392
Amherst College, 419
Amyda, 338
Anderson, F.M., 315, 327
Anderson, John A., 27-28
Andrews, Roy Chapman, vii, 231, 239, 445
Anisonchus cophater, 184
Aphelops fossiger (Teloceras), 78
Arapahoe Indians, 31, 116
Archaelurus debilis, 97
Aspiderites, 338
Aspidophyllum trilobatum, 19
Baena, 335, 338
Bailey, Prof., 407
Baily, Martin L., 288

Baird, Spencer F., 19
Balaenoptera borealis, 307
Balboa Park, 303-307, 322
Bannock Indian War, 33, 98, 100-101, 259
Barbour, Erwin, 175, 184
Barnitz, Lt. Albert, 274
Barsi, Carol, 306
Battle of Bull Run, 252
Bauer, C. Max, 331
Beaver Creek (KS), 79
Beeman, Lloyd, 431
Belly River Series, 111, 198, 222, 234, 337, 383, 403, 446
Bender, John, 331-334
Berkeley Daily Gazette, 281
Betulites Locality, 19
Betulites westii, 397
Bickmore, Albert S., 229-230
Big Horn Basin, 427
Bird, Roland T., 414-415, 445
Birdsall, Ralph, 250
Blackfoot Indian Reservation, 195-196, 422
Blue Mont College, 25
Boll, Jacob, 134, 143, 145
Borden, Lt. George P., 292
Bourne, W.O., 304-305
Brachyceratops (Monoclonius), 422
Brandenburg, Aliki, 421
Bridger Basin, 182, 424
British Museum (Natural History), 68-69, 152, 196, 203, 207, 229, 236, 241, 447
Brock, Reginald Walter, 211-212
Brodie Club of Toronto, 387
Broili, Dr. F., 149
Brooks, Rev. David 356
Brous, H.A., 29-30, 65
Brouse, Wilbur, 79
Brown, Barnum, iii, 167, 192, 199,

210, 212-214, 216-222, 234, 260, 331, 337, 413-415, 417, 445
Brown, John, 123
Brown, Kenneth, 324-325
Brown, Lilian, 45-46, 414
Buckeye Cemetery (Kanopolis), 285-286, 288, 295
Buffalo Park, KS, 30-31, 79-80, 107-109, 111, 281
Bumpus, Hermon C., 231
Buntline, Ned, 117
Butler's Pass, 332-333
Calamodon, 427
California Academy of Sciences, 314
California Institute of Technology, 315, 318
Calvin College, 364
Camelus americanus, 320
Camp, Charles L., 68
Canadian Geological Survey, 210-213, 216, 220, 225, 415, 440
Canyon City, OR, 91-92
Cardium cooperi, 327
Carlin, W.E., 110
Carnegie Museum, 201, 235, 320
Carnegie News Service, 319-320
Case, E.C., 145
Castle Rock (KS), 421, 438
Cather, Willa, 13
Cattell, J. Mark, 40
Centrosaurus longirostris, 441
Chaco Canyon, 333
Chasmosaurus, 50, 223, 307, 337
Chasmosaurus belli, 447
Chesnut, J.S., 148
Chicago Herald, 399
Chief Buffalo Horn, 98
Chief Egan, 98, 100
Chief Homely, 98, 100
Chief Joseph, 56
Chief Oytes, 98
Chief Red Cloud, 164
Chief Roman Nose, 279
Chief Sitting Bull, 50, 56

Chief U-Ma-Pine, 100
Chief Winnamucca, 98
Chisholm Trail, 117
Cholera, 253
Clark, F.R., 331
Clark, Robert D., 344
Clarno Formation, (John Day), 94
Cleveland, Grover, 260
Clidastes tortor, 36, 78, 155, 305
Coalinga, CA, 312-314, 322
Cody, Buffalo Bill, 34, 128, 180
Coelophysis, 232
Coffee Creek, 138, 143-145
Colbert, Edwin H., 65, 184 203, 232, 236
Colburn, A.E., 199
College of Physicians & Surgeons, 251
Como Bluffs, 110
Condon, Thomas, 85, 87, 174, 344
Constant, John T., 121, 240
Cook, Harold J., 64, 193-194, 236
Cook, James H., 64, 164-165, 258
Cooke, Edwin, 422, 433
Cooper, James Fenimore, 3, 251
Cooper, William, 3
Cooperstown, NY, 250-251
Cope, Edward Drinker, iii, viii, 26, 29-30 34-38, 41-56, 59-60, 64, 66, 73-76, 80, 85-88, 110, 112, 134, 146-147, death 148, 165, 167, 171, 173, 175-187, 232, 283, 301, 343, 348, 352-356, 375, 391, 394, 414, 427
Copeland, Mr., 89
Coralliochama orcutti, 327
Corythosaurus, 25
Corythosaurus casuarius, 447
Cosmos Club, 255
Cottonwood Ranch, (Kanopolis, KS), 274, 281
Cow Island, 49-50, 53-54, 56, 391
Coyote, KS, 108
Craddock Quarry, 112, 152-155

Crassatella tuscana, 326
Crater Lake, OR, 80
Crawford, Samuel J., 31
Creationist Science Institute, 364
Cronkite, Walter, 106
Crooked Creek (TX), 146, 241
Crooks, Ramsay, 93
Cummins, William Fletcher, 134, 136-137, 143, 185
Custer, Gen. George, 43-44, 190
Cutler, W.E., 203, 218
Cuvier, Georges, viii
Dalles, The, 85, 89, 93
Dana, James Dwight, 174, 200, 344, 351
Dane, OK, 289
Danforth, Clarence H., 309
Darwin, Charles, 344, 348, 353
Davies, D.C., 417
Davis, Leander S., 100
Dawson, John William, viii
Day, Bill, 94, 97, 99
Day, John, 94-95
Dayville, OR, 93-94, 99-101
Dead Lodge Canyon, (Red Deer), 214, 220, 416
deCamps, L. Sprague & Catherine, 173, 394
Deer, James, 49-50
Dentalium [Entalis] whiteavesi, 326
Deseado Formation, (Patagonia), 419
Diadectes, 138, 141
Diceratherium nanum, 97
Dick, Everett, 119, 128
Dimetrodon, 141, 144
Dimetrodon gigas, 153
Dinosaur Provincial Park, 214, 225, 446
Diospyros rotundifolia, 397
Diplocaulus magnicornis, 147
Diplodocus, 243
Disney, H.A. Patrick, 203
Dissorophus multicinctus, 138
Dix, The, 285

Dodge, Lt. Col. Richard Irving, 34, 257
Dog Creek, 42, 49, 161
Dolichorhynchops osborni, 397
Drevermann, Fritz, 308, 407
Drumheller, 214
Dryerson, Rev. Delmar L., 310, 367
Duncan, George, 80, 83, 85-86
Earp, Wyatt, 128
Ectoganus gliriformis, 427
Edaphosaurus, 141, 144-145
Edaphosaurus pogonias, 145
Edmonton Beds, 212-213, 221, 234
Eiseley, Loren, 346
El Marmol, (Baja), 324-325
Eldridge Hotel, (Lawrence), 121, 127
Elephas primigenius, 87-88
Ellsworth, KS, 115, 117-119, 249, 253, 270, 274-277, 280-293
English Lutheran Church, 304
Enhydrocyon stenocephalus, 97
Equus scotti, 156, 305
Ernst, John Frederick, 286
Eryops, 141
Escavada, The, (San Juan Basin), 335
Eschatius conidens, 88
Evangelical Quarterly Review, 272, 349
Falkenbach, C., 431
Field Museum, (Chicago, IL), 245, 416-419
First Lutheran Church, 304, 342
First Presbyterian Church (Ellsworth), 276-277
Flat Boat (Barge), 214-216
Ft. Benton, 41-43, 56
Ft. Bridger, 424
Ft. Claggett, 42, 47
Ft. Ellsworth, [Also Ft. Harker], 116, 279
Ft. Harker, 11-12, 18, 20, 115-117, 253-254, 261, 274-280, 285-286,

295
Ft. Hays State Teacher's College, 376, 420
Ft. Hays State Univ, 130, 376
Ft. Klamath, 80, 85
Ft. Pierre Group, 44
Ft. Riley, 127, 253, 290-293
Ft. Walla Walla, 90, 256,-257
Ft. Wallace, 256, 279, 295
Fossil Bandaging, 61, 68
Fossil Lake, OR, 85, 87
Free State Hotel, (Lawrence, KS), 121, 127
Frick, Childs, 377, 429-430
Fruitland, The, (San Juan Basin), 337, 339
Fryer, Dr., 254
Galileo, 348
Galyean, Frank, 142-143
Garbutt, Frank A., 316
Gazin, C. Lewis, 427-431
Geismar, Jacob, 74, 375
Gibson, John M., 255, 262
Gidley, James W., 109, 219, 260, 398
Gidley's Horse Quarry, 156
Gillicus arcuatus, 421
Gillispie, Charles Coulston, 353
Gilmore, C.W., 195-198, 219, 235, 241, 260, 308, 335, 338, 376, 387, 398-399, 402, 416, 422-428, 432
Goldbaum, David, 324
Gorgosaurus sternbergi (Albertosaurus libratus), 220, 441, 444
Gould, Charles, 137, 166, 206-207
Gove County, KS, 107
Granger, Walter, 162-163, 220, 232
Gray, Asa, viii
Green, J. Elton, 307
Greene, Alfred A., 294, 410
Greene, Susan A., 294, 410
Gregory, William King, 232, 390-391, 401
Grinnell, George Bird, 174, 181,

258
Gross, Renie, 217-218, 415
Gryphaea, 326
Gyrodes conradiana, 326
Hackberry Creek, 155
Haddonfield, PA, 73-74, 178
Hagerman Horse Quarry, 430
Halemenites major, 334
Hall, Steve, 216-217
Hamites duocostatus, 325
Hamman, George, 139
Hancock, Gen. Winfield S., 31
Hanna, G. Dallas, 68, 314-316, 323, 326-327
Happy Jack's Ferry, (Red Deer), 224
Harker, Gen. Charles G., 116
Harper's Magazine, 108
Hartwick Coll Seminary, 2, 3, 8-11, 249, 273, 281, 284
Hartwick, John Christopher, 3, 10
Haskell Institute, (Lawrence), 410
Hatcher, John Bell, iii, 2, 173, 175-177, 186, 192, 199-202, 207, 245, 260, 414, 417-420
Hayden Survey, 59-60, 77, 134
Hazelius, Ernst L., 10, 282
Hedra ovalis, 397
Helaletes, 425
Hertlein, L.G., 315
Hesperornis, 78
Hickok, Wild Bill, 128
Hickcox, Rev. W.H., 291
Hill, Russell T., 65, 79
Hill, Rev. Timothy, 276
Holomeniscus hesternus, 88
Holomeniscus vitakerianus, 88
Holt & Company, Henry, 390
Holyrood, KS, 288
Homestead Act of 1862, 120
Horner, John (Jack), 194, 394, 445-446
Howard, Gen. Oliver O., 98, 100
Huene, Friedrich von, 154, 417

489

Huff, Joe, 90
Humlong, Francis, 290
Humlong, George Arthur, 290
Humlong, Margaret Louise, 290
Humlong, Robert Sternberg, 290
Humboldt, Alexander, 181
Hunter's Wash (NM), 335
Hylodon sodalis, 88
Hypacrosaurus, 423
Hyrachyus, 425
Indian Creek, 148
Inman, Maj. Henry, 117
Inoceramus, 67-68, 334
Inoceramus pacificus, 326
Interior, The, 400
Iowa Lutheran College, (Albion, IA), 11
Isaac, J.C., 37-38, 41, 44, 48, 51, 56, 13
Jackson, Hansel Gordon ("Happy Jack"), 224
James, Frank, 124
James, Jesse, 124
Janeway, Dr. J.H., 256
Jenkins, Olaf P., 164, 313
Jennison, Charles, 125
Johanson, Donald, 106
John Day, The, 91-93, 96-97, 301, 352
Johnson, Col. Henry, 275
Johnson, Luther, 276
Johnson, Phillip E., 366-367
Jones, Harold S. "Corky", 67, 166
Josephine, The, viii, 53-56, 66
Journal of Geology, 401
Judd, Neil M., 333
Judith River Group, 44, 59, 66, 337, 403, 427
Kaisen, Peter, 218, 337, 413, 415
Kanopolis, KS, 12-13, 274-275, 280, 285, 288
Kansas Academy of Science, 75, 129, 175, 186, 208, 234, 374, 392
Kansas chalk, 14, 76

Kansas City Review of Science & Industry, 30, 66, 106, 350
Kansas State Agricultural College, 25, 30, 171, 292, 311
Kansas State Teacher's College Leader, 311, 320
Kimbeto(h), The, (San Juan Basin), 335-336, 338
King, Clarence, viii, 185, 344
King, William S., 252
Kirtland Formation, The, (San Juan Basin), 335, 338-339
La Brea Tar Pits, 314, 316-321
Labidosaurus hamatus, 138
Lambe, Lawrence M., 211-212, 220, 223, 415-416
Lambeosaurus lambei, 220, 441
Lane, Sen. James, 125-126
Lanham, Url, 200
Lantz, D.E., 409
Lawrence, Amos A., 123
Lawrence Gazette, The, 400
Lawrence Journal World, 117
Leakey, L.S.B., 218
Leakey, Richard, 106, 193-194, 356
LeConte, Joseph, 125, 344, 354-356
Leidy, Joseph, 174, 256, 375
Lesquereux, Leo, 19-21, 65
Lewin, Roger, 356
Lewis, William A., 428
Lincoln, Abraham, 8, 123, 142
Lincoln, Robert T., 142
Lindblad, Ann, 386
Lindblad, Gustav, 386
Little Sandhill Creek (Red Deer), 440
Little Wichita River, 148
Livingston, William, 288
Llano Pliocene Beds, 185
Long, David Burton, 117, 253, 273-274, 276, 278-280, 288, 295
Long, Harriet Sage, 279, 295
Long, Reub A., 85, 87
Long Island, KS, 175, 199, 201,

251, 317, 374
Loomis, Frederic B., 245, 417-419
Loosely, George, 85
Lord Beresford, 224
Louis XIV, 268
Loup Fork Beds, 77, 91, 202, 406
Love, David, 347
Loveland Ferry, 220
Lucas, F.A., 255
Lull, Richard Swann, 27
Lyon, Charles J., 273
Maclain Hotel (Seymour, TX), 150
Malheur Indian Agency, (OR), 98
Marsh, Capt. Grant, 54
Marsh, Othniel C., iii, 26, 30, 32, 51, 54, 61, 66, 75, 78-79, 94, 100, 106, 110-112, 164-168, 171-187, 199-200, 202, 256, 343, 375, 407, 414
"Marsh" (Sternberg) Quarry, 78, 175-176, 199, 247, 317, 321, 374
Martin (Sternberg), Glenn, 440
Martin, Handel T., 145, 154
Martin, Myrtle, 438-441
Martin (Sternberg), Raymond McKee, 366, 386, 391, 439-440, 444-445
Martin (Sternberg), Stan, 440
Mascall, Bill, 99, 101
Mascall Formation, (John Day), 94, 101
Mascall, Lillian, 101
Masterson, Bat, 128
Mastodon, 77
Mather, Stephen Tyng, 196, 198, 422
Matthew, William Diller, 89, 144-145, 154, 160-163, 177, 232, 258, 315, 336-337, 397, 406
McClung, C.E., 96
McGee, Jack, 216
McKinley, William, 260
McKittrick, CA, 303, 311, 313-320, 327

McPhee, John, 346
Merriam, John C., 87, 100-101, 314-315
Merrill, Austin, 49-50
Mexico Ranch, 224
Middleburgh, NY, 1, 4-5, 8, 73, 249, 282
Midland College, 191, 429
Midway Royal Oil Company, 315-316
Miles, Col. Dixon S., 52
Milk River, 422-423
Miller, Dehlia B. Snyder, 10
Miller, George B., 10-11, 444
Miller, Loye, 96-97, 101
Miller, Margaret Levering *see* Margaret L. Miller Sternberg
Miller, Max, 325
Mongolian Gobi Desert Expedition, 163, 231
Monoclonius, 50, 63, 422-423
Monoclonius crassus, 50
Monoclonius fissus, 50-51
Monoclonius sphenocerus, 50, 65
Montgomery, James, 125
Monument Station, 106
Moody, Frank, 323
Mook, Charles C., 337
Mormon Trail, 120
Mosasaurus copeanus, 178
Mount Temple, The, 241
Mudge, Benjamin F., iii, 6-30, 53, 110, 129, 171-172, 348
Muhlenberg, Henry Melchior, 282
Muhlenberg, William Augustus, 282
Museum of Comparative Zoology, 134-135, 142
Museum of Man (San Diego), 331
Naosaurus, 144-146
Nation, The, 400
National Museum of Canada, 225
National Museum of France (Paris), 447
National Museum of Germany

(Berlin), 447
"Nauchville", 117
Nautilus sternbergi, 325
Navaho Indians, 92, 332
Nelkin, Dorothy, 362, 365
Nesodont, 245
Neurankylus, 338
New England Emigrant Aid Company, 123
New York Times, 168, 318
Newberry, John Strong, 19
Newton, Sir Isaac, 344
Nichols, Alice, 125
Niobrara Formation, 152, 374, 438
North, Frank, 179
North, Luther, 179-180
Oak Hill Cemetery (Lawrence), 375, 383
Oakley, KS, 420
Ohio University, 407
Ojo Alamo (San Juan Basin), 335
Old Woman Creek (WY), 438
Omaha, NB, 41
Opdycke, Wilda, 432
O'Rourke, Patrick F., 307
O'Rourke Zoological Institute, 307-308
Orohippus, 425
Orton Jr., Edward, 407
Osborn, Henry Fairfield, 60, 74, 76, 90, 111-112, 137, 145, 151, 180, 184, 193-194, 220, 233, 236, 254-255, 283, 289, 336-337, 390, 406, 413, 447
Ostrea, 326
Otis, D.H., 409
Otsego Herald, 250
Overton, Anthony, 176-177
Overton, William, 283
Pachydiscus catarinae (Parapachdiscus catarinae), 323, 325
Pacific Beach (San Diego), 307
Pacific Fur Company, 93
Pacific Oil Company, 315-316

Padilla, Dan, 334-335
Palatines, 267-269
Paleontological Museum of Munich, 149
Paleontological Society, 403
Panoplosaurus, 423
Panoplosaurus mirus, 441
Pap Burns Oil Well, 323
Parallelodon brewerianus, 326
Parallelodon vancouverensis, 326
Parasaurolophus tubicans, 336, 338
Parker, Ronald B., 345
Parkman, Frances, 17, 32
Parks, Thomas, 107
Parks, William Arthur, 386
Patagonia, 2, 199, 201, 245, 376, 418-419
Patterson, Bryan, 418
Pawnee Indians, 179-180
Peabody Museum (Yale), 174, 184
Pearce, F.L., 431
Pearce, George B., 424, 430, 433
Peck, The C.K., 54-55
Pentaceratops, 335
Pentaceratops fenestratus, 336-337
Pentaceratops sternbergii, 336-338, 447
Peterson, O.A., 2, 199
Petrified Forest (Arizona), 332
Phelps, Ira Warner, 284
Philadelphia Academy of Sciences, 79, 229
Philophropean Society (Hartwick), 7
Phinney, Elihu, 250, 251
Phinney, Henry, 250, 251
Pierce Hotel (Tacoma), 282
Piltdown "man", 183
Pine Creek, WA, 88, 90-91
Pioneer Cemetery (Ellsworth, KS), 280, 283
Plastomenus, 338
Plate, Robert, 178
Platecarpus (Platycarpus), 36, 78, 155

Platecarpus coryphaeus, 155
Plesiosaur, 155
Plesippus, 429
Popular Science Monthly, 243, 303, 392, 401
Popular Science News, 86
Port Neuf Canon, OR, 42
Portheus molossus (Xiphactinus audax), 36, 69, 234, 240, 244, 305, 373, 421-422, 446
Powell, John Wesley, 134, 185
Preston, Douglas J., vii, 18, 233
Pride W.F., 293
Princeton Univ, 202, 447
Proadiantus ameghino, 419
Productus, 332
Prosaurolophus maximus, 446
Protophyllum sternbergii, 19
Protostega gigas, 75, 186
Pterodactyl, 79
Pteranodon, 155
Pueblo Bonito, 333
Punam, Irene, 305
Quantrill, William Clarke, 121-128, 293
Rattlesnake Formation, 94
Raymenton, W.H., 307
Red Deer, The, 111, 187, 213-214, 217, 241, 334, 413, 440, 445
Reed, T.G., 425
Reed, Walter, 263
Reed, William Harlow, 110, 263
Reeside, J.B., 331, 335
Reynolds, Anna Frieda, 297
Reynolds, Anna Musgrove *see* Anna Musgrove Reynolds Sternberg
Reynolds, Rev. Charles, 126-127, 291-293, 296, 444
Reynolds, Elizabeth (Bessie), 292, 296
Reynolds, Herbert E., 296-297, 307
Reynolds, Mary Braine, 292, 296
Reynolds, Theodore B., 296
Riggs, Elmer S., 245, 418-419

Rinehart, Maj. W.V., 98
Robidoux, Peter, 118
Robinson, Gov. Charles, 123
Rock Creek Horse Quarry, 155
Rocky Mountain News, 84
Rodriquez, A.L., 324
Roemer Museum (Hildensheim), 447
Roessler, A.R., 136
Rogers, Katherine, 48, 282, 289-290, 294, 410
Roman Catholic Church, 268, 348
Romer, Al, 342
Roosevelt, Teddy, 159
Royal Canadian Navy, 439
Royal Ontario Museum, 385, 387
Rudwick, Martin J.S., 347
Russell, Loris, 212, 433
Russell, Maria Louisa, 253
Samoa, The, 285
San Diego Museum of Natural History, 304-306, 308, 336
San Diego Union, 07, 31
San Francisco Argonaut, The, 399
Santa Catarina, Baja, 324-325
Santillian, Manuel, 324
Sappa Creek (KS), 79
Saskatchewan Museum, 247
Sassafras dissectum, 19
Sassafras Hollow, 17 19
Scates, Elisha, 288
Scates, Merle Ione, 296
Schoharie, NY, 4-5, 268-270, 280
Schuchert, Charles E., 174
Science, 392
Scott, William Berryman, 61, 100, 112, 174, 180, 219, 445
Scripps, E.W., 304
Scripps, Ellen, 304-305
Scripps Institute of Oceanographic Research, 304
Seaman's Ranch, 439
Senckenberg Natural History Museum (Frankfurt), 308, 407, 447
Seymour, TX, 139, 142, 150

Seymouria, 138
Sherman, French & Company, 401
Shor, Elizabeth Noble, 202
Shouver, Ned, 334-335
Shumard, Benjamin F., 136
Silliman, Benjamin, viii, 344
Silver Lake, OR, 79, 83-84, 86
Simpson, George Gaylord, 232, 445
Smith Woodward, A., 68-69, 152, 203, 207, 236, 242, 413
Smith, Zoward B., 307
Smithsonian Museum *see* U.S. National Museum
Smoky Hill Trail, 255
Snake River Piutes, 98
Southwestern Onyx & Marble Company, 324-326
Spielhagen, Frederick, 270
Spier, F., 180
Spondylus, 326
Stegosaur, 216
Stephanosaurus marginatus (Lambeosaurus), 220
Stephenson, Mr., 156
Sternberg, A. Irving, 409
Sternberg, Albert, 115, 270
Sternberg, Anna Musgrove Reynolds, 121, 177, 211, 291, 294-296, 302, 308-310, 333, birth 296, death, 296, 310
Sternberg, Anna Ziegler, 425-426
Sternberg, Bertha, 285, 288, 295-296
Sternberg, Charles Hazelius, ailments/accidents, 41, 146, 161-163, 217, 254, 273, 309, Birth, 1, California move, 301-303, camp life, 45-47, Canadian move, 206, 210-211, Christianity, 48-49, 272, 341-367, education, 7-9, 25-26, farming, 288, financial woes, 150, 152, 211, 405-406, fossil bandaging, 62-68, fossil leaves, 17-20, heritage, 267-270, Iowa move, 11, Kansas move, 12-13, 115, 262, laboratory, 121-122, 239-240, Lawrence move, 119, lecturing, 403-403, marriage, 290-291, 302, 309, other businesses, 121, 406-411, poetry, 359-362, 384, 395-396, siblings, 1-2, working for Cope, 29-30, 35-38, 41-56, 172, working for Marsh, Sternberg, Charles Mortram, iii, 166, 208, 212, 235, 288, 291, 349, 366, 372, 380-383, 398, 406, 410-411, 416, 437-441
Sternberg, Charles W., 338, 374, 417, 425-426
Sternberg, Charlotte (Lottie), 285-286, 288, 295
Sternberg, Clarence Denby, 283, 296
Sternberg, David, 270
Sternberg, Edward Endress, 2, 115, 249, 270, 288-289
Sternberg, Emily, 290
Sternberg (Beeman), Ethel, 417, 425, 431
Sternberg, Florence Ethel, 296
Sternberg, Francis (Frank), viii, 30, 115, 270, 283-284, 287, 296
Sternberg Geology Club, 421
Sternberg, George Fryer, iii, 122, 129-130, 149, 154, 161-162, 167, 210, 216-218, 232-233, 259, 294, 338-339, 371-377, 379-380, 385, 397-398, 409, 411, 413-432, death 433
Sternberg, George Miller, viii, 1, 2, 6, 9, 12, 90-91, 249-264, 273, 278, 284, 289, 294, 371
Sternberg, John Frederick, 52, 62, 115, 270, 284, 286-287
Sternberg, John Levi, 295
Sternberg, Lambert (Lampert), 269-270
Sternberg, Levi, iii, 6-8, 154, 212,

223, 294, 338, 358-359, 372, 380-381, 385-386, 392, 437-441
Sternberg, Levi (Sr), 1, 2, 4, 11-12, 119, 262, 270, 272-273, 275-280, 291-292, 295, 349, 444
Sternberg, Mabel Smith, 333, 374, 417
Sternberg, Margaret Levering Miller, 1, 6, 261, 270, 275, 281-282, 295
Sternberg, Margaret Pattison, 6, 12, 250, 252, 254, 258-259, 262, 269, 277
Sternberg, Margery Anna, 374-375
Sternberg, Maude (Maud), 294, 375, 382-385, 396
Sternberg Memorial Museum, 376, 416-417, 421, 446
Sternberg, Nellie, 283-284, 296
Sternberg, Robert, 290
Sternberg, Rosina, 2, 289
Sternberg, Theodore, 2, 115, 117, 252, 270, 284-285, 288, 295-296
Sternberg, William Augustus, viii, 107, 115, 270, 281-282
Sternbergh, James Hervey, 409
Sternberg's Silicious Soap, 121, 398, 405-407
Steve Hall's Hotel, 216
Stevens, Frank, 305
Stevens, Kate, 305 (see also 315)
Steveville, 216, 224, 447
Stock, Chester, 315, 317, 319-320
Stockman/Grower's Associations, 280
Styracosaurus albertensis, 247, 423
Sullivan, Bob, 305
Swinton, W.E., 61
Sykes, Gen. George, 251
Tacoma News Tribune, 281
Tambiago, 98
Taeniodont, 427
Tanupolama stevensi, 320
Tessarolax incrustata, 326

Testudo (Xerobates) orthopygia, 77
Texas Almanac, 136
Thompson, Billy, 277
Thompson Creek, 19, 270-271, 276
Thompson, P.M., 271
Thornton, Mae, 286
Tokaryk, Tim, 247
Toronto Star Weekly, 386
Toxochelys, 78
Trachodon (Anatosaurus?), 186, 233, 245, 416
Triceratops, 240-241, 307
Trinity Episcopalian, 293
Troodon (Stegoceras), 426
Truman, Harry, 224
Turritella peninsularis, 326
Turtle Cove (John Day), 96-97, 101
Two Medicine Formation, 422-423
Tyler, Mr., 336
Tylosaurus, 155
Tylosaurus dyspelor, 36, 245
Tyrannosaurus rex, 220, 244, 307, 441
Tyrrell, Joseph Burr, 212
Uintacrinus socialis, 155
Uncle James (from Ames), 73
Union Pacific Railroad, 117, 281
Univ of Calif @ Berkeley, 87, 96, 101, 259, 296, 447
Univ of Kansas, 96, 128
Univ of Tubingen, 447
Univ of Toronto, 86, 447
U.S. National Museum (Smithonian),19, 91, 153, 247, 255, 258-259, 308, 314, 333, 335, 338, 377, 398, 402, 422-423, 426, 429, 432
Ussher, Bishop James, 351
Utamilla Indians, 100
Vail, Bishop Thomas H., 291
Valley Fever, 312-313
Victoria Memorial Museum, 240
Wagenhalls, Margaret, 390
Wagner, J.L., 214
Wakeeney, KS, 108

Walker, Myrl V., 432
Warwick, Mr., 94
Washtucna, Lake, 258
Wegeforth, Harry, 304
Weiser, Conrad 269, 282
Weiser's Dorp, 269, 282
Wetmore, Alexander, 429
Whelan, Pat, 140-141
Whiskey Jack, 42
White, Andrew, 345
White, Dr. Charles A., 134
Whitaker, George, 232
Whiteaker, Charles, 85
Whiteaker, Gov. John, 85
Whitney, Sheriff C.B., 117, 277
Wiest, Rev. C.F., 425
William, John, 268
Williams, Capt., 44
Willis, Bailey, 140
Williston, Frank, 173
Williston, Samuel W., iii, 25-27, 29-30, 53, 64-65, 75, 79, 106-107, 110-111, 129, 145, 152-154, 166-167, 173, 202, 213, 260, 283, 348, 356-357, 414
Willow Creek (TX), 148
Wilson, Francis L., 283
Wilson, Woodrow, 198
Wiman, Dr., 336
Wortman, Jake L., iii, 90-91, 97, 99-100, 212-213, 254, 320, 414, 427
Wright, Mr., 141-142
Wright, Charles D., 284
Yale University, 168, 174, 180, 447
Young, Davis A., 364, 366
Younger, Cole, 124
Zakrewski, Richard, 416, 433
Ziegler, Anna *see* Anna Ziegler Sternberg
Zittel, Karl von, 149-150